Challenges for Smart City Infrastructure, Technologies, and Their Future

Mohammed Balfaqih
University of Jeddah, Saudi Arabia

Akram A. Almohammedi
Karabük University, Turkey

Malek Alsmadi
Confederation College, Canada

IGI Global
Scientific Publishing
Publishing Tomorrow's Research Today

Vice President of Editorial | Melissa Wagner
Director of Acquisitions | Mikaela Felty
Director of Book Development | Jocelynn Hessler
Production Manager | Mike Brehm
Cover Design | Jose Rosado

Published in the United States of America by
IGI Global Scientific Publishing
701 East Chocolate Avenue
Hershey, PA, 17033, USA
Tel: 717-533-8845 | Fax: 717-533-7115
Website: https://www.igi-global.com E-mail: cust@igi-global.com

Library of Congress Cataloging-in-Publication Data

Names: Balfaqih, Mohammed, 1985- editor http://id.loc.gov/rwo/agents/n2025014216 | Almohammedi, Akram, 1986- editor http://id.loc.gov/rwo/agents/n2025014219 | Alsmadi, Malek, 1981- editor http://id.loc.gov/rwo/agents/n2025014221
Title: Challenges for smart city infrastructure, technologies, and their future / edited by Mohammed Balfaqih, Akram Almohammedi, Malek Alsmadi.
Description: Hershey, PA : IGI Global Scientific Publishing, [2026] | Includes bibliographical references and index. | Summary: "This book provides a comprehensive exploration of the rapidly evolving domain of smart cities, covering the critical technologies, practical applications, key challenges, and future outlooks. It delves into the technological foundations that enable smart city initiatives, highlights innovative applications transforming urban life, and addresses the complex challenges that cities face in becoming smarter and more connected"-- Provided by publisher.
Identifiers: LCCN 2024055564 (print) | LCCN 2024055565 (ebook) | ISBN 9798337301549 h/c | ISBN 9798337301556 s/c | ISBN 9798337301563 ebook
Subjects: LCSH: Smart cities--Technological innovations | Smart cities--Forecasting | Sustainability http://id.loc.gov/authorities/subjects/sh2009000375
Classification: LCC TD159.4 .C43 2025 (print) | LCC TD159.4 (ebook) | DDC 307.1/416--dc23/eng/20250331
LC record available at https://lccn.loc.gov/2024055564
LC ebook record available at https://lccn.loc.gov/2024055565

British Cataloguing in Publication Data
A Cataloguing in Publication record for this book is available from the British Library.

Table of Contents

Detailed Table of Contents

This chapter provides a comprehensive exploration of foundational technologies that underpin the development of smart cities. It discusses the pivotal role of the Internet of Things (IoT) as a fundamental component enabling intelligent urban infrastructure. Key IoT architectures, including both functional and sensing-to-cloud frameworks, are analyzed to highlight their significance in enhancing urban management, sustainability, and quality of life. The chapter further examines practical IoT applications across various sectors such as industrial and manufacturing, consumer technology, retail, healthcare, transportation, agriculture, energy, and government operations. Additionally, critical technological challenges and security considerations associated with large-scale IoT deployments are reviewed, emphasizing the importance of robust standards and protocols to ensure data integrity, reliability, and overall system efficiency.

The convergence of cloud computing with emerging technologies is reshaping industries by enhancing efficiency, scalability, and innovation. This chapter examines how cloud platforms integrate with blockchain, serverless systems, artificial intelligence, machine learning, the Internet of Things, edge computing, and cybersecurity to deliver real-world benefits across sectors such as healthcare, manufacturing, retail, and supply chain management. It highlights the transformative potential of cloud-enabled innovations in reducing costs, fostering collaboration, and accelerating digital transformation, while also addressing key challenges including legacy system migration, regulatory compliance, and data privacy. By combining technological foundations with practical applications, the chapter offers valuable insights for researchers and practitioners, demonstrating how the fusion of cloud computing and emerging technologies is driving industrial transformation and shaping the future of business.

This chapter explores the role of blockchain technology in enabling sustainable and inclusive smart cities amid rapid urbanization and resource constraints. It examines how blockchain's core features—decentralization, transparency, and immutability—integrate with smart city frameworks to enhance governance, energy management, urban planning, and transportation. Case studies from Dubai, Zug, and Telangana demonstrate practical applications in healthcare, public services, and land registries, showcasing blockchain's transformative potential. The chapter also discusses key challenges such as scalability and energy consumption, while highlighting emerging trends like decentralized finance, DAOs, and integration with AI, IoT, and 5G. By combining technological foundations with real-world implementations, the chapter underscores blockchain's capacity to strengthen transparency, citizen engagement, and resilience in smart cities, paving the way for more efficient, secure, and sustainable urban futures.

This chapter explores how Artificial Intelligence (AI) works together with the Medical Internet of Things (MIoT) in modern healthcare. It focuses on smart hospitals. AI helps MIoT by making it possible to predict health problems, track disease progress, and make better decisions using data. IoT tools like asset tracking, energy monitoring, and staff management help reduce hospital costs and save resources. A cost analysis shows that MIoT can reduce hospital readmissions by 30% and cut healthcare costs by 20%. It also offers a return on investment within 2–3 years. There are some challenges, such as device compatibility, system expansion, and data security. The chapter provides simple solutions like using cloud or edge computing, blockchain, and common international standards. Real-life examples from South Korea and the UK show that MIoT improves patient care and hospital performance. The chapter also covers new trends like 5G, AI-based automation, and robots doing routine tasks. These trends point toward a future of smart, patient-focused, and technology-driven healthcare.

This chapter examines the impact of Industry 4.0 on the digitalization and automation of logistics and supply chain management (SCM). It explores how technologies such as the Internet of Things (IoT), big data analytics, artificial intelligence (AI), blockchain, and automation are transforming supply chain processes by improving efficiency, transparency, and decision-making. Particular emphasis is given to data-driven models, predictive analytics, and digital twins, which enable real-time optimization, operational agility, and resilience in global supply networks. The findings demonstrate that digital transformation streamlines logistics operations, reduces inefficiencies, and supports proactive disruption management. By integrating these technologies, organizations can build smarter, adaptive, and sustainable supply chains. This study provides insights into how Industry 4.0 is reshaping logistics and SCM, offering practical guidance for academics, practitioners, and policymakers.

The rising demand for sustainable and energy-efficient solutions in urban areas has driven interest in renewable systems for smart cities. This chapter presents a solar-assisted combined cooling, heating, and power (SA-CCHP) system designed for Doha, Qatar, where high solar radiation and cooling needs prevail. Powered solely by a parabolic trough collector (PTC) field, the system delivers net power from 120 kW in winter to 195 kW in summer, with cooling loads of ~210–340 kW and heating loads of ~900–1450 kW. Increasing the superheating degree at the ORC turbine inlet enhances power and heating but reduces cooling, while raising the pressure ratio (α) from 0.5 to 0.8 boosts net output and efficiency, cutting CO_2 emissions from 0.22 to 0.13 kg/kWh. Overall energy efficiency rises from 85% to 90% and exergy efficiency from 76% to 78.5%, while costs decline from $40/hr to $36/hr,

of technological innovation and sustainable water management, the paper offers policy prescriptions to climate proof agriculture. With means of interdisciplinary views, practical applications and policy suggestions, this book snapshots the agro-environmental management policy and the smart irrigation systems towards the protection of water resources, and future generations of smart farming applications worldwide.

Despite the growing interest in smart cities development and their potential benefits, the existing research lacks empirical studies into the nuanced factors shaping purchase intentions of homeowners in the 21st century. Adopting Value-based Adoption Model (VAM), hypotheses were developed and tested. This study addresses the gap and is among the first to empirically utilize perceived price value as a mediator; and provides insights into how perceptions of affordability and value-for-money influence consumers in developing countries. By utilizing PLS-SEM the findings indicate that the relationship among perceived environmental initiatives, convenience, IT resources and behavioral intention to purchase homes in smart cities are mediated by perceived price value, while perceived privacy and perceived security did not. These findings offer city officials an approach to gauging Malaysian' intentions to purchase homes in smart cities as well as identify those factors critical to developing a successful smart-city strategy effectively cater to the evolving needs of future homeowners.

The evolution of smart cities, driven by 5G and B5G, enhances quality of life and sustainability but faces challenges in disaster management. This chapter explores solutions using advanced communication technologies to ensure operational continuity post-disaster. Device-to-device (D2D) communication facilitates crucial message exchange between victims and relief teams without relying on base stations. Drone-assisted D2D cooperative communication (DA-DD-CC) improves reliability by incorporating drones as relays, enabling direct, drone-assisted, or combined links. Rate-splitting multi-user cooperation by drone (RSMUCD) integrates rate-splitting

This paper aims to describe the relationships between data privacy and security issues and smart cities, discussing data as the driver of the city development. The author explains what smart cities are, why they are important and what the research is going to focus on. This work explores various aspects of data privacy including the type of gathered data, privacy issues, case studies where breaches occurred and the legal and ethical analysis. It continues with a discourse of security concerns, which gives information about typical risks, hostile elements of structures, actual events, and legislations. Proposed interventions are data protection/encryption, distributed ledger technologies such as blockchain, privacy by design, security roles of Artificial Intelligence (AI) etc. The future direction for this paper discusses future advancement in technology and policy suggestions to comprehensively establish a smart city solutions framework. A call to action encourages stakeholders to focus on privacy and security to make smart cities the secure, efficient, and trustworthy part of the future city.

As urban environments become increasingly digitized, the evolution from smart cities to cognitive cities marks a transformative shift in how cities sense, learn, and adapt. While smart cities rely on data collection and automation to optimize urban systems, cognitive cities integrate artificial intelligence, machine learning, and cognitive computing to create self-aware, context-sensitive, and continuously learning ecosystems. This chapter explores the theoretical underpinnings and enabling technologies of cognitive cities, distinguishing them from their smart city predecessors. It examines key applications in mobility, public services, and citizen engagement, and highlights real-world case studies from leading global cities. Ethical, social, and governance implications are discussed, including issues of data privacy, algorithmic transparency, and participatory design. The chapter concludes by identifying current challenges and outlining future research directions.

Preface

In recent years, the concept of the smart city has moved from being a futuristic vision to a pressing reality that is reshaping urban life across the globe. With the rapid pace of urbanization, the growing demands on infrastructure, and the urgent need for sustainable development, smart cities have emerged as a cornerstone for the future of human society. By integrating advanced technologies such as the Internet of Things (IoT), artificial intelligence (AI), cloud computing, blockchain, and next-generation wireless networks, smart cities promise to deliver safer, more efficient, and more sustainable urban environments.

This edited volume, Challenges for Smart City Infrastructure, Technologies, and Their Future, is a comprehensive collection of scholarly contributions that examine the technological foundations, innovative applications, and critical challenges associated with smart cities. The book is organized into four interconnected sections: technologies, applications, challenges, and future perspectives, ensuring holistic coverage of this complex and rapidly evolving field.

The goal of this book is twofold: first, to provide readers with fundamental and technical insights into the enabling technologies that drive smart city initiatives; and second, to critically examine the practical applications, infrastructure complexities, and societal challenges that must be addressed for their sustainable development. Each chapter contributes to this vision by exploring distinct aspects of smart cities—from IoT architecture and cloud-enabled innovations to medical IoT, logistics automation, renewable energy integration, urban planning, and sustainable farming. Further, the book highlights the challenges of data privacy, disaster management, and geospatial complexities, while also offering future perspectives on global strategies and sustainable development.

This volume brings together experts, researchers, and practitioners from a wide range of disciplines and regions, reflecting the inherently interdisciplinary nature of smart city research. By bridging theory and practice, and by linking current technologies to future trends, the book provides both a reference guide for today and a roadmap for the next generation of smart cities. The content of the book is as follows:

SECTION 1: SMART CITY TECHNOLOGIES

This section presents key technological pillars—IoT, cloud computing, and blockchain—that collectively drive the innovation, efficiency, and resilience of future smart cities.

Chapter 1: Internet of Things as a Core Enabler for Smart Cities: Architectures, Technologies, and Applications

The chapter examines the Internet of Things (IoT) as a cornerstone of smart city development. It reviews core IoT architectures and sensing-to-cloud frameworks, highlighting their role in enhancing urban management, sustainability, and quality of life. Practical applications are illustrated across sectors including healthcare, transportation, agriculture, energy, and government, while critical challenges such as scalability, security, and standardization are also addressed, underscoring the need for reliable and efficient IoT systems in future cities.

Chapter 2: Cloud Computing as a Catalyst for Technological Innovation and Industry Transformation

The convergence of cloud computing with emerging technologies is transforming industries by driving efficiency, scalability, and innovation. This chapter explores how cloud platforms integrate with blockchain, AI, ML, IoT, edge computing, serverless systems, and cybersecurity to deliver practical benefits across sectors such as healthcare, manufacturing, retail, and supply chain management. It highlights the potential of cloud-enabled innovations to reduce costs, enhance collaboration, and accelerate digital transformation, while also addressing challenges like legacy migration, compliance, and data privacy. By linking technological foundations with real-world applications, the chapter provides valuable insights into how cloud computing and emerging technologies together are reshaping the future of business.

Chapter 3: Blockchain for Smart Cities: Enabling Sustainable and Inclusive Urban Development

This chapter explores how blockchain can enable sustainable and inclusive smart cities amid rapid urbanization and limited resources. It highlights blockchain's core features—decentralization, transparency, and immutability—and their applications in governance, energy, urban planning, and transportation. Case studies from Dubai, Zug, and Telangana showcase real-world impacts in healthcare, public services, and land registries. The chapter also discusses challenges such as scalability and energy consumption, alongside emerging trends like DeFi, DAOs, and integration with AI, IoT, and 5G. Together, these insights demonstrate blockchain's potential to enhance transparency, citizen engagement, and resilience in future urban environments.

SECTION 2: SMART CITY APPLICATIONS

This section highlights how AI, IoT, blockchain, and renewable energy are transforming healthcare, logistics, energy, urban planning, and agriculture. These chapters showcase practical models for building efficient, sustainable, and resilient cities.

Chapter 4: AI-Enabled Applications of Medical Internet of Things in Smart Hospitals

This chapter examines the integration of AI and the Medical Internet of Things (MIoT) in advancing smart hospitals. It highlights how AI enables predictive care, disease tracking, and data-driven decision-making, while IoT tools enhance efficiency through asset tracking, energy monitoring, and workforce management. Case studies from South Korea and the UK demonstrate reduced readmissions, lower healthcare costs, and improved patient outcomes. The chapter also addresses challenges such as interoperability, scalability, and data security, proposing solutions like cloud/edge computing, blockchain, and global standards. Finally, it explores future trends, including 5G connectivity, AI-driven automation, and robotic support, pointing toward a more sustainable, patient-centered, and technology-enabled healthcare system.

Chapter 5: Digitalization and Automation of Logistics and Supply Chain Management in the Industry 4.0 Era

This chapter explores how Industry 4.0 is driving the digitalization and automation of logistics and supply chain management (SCM). It highlights the role of IoT, big data analytics, artificial intelligence, blockchain, and automation in enhancing efficiency, transparency, and decision-making. By focusing on predictive analytics, digital twins, and data-driven models, the chapter shows how digital transformation enables real-time optimization, operational agility, and sustainable supply chains, offering practical insights for researchers, professionals, and policymakers.

Chapter 6: Design and Analysis of a Solar-Assisted Combined Cooling, Heating, and Power System for Smart Cities: Case Study from Doha

This chapter explores the design and evaluation of a solar-assisted combined cooling, heating, and power (SA-CCHP) system tailored for Doha, Qatar, a city with high solar potential and significant cooling demand. Using a parabolic trough collector (PTC) field as the sole energy source, the system demonstrates strong technical and economic performance, achieving seasonal variations in power, cooling, and heating output while improving overall efficiency and reducing CO_2 emissions. The findings highlight the feasibility of solar-powered CCHP systems as scalable, sustainable solutions for smart cities aiming to meet clean energy targets.

Chapter 7: Utilizing IoT and AI to Enhance Urban Planning and Development

This chapter explores the role of the IoT and machine learning (ML) in advancing the vision of smart cities. It highlights how IoT devices collect real-time data that, when combined with ML analytics, enable predictive maintenance, anomaly detection, and dynamic decision-making. The chapter focuses on practical applications in smart transportation, energy management, sustainability, public safety, and building management, while also emphasizing the importance of citizen engagement. By integrating IoT and ML, the study illustrates future trends in smart urban planning and the creation of more efficient, resilient, and sustainable cities.

Chapter 8: Centering Ground Water Scarcity Ensuring Futuristic Sustainable Farming with Blockchain and IoT

This chapter addresses the critical issue of groundwater depletion and its implications for agricultural sustainability, particularly in irrigated regions. It explores how IoT-powered sensors enable real-time monitoring of water use, soil moisture, and environmental conditions, supporting more efficient irrigation management. The discussion also highlights the role of decentralized smart contracts in promoting fair water use rights and incentivizing conservation among farmers. By combining technological innovation with sustainable practices, the chapter offers policy recommendations and practical insights for advancing smart irrigation systems and safeguarding water resources for future generations.

SECTION 3: CHALLENGES IN SMART CITIES

This section addresses the multifaceted challenges smart cities face as they evolve toward greater connectivity, sustainability, and resilience. It examines issues ranging from homeowners' adoption of smart housing and the affordability-value tradeoff to the critical role of next-generation wireless systems in disaster management. It further explores the complexities of integrating geospatial and computing technologies, as well as the governance of environmental data for regulatory compliance. Together, these chapters highlight the technological, social, and policy hurdles that must be addressed to build secure, inclusive, and future-ready smart cities.

Chapter 9: Are you Ready for your Future Home?: Modeling the Intention to Purchase Homes in Smart Cities

This chapter investigates the factors influencing homeowners' intentions to adopt smart city housing, applying the Value-based Adoption Model (VAM) to explore the role of perceived price value. Using Malaysia as a case study, the findings show that environmental initiatives, convenience, and IT resources positively affect adoption through value perceptions, while privacy and security concerns have little impact. The study provides early empirical evidence on how affordability and value-for-money drive smart city home adoption, offering practical guidance for policymakers and city planners.

Chapter 10: Role of Next Generation Advanced Wireless Techniques for Disaster Management in Smart Cities

This chapter examines how advanced communication technologies, including 5G and B5G, can strengthen disaster management in smart cities. It highlights drone-assisted device-to-device (D2D) communication and rate-splitting multi-user cooperation as effective approaches to ensure connectivity, optimize bandwidth, and enhance reliability during crises. The findings demonstrate how these innovations improve outage probability, data rates, and throughput, offering practical solutions for resilient and sustainable smart cities.

Chapter 11: Geospatial Data Complexities and Technology Challenges in Smart Cities

This chapter explores the integration of geospatial and computing technologies in smart cities, focusing on their role in improving urban services and sustainability. It highlights the potential of Geospatial AI while addressing challenges related to data quality, scalability, privacy, interoperability, workforce skills, and legal-ethical concerns. By outlining these barriers, the chapter provides actionable insights for researchers, policymakers, and administrators working toward future-ready urban systems.

SECTION 4: FUTURE PERSPECTIVES

This section explores the future trajectories of smart cities, focusing on global strategies, sustainability, security, and the next stage of urban intelligence. The chapters examine lessons from leading cities, and the critical role of data privacy and security. They also introduce the emerging concept of cognitive cities, where AI and advanced computing enable adaptive, self-learning urban ecosystems. Together, these perspectives provide guidance for shaping resilient, secure, and future-ready cities.

Chapter 12: Data Privacy and Security in Smart Cities: Avoiding Problems and Resolution

This chapter examines the critical relationship between data privacy, security, and the development of smart cities, recognizing data as the key driver of urban innovation. It explores the nature of collected data, the privacy challenges it raises, and lessons from real-world breaches, supported by legal and ethical analysis. Security concerns are addressed through discussions of risks, vulnerabilities, and regulatory frameworks, alongside proposed interventions such as encryption, blockchain, privacy-by-design, and AI-driven safeguards. By outlining both current challenges and future directions, the chapter provides practical insights and policy recommendations for building secure, efficient, and trustworthy smart city frameworks.

Chapter 13: Global Perspectives on Smart Cities: Strategies, Technologies, and Lessons Learned

This chapter traces the evolution of smart cities and their role in addressing global urban challenges through technology and sustainability-focused frameworks. It analyzes technological, economic, social, and governance dimensions while comparing five leading cities—Zurich, Oslo, Canberra, Geneva, and Singapore. By aligning initiatives with key UN Sustainable Development Goals, it highlights impacts on healthcare, energy, mobility, inclusion, and environmental resilience. The chapter concludes with lessons on policy, ethics, and collaboration, offering practical insights for researchers, policymakers, and urban planners dedicated to building intelligent and equitable cities.

Chapter 14: From Smart to Cognitive Cities: The Next Evolution in Urban Intelligence

This chapter examines the transition from smart cities to cognitive cities, where artificial intelligence, machine learning, and cognitive computing enable self-aware and adaptive urban ecosystems. It distinguishes cognitive cities from traditional smart cities, highlighting applications in mobility, public services, and citizen engagement, supported by global case studies. Ethical and governance issues such as data privacy and transparency are addressed, alongside current challenges and future research directions.

Mohammed Balfaqih
University of Jeddah, Saudi Arabia
Akram Almohammedi
Karabük University, Turkey
Malek Alsmadi
Lakehead University, Canada

Chapter 1
Internet of Things as a Core Enabler for Smart Cities:
Architectures, Technologies, and Applications

Mohammed Balfaqih
https://orcid.org/0000-0002-6498-2408
University of Jeddah, Saudi Arabia

ABSTRACT

This chapter provides a comprehensive exploration of foundational technologies that underpin the development of smart cities. It discusses the pivotal role of the Internet of Things (IoT) as a fundamental component enabling intelligent urban infrastructure. Key IoT architectures, including both functional and sensing-to-cloud frameworks, are analyzed to highlight their significance in enhancing urban management, sustainability, and quality of life. The chapter further examines practical IoT applications across various sectors such as industrial and manufacturing, consumer technology, retail, healthcare, transportation, agriculture, energy, and government operations. Additionally, critical technological challenges and security considerations associated with large-scale IoT deployments are reviewed, emphasizing the importance of robust standards and protocols to ensure data integrity, reliability, and overall system efficiency.

DOI: 10.4018/979-8-3373-0154-9.ch001

INTRODUCTION

The notion of the smart city emerged prominently during the 1990s as urban planning increasingly integrated technology, innovation, and global interconnectedness (Gibson et al., 1992). This concept gained significant traction following IBM's introduction of the Smarter Planet initiative in 2008 (Palmisano, 2008). Although extensively discussed across numerous publications, scholars and practitioners have yet to agree upon a singular, universally accepted definition of what constitutes a smart city. Most interpretations, however, emphasize leveraging information and communication technologies (ICT) as essential tools to enhance urban services and decision-making capabilities, ultimately aiming to improve residents' quality of life (Law & Lynch, 2019).

For example, one viewpoint describes a smart city as one that employs ICT to systematically gather, analyze, and utilize essential information from fundamental urban infrastructures to deliver intelligent, efficient, and responsive everyday services and automation (Su, Li, & Fu, 2011). Another perspective suggests that a city attains smart status when diverse ICT applications effectively enhance the daily experiences and overall wellbeing of its inhabitants (Syed et al., 2021).

Achieving the status of a smart city requires cohesive integration and communication among various urban systems to holistically address city-wide needs and challenges (Monzon, 2015). Smart cities strategically allocate investments toward sustainable economic development, improving living standards, and responsibly managing environmental resources, guided by inclusive and participatory governance practices (Caragliu et al., 2011). In general, smart city concepts typically fall into two distinct frameworks. The first emphasizes technological advancement, infrastructure efficiency, energy management, environmental sustainability, and mobility solutions. The second focuses primarily on enhancing citizen engagement, educational opportunities, social cohesion, effective governance, and cultural creativity. Thus, reaching a unified and comprehensive definition of a smart city remains a critical step forward for urban development discourse (Abadía et al., 2022).

Currently, professionals including technologists, engineers, and developers are actively pursuing advanced IoT-based solutions aimed at enhancing the structural design and functionality of smart cities. Figure 1 shows the development of smart city frameworks involving three principal stages: firstly, the collection and extraction of valuable insights from urban data; secondly, the systematic storage, management, and analysis of this information to facilitate informed, autonomous decision-making; and thirdly, the practical implementation of these decisions through efficient urban services (Nathali Silva et al., 2017).

IoT infrastructures are rapidly advancing due to their increasing significance in both academic research and industrial applications focused on smart city initiatives

(Petrolo et al., 2017). With the steady rise in urban populations, significant challenges are emerging for urban governance and infrastructure. According to projections by the United Nations Population Fund, the proportion of people residing in urban areas, which was 50% in 2007, is expected to experience substantial growth by 2030 (Naphade et al., 2011). Consequently, IoT-based infrastructures are anticipated to become integral components across various smart city applications, benefiting both businesses and city residents.

Figure 1. The development of smart city frameworks.

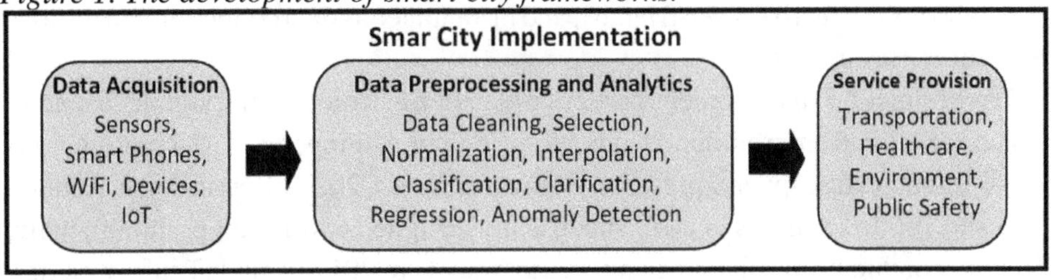

This chapter is designed to offer an in-depth exploration of the foundational role of IoT technologies in the development of smart cities. The primary objectives addressed in this research include:

- Evaluating current IoT architectures, highlighting functional and sensing-to-cloud frameworks and their impact on urban sustainability and quality of life.
- Exploring various enabling technologies that facilitate the successful integration of IoT within smart city infrastructures.
- Analyzing diverse IoT applications across critical sectors such as industrial and manufacturing, consumer products, retail, healthcare, transportation, agriculture, energy, military, and government operations.

The chapter begins by contextualizing the significance of IoT in smart city initiatives, followed by detailed discussions on key architectural designs, including far-edge, pan-LAN, WAN, and cloud layers. Essential enabling technologies, such as RFID, smart sensors, communication protocols, and data management tools, are elaborated upon. Practical applications within critical sectors are thoroughly reviewed, accompanied by analyses of real-world implementation examples. Additionally, significant challenges, including security considerations and data integrity, are assessed, emphasizing the importance of adhering to international standards and protocols. The chapter concludes with summarizing insights and highlighting the implications for future smart city developments.

INTERNET OF THINGS (IOT)

The IoT significantly contributes to implementing and enhancing smart city initiatives worldwide. Prominent cities such as Singapore, London, and Helsinki heavily depend on IoT technologies to improve urban living standards and sustainability (Balfaqih & Alharbi, 2022). IoT enables cities to monitor essential factors such as energy usage, water quality, and traffic conditions, facilitating early identification of potential issues and proactive mitigation strategies (Sumi & Ranga, 2016). By integrating countless connected sensors and devices online, IoT solutions provide cities with cost-effective, user-friendly, and instantaneous data collection capabilities (Salman & Hasar, 2023).

Environmental management particularly benefits from IoT through its capacity to monitor critical environmental indicators, including temperature fluctuations, gas emissions, and atmospheric pressure (Salman & Hasar, 2023). For instance, London deployed 100 IoT sensors in 2019 specifically designed to track air pollution, while Helsinki conducted similar air-quality assessments in 2017 using IoT infrastructure (Vodák et al., 2021). Such real-time data collection enables optimized energy usage and more efficient allocation of public services. In Singapore, IoT systems are integral to waste management processes, enhancing the effectiveness of waste treatment and significantly cutting down greenhouse gas emissions (Huseien & Shah, 2021).

Moreover, IoT technology strengthens urban disaster preparedness and risk management. Practical deployments illustrate IoT's effectiveness in detecting environmental hazards, issuing timely alerts, and informing relevant disaster response agencies (Munawar et al., 2022). By 2020, effective IoT-driven early warning systems were operational in 23 United Nations member states, safeguarding approximately 93.63% of the vulnerable populations against natural disasters (Esposito et al., 2022). Consequently, IoT contributes substantially to environmental sustainability in smart cities.

IoT also significantly supports social sustainability, fostering greater transparency and trust between cities and their residents (Vodák et al., 2021). In cities like Singapore, London, and Helsinki, IoT platforms are citizen-centered, allowing transparent and reliable information dissemination through public digital channels, thereby improving civic trust and engagement (Vodák et al., 2021). Additionally, IoT infrastructure contributes to enhancing urban safety. In Singapore, sensor-equipped cameras deployed on streets for facial recognition significantly reduced theft and robbery incidents, achieving 322 consecutive crime-free days in 2018 (Vodák et al., 2021).

IoT further extends its benefits to healthcare by allowing real-time monitoring of patients' physiological data, thus facilitating early detection of medical emergencies and clinical deterioration (Catarinucci et al., 2015). IoT also proves valuable in disaster relief efforts, where rapid and accurate information is crucial for effective response operations (Sinha et al., 2019). Specifically, the initial 72 hours following a disaster, often termed the "golden hour" for rescue operations, are critical due to diminishing survival probabilities thereafter (Ochoa & Santos, 2015). IoT capabilities greatly enhance search and rescue effectiveness during this crucial period by detecting vital signs and accurately pinpointing locations of individuals in distress (Zeng et al., 2023).

IOT ARCHITECTURE

The IoT constitutes an integrated network composed of diverse interconnected devices, including RFID tags, NFC technology, sensors, actuators, among others. Each device within this network possesses a unique identifier (UID), facilitating efficient and precise data exchange and communication between devices. Designed to accommodate various industrial applications and operational requirements, IoT typically follows a multi-tiered architecture. Broadly, IoT architectures can be classified into two main types: functional-based architectures, focusing on specific operational roles, and sensing-to-cloud architectures, encompassing the entire process from initial data collection to cloud-based data processing and analysis.

Sensing-to-Cloud Stages Architectures

Architects exploring IoT solutions must carefully evaluate numerous potential configurations. Figure 2 illustrates communication pathways between sensors and cloud infrastructures, depicting both direct sensor-to-cloud connections and those utilizing edge gateways (Lea, 2020).

Figure 2. Sensing-to-cloud stages architecture.

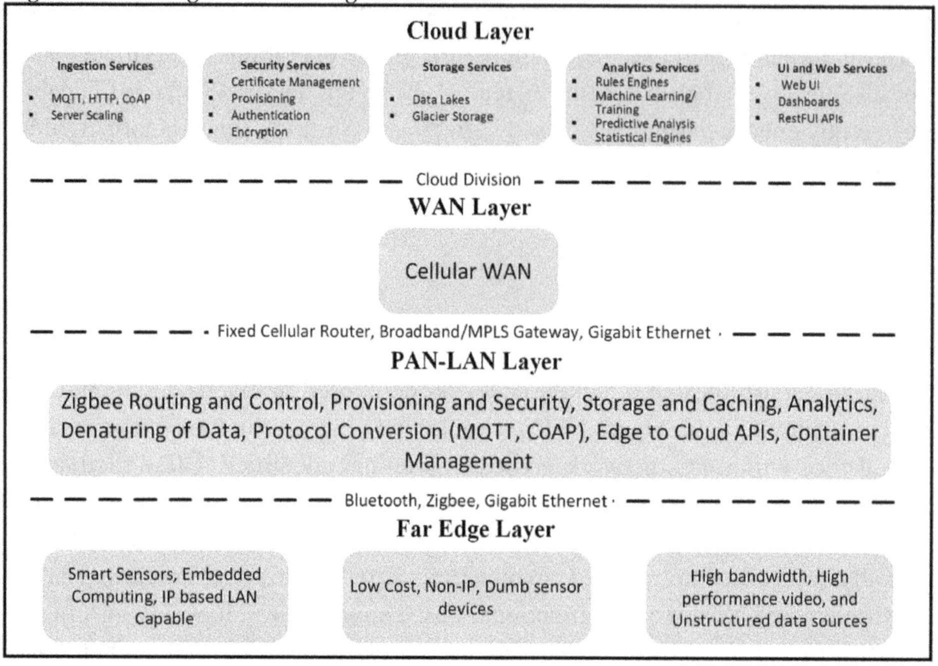

The diagram further emphasizes the distinct functionalities offered by edge computing nodes and cloud-based components. Key elements within a comprehensive IoT architecture that require thorough consideration include (Lea, P., 2020):

Far Edge Layer

The Far Edge layer in IoT architecture consists primarily of sensors, actuators, and physical systems that include embedded devices, real-time operating systems, energy-efficient components, and micro-electro-mechanical systems (MEMS). These components play a crucial role in capturing and responding to real-world data and interactions.

Communication within the Far Edge layer typically occurs through sensor communication networks, notably Wireless Personal Area Networks (WPANs). These networks operate over relatively short distances, often up to approximately 100 meters, and use low-power, bandwidth-efficient, and frequently non-IP-based communication protocols optimized for sensor data transmission.

Additionally, local area networks (LANs), predominantly IP-based solutions like Wi-Fi (IEEE 802.11), facilitate rapid and reliable data exchange in peer-to-peer or centralized network topologies. Essential to the functionality of the Far Edge layer are aggregation devices, routers, and gateways, which are embedded systems commonly supplied by economical hardware providers. These components aggregate and route data effectively, bridging sensor networks to broader IoT infrastructures.

PAN-LAN Layer

Decentralized processing infrastructure positioned closer to the data sources to reduce latency, enhance real-time responsiveness, improve redundancy, and address connectivity challenges. This includes analyzing hardware such as processors, memory, and storage; reviewing manufacturers for modules, passive components, thin clients, wireless radios, and cellular equipment; evaluating middleware platforms, fog computing solutions, edge analytics, edge security frameworks, certification management systems, WPAN-to-WAN protocol conversion, advanced routing methodologies, and software-defined network architectures.

Wide Area Networks (WANs)

Utilize cellular technologies (LTE, Cat M1), satellite communications, and low-power wide-area network (LPWAN) technologies like Sigfox or LoRa. WAN communication often employs IoT-specific protocols such as MQTT, CoAP, or even HTTP for constrained device environments.

Cloud Layer

The cloud layer in IoT architecture encompasses the infrastructure and services provided by cloud platforms, such as Infrastructure as a Service (IaaS) and Platform as a Service (PaaS). These services include database management systems, tools for both real-time and batch data processing, analytical software suites, Software as a Service (SaaS) applications, comprehensive data storage solutions like data lakes, and advanced machine learning capabilities. This layer serves as the central hub where extensive IoT data is stored, processed, and analyzed to deliver actionable insights.

Data analytics within the cloud layer is essential for deriving valuable insights from the massive quantities of IoT-generated data. Techniques used in this layer include complex event processing, statistical analysis, rules-based decision engines, and machine learning algorithms. These methodologies enable organizations to understand patterns, predict future trends, and make informed, data-driven decisions.

Security represents a critical aspect of the cloud layer and extends across the entire IoT architecture. Ensuring robust, end-to-end security involves securing hardware components, protecting physical devices, encrypting data communications, and employing secure protocols. Each level within the IoT system must prioritize authenticity, integrity, and confidentiality to safeguard against vulnerabilities that could lead to significant risks, particularly due to IoT's extensive and interconnected global nature.

Functional-based Architectures

Numerous IoT architectural frameworks have been proposed in academic research. Figure 3 highlights various structural configurations ranging from three to six distinct layers (Khalifeh et al., 2019, Lee et al., 2017, Atzori et al., 2012, Hassan et al., 2009). A six-layer architecture typically includes a coding layer dedicated to uniquely identifying smart objects; however, this layer can often be effectively integrated into the perception layer, particularly when advanced identification technologies such as blockchain are employed (Goyal et al., 2021, Younan et al., 2021).

Figure 3. Numerous IoT architectural frameworks

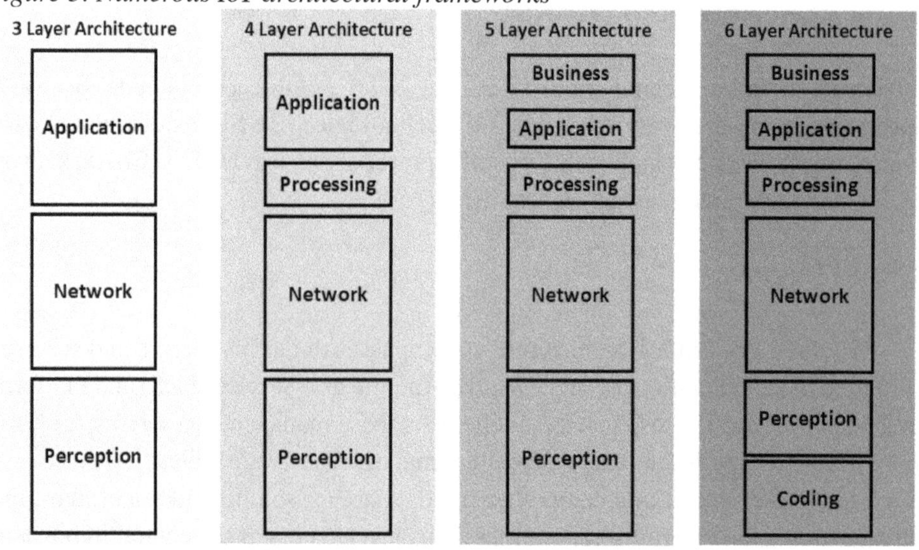

Consequently, five-layer architecture is frequently preferred due to its balanced approach, effectively addressing critical requirements for data security and privacy (Goyal, P., et al., 2021). These layers typically include perception, network, processing (cloud), application, and business layers. A brief overview of each layer is as follows:

Perception Layer

Also referred to as the sensor layer, this foundational layer facilitates interactions with physical devices through smart technologies such as RFID tags, sensors, and actuators (Syed et al., 2021). The primary responsibilities of the perception layer include connecting physical objects to the IoT ecosystem, capturing and preliminary processing of relevant data from these objects, and transferring this processed data upward through interfaces for further analysis.

Network Layer

Positioned centrally within the IoT architecture, this layer is responsible for receiving processed information from the perception layer and managing its routing and transmission to higher layers, including IoT hubs, devices, and applications (Abadía et al., 2022). The network layer is particularly crucial as it integrates diverse hardware components (such as gateways, switches, and cloud interfaces) and employs various communication technologies (e.g., Bluetooth, Wi-Fi, LTE) to facilitate seamless and reliable data transfer across heterogeneous networks.

Processing Layer

The processing layer plays a critical role in IoT-based smart city architectures, enabling meaningful insights to be drawn from large datasets, thus promoting informed decision-making and sustainable urban living. For example, smart meters installed across residential areas can provide valuable data on energy consumption patterns, enabling homeowners, utility companies, and service providers to dynamically manage energy usage, thereby improving overall efficiency.

Traditional computing architectures often rely on vertical scaling, which involves upgrading a single system with enhanced hardware components like increased memory, additional processors, or expanded storage. Although this approach can increase computational capabilities, it typically incurs high costs, introduces a single point of failure, and creates complex dependencies among software modules, negatively impacting system flexibility and maintainability.

Alternatively, distributed computing architectures offer a cost-effective and efficient solution by leveraging multiple moderate-capacity systems operating concurrently.

Several distributed computing frameworks have been designed to efficiently manage extensive data processing tasks (Noaman et al., 2022). Apache Hadoop, a prominent framework, employs batch processing through its MapReduce model (Hong & Shi, 2023). MapReduce distributes tasks across Hadoop Distributed File

System (HDFS) nodes, handling large-scale datasets in batches (Phuttharak & Loke, 2023). Its operations include mapping data into key-value pairs, grouping similar keys through shuffling, and reducing these groups into finalized outputs. Although effective, MapReduce often requires specialized programming expertise; however, SQL-based interfaces like Hive and SparkSQL simplify this by converting SQL queries into executable MapReduce operations (Fortino et al., 2020).

In contrast, stream processing frameworks, such as Apache Spark Streaming, perform continuous, real-time data analysis. Spark Streaming utilizes Resilient Distributed Datasets (RDDs) for efficient in-memory processing, represented via Directed Acyclic Graphs (DAGs) for rapid execution (Karunarathne et al., 2018, Chen et al., 2019). Unlike MapReduce, Spark avoids storing intermediate results, significantly reducing query response times. Additionally, Spark supports multiple programming languages, including Scala, Java, Python, and R, (Yaqoob et al., 2017).

Hybrid processing frameworks, combining both batch and real-time processing capabilities, offer versatile data management solutions. Apache Spark provides robust hybrid functionality through integrated batch and stream processing methods (Al-Ali et al., 2024). Another notable hybrid framework, Apache Flink, similarly manages simultaneous batch and streaming operations directly from HDFS (Al-Ali et al., 2024).

Online Analytical Processing (OLAP) supports complex queries and comprehensive data analysis through multidimensional data structures. OLAP arranges data into multidimensional cubes defined by measures (quantitative values) and dimensions (attributes for categorization), structured hierarchically to enhance query efficiency. Utilizing star schema designs, OLAP features a central fact table linked to multiple dimension tables, facilitating detailed analysis based on parameters like time or geographic location (Al-Ali et al., 2024).

Application Layer

The application layer occupies the topmost position in IoT architectures, primarily handling data received from the network layer (Al-Fuqaha et al., 2015). This layer leverages incoming data to execute specific services and functionalities. Tasks performed at this level include securely storing data in databases and performing advanced analytical functions to forecast device states and conditions. The diversity of IoT applications within this layer is significant, encompassing sectors such as smart grids, intelligent transportation systems, and smart city infrastructures, each with unique operational requirements (Tan & Wang, 2010, Wu et al., 2010).

Business Layer

The business layer manages the overall IoT ecosystem, including both services and applications. It utilizes data provided by the application layer to create strategic insights and operational models, such as business frameworks, visual flowcharts, and process diagrams. The efficacy and sustainability of IoT solutions greatly depend on the strategic decisions and business models crafted at this layer. By analyzing operational outcomes, the business layer guides future planning, informs strategic business decisions, and supports the long-term success and optimization of IoT implementations (Lin, J., et al., 2017).

ENABLING TECHNOLOGIES AND CHALLENGES

According to the structure illustrated in Figure 2, the implementation of IoT solutions relies on numerous essential enabling technologies. This section will comprehensively explore and elaborate on the significant technological advancements applicable to each of the five distinct layers within the IoT architectural framework. By examining these technologies, we aim to highlight their critical roles in effectively realizing and supporting the functionalities required for robust IoT ecosystems.

Perception Layer

Physical devices equipped with sensors, commonly referred to as smart devices or "things," include smartphones, laptops, and various electronic components capable of autonomously gathering environmental data without direct human involvement. These devices primarily utilize enabling technologies such as Radio Frequency Identification (RFID), Quick Response (QR) codes, and Infrared (IR) readers. RFID technology, in particular, can be classified into passive, semi-passive, and active categories. Passive RFID tags operate without an internal power source, relying instead on external energy and offering limited communication range. Semi-passive tags include batteries, enhancing their data read-write capabilities. In contrast, active RFID tags are equipped with robust batteries, enabling long-range communication suitable for real-time tracking applications (Wang et al., 2020).

Extending the functionality further, smart sensors represent devices specifically designed to perform continuous data collection autonomously, without human intervention. These sensors can be embedded in various platforms, ranging from computer systems to wearable health devices, such as fitness trackers that measure physiological signals like heart rate (Vinodhini et al., 2023). Smart sensors collaborate closely with smart devices, forming integrated systems designed to fulfill complex operational requirements (Astill et al., 2020).

Completing the IoT ecosystem are smart actuators—automated physical machines that act based on sensor data without human oversight. The integration of smart actuators significantly enhances operational efficiency, reduces labor costs, and optimizes resource utilization. In advanced manufacturing frameworks like Industry 4.0, smart actuators enable comprehensive automation throughout production facilities, minimizing the necessity for human intervention and increasing productivity and efficiency (Vinodhini et al., 2023).

Network Layer

The evolution of electronic devices has led to enhanced functionalities, primarily facilitated through IoT technologies and their dedicated wireless communication standards. IoT fundamentally relies on establishing seamless network connectivity between applications, enabling efficient data sharing. Communication protocols within IoT define essential functions such as data formatting, encoding, addressing, and routing, along with flow control, sequence management, and retransmission capabilities to recover lost data. One of the foundational standards supporting these functions is the IEEE 802.11, developed by the Institute of Electrical and Electronics Engineers, forming the basis for Wireless Local Area Networks (WLANs) (Beard & Stallings, 2002).

WiFi, adhering to IEEE 802.11 standards, comprises several versions, including 802.11a, b, g, and n, each providing varying operational characteristics. Utilizing orthogonal frequency-division multiplexing (OFDM), WiFi typically operates in frequency bands around 2.3 GHz and 4 GHz, with indoor ranges from approximately 125 feet (802.11g) to 235 feet (802.11n) (Beard & Stallings, 2002).

In contrast, Wi-Max technology, based on the IEEE 802.16 standards, supports wireless metropolitan area networks (WMANs) with enhanced capabilities in static, portable, and high-speed internet access. Wi-Max utilizes both licensed and unlicensed frequency bands ranging from 700 MHz to 5.8 GHz, also employing OFDM for effective communication (Beard & Stallings, 2002).

The IEEE 802.15.4 standard specifically addresses Low-Rate Wireless Personal Area Networks (LR-WPANs), designed to efficiently connect numerous low-power and low-storage devices. This standard is particularly useful for applications such as smart home automation and smart city implementations (Lea, 2020).

Cellular network standards have evolved significantly through generations, from GSM and CDMA (second generation) to UMTS and CDMA 2000 (third generation), and currently advanced 4G technologies including LTE. Emerging 5G technology utilizes advanced radio access technologies (RAT) optimized for IoT applications, operating in both sub-6 GHz and mm-wave bands (20-100 GHz). The introduction of 5G NR (New Radio) by 3GPP releases 15 and 16 supports advanced features such as massive MIMO and sophisticated waveforms, significantly enhancing IoT applications across various industries, including smart manufacturing and critical services (Wang et al., 2017, Almohammedi et al., 2023a).

Bluetooth technology, based on the IEEE 802.15.1 standard, facilitates short-range wireless communication within Personal Area Networks (PANs), typically offering ranges up to approximately 10 meters and bandwidth around 1-3 Mbps, making it suitable for low-power handheld devices (Lea, 2020).

Long Range (LoRa) is another vital technology leveraging spread-spectrum modulation within the 868 and 900 MHz ISM bands, catering to IoT applications in smart logistics and supply chains. LoRa ensures efficient data communication with a low data rate but significantly extends battery life, making it suitable for various long-term deployments.

Light Fidelity (LiFi) technology utilizes Visible Light Communication (VLC) through LEDs, offering high-speed data transmission while minimizing energy consumption. Although LiFi delivers efficient data rates, its effectiveness can be limited by interference and signal loss, restricting its use to specific loss-tolerant applications (Lea, 2020).

Effective management of extensive data generated by IoT networks necessitates robust application-layer protocols. Protocols such as CoAP, MQTT, XMPP, and AMQP facilitate data management using publish/subscribe or request/response mechanisms, depending on system requirements. CoAP, built for constrained devices with low power consumption, operates over UDP and supports multicast messaging. MQTT, designed by IBM, reduces bandwidth consumption and network traffic overhead, using TCP/IP to guarantee reliable packet delivery across varying quality-of-service levels. XMPP, a lightweight and extensible protocol, excels in real-time communication scenarios, including instant messaging and multimedia conferencing. Lastly, AMQP provides secure, interoperable communication with guaranteed message delivery, primarily designed for asynchronous messaging and management of IoT applications, enabling reliable client-server interactions (Lea, 2020).

Collectively, these protocols and standards form a comprehensive IoT communication infrastructure, ensuring efficient, reliability, and secure connectivity across diverse applications and environments.

Processing Layer

The global infrastructure layer represents the third tier in IoT architecture, focusing on enhancing performance through centralized data processing. Depending on the required performance metrics, nodes can be deployed either in cloud or fog computing environments. Cloud computing is commonly employed when data sources and applications can easily communicate with centralized services. However, when nodes are significantly distant from cloud data centers, fog computing offers a more suitable alternative by reducing latency and providing localized computational resources (Lea, 2020).

Cloud computing serves as the backbone for many IoT applications, integrating a diverse set of digital tools and services. It efficiently collects and analyzes sensor data originating from dispersed geographical locations. The flexibility and scalability of cloud infrastructure allow for on-demand data processing, storage, and analysis, facilitating comprehensive insights into customer and system behavior. Cloud providers such as Kaa and DeviceHive offer Platform as a Service (PaaS) tailored for IoT, enabling IoT service providers to leverage Infrastructure as a Service (IaaS) to deliver Software as a Service (SaaS) solutions (Lea, 2020).

Fog computing complements cloud computing by bringing computational capabilities closer to the data sources, significantly reducing data transmission latency and enhancing scalability. Fog computing is particularly advantageous in scenarios involving extensive geographical distribution or mobility, connecting numerous devices efficiently and maintaining consistent service delivery (Al-Ali et al., 2024).

In managing vast data volumes, IoT implementations utilize robust big data ingestion and streaming tools such as Apache Flume and Apache Nifi. Apache Flume excels at reliably collecting, aggregating, and transferring extensive data streams to Hadoop for further storage and analysis, ensuring fault tolerance through its distributed architecture. Apache Nifi provides an intuitive and reliable platform for secure data distribution and processing, offering advanced monitoring, control, and secure data encryption via SSL, HTTPS, and SSH (Al-Ali et al., 2024).

Hadoop Distributed File System (HDFS), a Java-based storage solution, provides a secure and scalable environment for handling extensive datasets. By employing a master/slave architecture, HDFS facilitates parallel data processing across multiple nodes, ensuring redundancy and reliability through distributed storage. Gluster File System (GlusterFS) is another open-source, scalable, distributed storage solution offering high performance for unstructured data management. Its ability to handle large-scale deployments makes it suitable for cloud computing, media streaming, and content distribution applications (Al-Ali et al., 2024).

Amazon Simple Storage Service (Amazon S3) provides a reliable, scalable object storage web service suitable for cloud-based applications. Known for high durability, ease of data migration, and cost-efficiency, Amazon S3 efficiently manages large data repositories while ensuring robust fault tolerance and recovery mechanisms (Al-Ali et al., 2024).

Application Layer

Application layer forms the foundational support for IoT applications, providing essential services required by end-users through various smart devices such as smartphones, wearables, and smart home gadgets. Key functionalities delivered at this layer include identity management, data aggregation, and collaborative interactions among IoT devices (Houssein et al., 2024).

Identity management services at this layer continuously track and maintain user or device identities. For instance, medical records can be digitized using RFID technology, assigning a unique digital identity to each patient. This information is securely captured, stored, and updated within databases, accessible through web or mobile platforms for real-time health record management (Houssein et al., 2024).

Data aggregation services facilitate the consolidation and analysis of information received from multiple IoT sensors. An example of this service is the management of electrical load distribution within smart grid systems, where sensor data from diverse sources is analyzed to optimize energy distribution efficiently (Lea, P., 2020). Collaborative services enable seamless interaction and data sharing between multiple smart devices within a specific environment. Smart home automation exemplifies this functionality, allowing various household devices to communicate and coordinate tasks, offering users complete, real-time control and accessibility (Jabbar et al., 2019).

Business Layer

Implementing IoT in business environments involves strategic interactions and critical decision-making processes that leverage advanced computing capabilities. Ensuring security and maintaining data integrity become paramount, especially when IoT solutions are deployed extensively across various sectors. Effective security measures must be integrated throughout all technological components to mitigate vulnerabilities and reduce risks associated with potential security breaches (Houssein et al., 2024).

To guarantee adherence to security standards, IoT deployments should comply with international guidelines such as those established by the International Electrotechnical Commission (IEC). The IEC provides comprehensive standards and conformity assessments to ensure reliability and safety in electrical and electronic technologies. Additionally, the National Institute of Standards and Technology (NIST) continuously works towards enhancing security by formulating and promoting updated standards. This institution plays a crucial role in developing robust practices to protect IoT environments from emerging security threats (Houssein et al., 2024).

Security professionals are tasked with continually monitoring systems, swiftly identifying vulnerabilities, and evaluating their potential impacts. Upon discovering vulnerabilities, these security teams must promptly implement corrective actions, including debugging, patching, and updating the systems to minimize any negative impacts on end users. Therefore, establishing a comprehensive security framework is essential before deploying any IoT device in real-world scenarios. This structured approach ensures immediate action and effective risk management once vulnerabilities are identified.

IOT APPLICATIONS FOR SAMR CITIES

This section explores the diverse applications of IoT within smart city contexts, highlighting their transformative roles across various sectors. The areas covered include industrial and manufacturing processes, consumer products, retail, finance and marketing strategies, healthcare systems, transportation and logistics, agricultural and environmental management, energy utilization, as well as military and governmental functions.

Industrial and Manufacturing

The Industrial Internet of Things (IIoT) represents a rapidly expanding segment within the broader IoT ecosystem, significantly impacting manufacturing efficiency and factory automation. Traditionally rooted in operational technology (OT), IIoT integrates specialized hardware and software solutions designed to monitor and manage physical industrial processes in real time. Historically, these tasks have relied on on-site computing systems, typically managed through supervisory control and data acquisition (SCADA) frameworks (Lea, 2020).

Distinct differences have traditionally separated OT from information technology (IT). Operational technology primarily emphasizes real-time metrics such as production yields, equipment uptime, and operational safety, whereas IT focuses predominantly on data security, structured data management, and service delivery. However, with the increasing adoption of IoT solutions in industrial environments, these previously separate domains are merging, notably through predictive maintenance applications that generate vast quantities of operational data, subsequently managed via cloud-based infrastructures (Lea, 2020).

Critical to the IIoT sector is the requirement for near-instantaneous or real-time decision-making capabilities, underscoring latency reduction as a central priority on the production floor. Furthermore, mitigating downtime and ensuring robust cybersecurity are essential, leading to investments in redundant systems and potentially private cloud solutions tailored for industrial contexts. A notable challenge unique to this sector is integrating legacy or brownfield equipment, often employing outdated communication protocols such as RS485 serial connections, rather than contemporary wireless technologies. Several key applications illustrate IIoT's profound impact on industrial efficiency and safety, including:

- Predictive and preventative maintenance for both new and legacy manufacturing equipment
- Enhanced production throughput via real-time demand analysis
- Significant reductions in energy consumption
- Implementation of safety measures including thermal detection, pressure monitoring, and gas leak identification
- Deployment of expert systems to optimize factory floor operations

Consumer

Consumer-oriented devices were among the earliest adopters of internet connectivity, marking the inception of consumer IoT with innovations such as an internet-connected coffee machine (Alzhrani et al., 2024). The proliferation of consumer IoT accelerated notably with the widespread adoption of Bluetooth technology in the early 2000s.

Today, IoT-enabled products such as Nest thermostats, Hue lighting systems, Alexa virtual assistants, and Roku streaming devices are commonplace in millions of homes. Additionally, wearable devices like Fitbit have significantly connected people personally and continuously. The consumer market typically acts as an early adopter of emerging IoT technologies, popularizing user-friendly, straightforward devices designed for immediate use (Almohammedi et al., 2023b).

One significant challenge within the consumer IoT market is the existence of multiple competing communication standards, including Bluetooth, Zigbee, and Z-wave, each of which lacks interoperability.

Though consumer IoT devices share certain similarities with healthcare-oriented IoT solutions—such as wearable fitness trackers and home medical monitors—the two sectors are typically viewed separately, given that healthcare IoT extends beyond basic fitness tracking and personal health devices. Prominent consumer IoT applications include:

- Smart home technology: intelligent irrigation systems, automated garage doors, smart locks, lighting controls, thermostats, and security systems.
- Wearable technology: fitness trackers, health monitoring devices, and advanced smart clothing.

- Pet care solutions: GPS-enabled pet tracking systems and automated pet entry devices.

Retail, Finance, and Marketing

This sector encompasses environments dedicated to consumer-focused commerce, ranging from traditional retail establishments to temporary retail setups such as pop-up shops. It also extends to financial institutions like banks and insurance companies, along with leisure and hospitality services. IoT adoption in the retail industry primarily aims to enhance customer experiences while simultaneously reducing operational costs (Balfaqih et al., 2023). To maintain clarity and consistency, marketing and advertising applications are included within this sector.

The effectiveness of IoT solutions within retail, finance, and marketing is often evaluated based on their direct financial impact. IoT implementations are rigorously assessed to ensure they effectively reduce expenses or significantly increase revenue. By enhancing operational efficiency and customer engagement, IoT technologies enable businesses to offer superior service, improve consumer satisfaction, and optimize cost management (Khayyat et al., 2024). Key IoT applications within this sector include:

- Targeted marketing through proximity detection of current or potential customers, enabling personalized promotions.
- Beacon technologies for customer proximity detection, analyzing customer flow patterns, and improving marketing strategies.
- Asset tracking systems designed to manage inventory efficiently, minimize loss, and streamline supply chain operations.
- Monitoring systems for cold storage, employing predictive analytics to optimize the management and preservation of perishable goods.
- Insurance applications for tracking and managing assets, including assessing driver-related risk profiles.
- Digital signage solutions for enhancing consumer interactions within retail, hospitality venues, or public spaces.
- Beacon-based solutions within entertainment venues, conferences, amusement parks, concerts, and museums for improved customer engagement and navigation.

Healthcare

The healthcare sector is emerging as one of the leading industries in terms of IoT-generated revenue and overall societal impact, closely rivaling manufacturing and logistics (Islam et al., 2024). Enhancing patient outcomes and reducing healthcare expenditures remain paramount objectives in developed countries, driving the widespread adoption of IoT solutions. IoT technologies facilitate remote, continuous patient monitoring, significantly expanding healthcare accessibility and flexibility (Hafiz et al., 2024).

Integration of advanced analytics and machine learning algorithms enables continuous monitoring and precise diagnosis, ensuring timely and effective treatment recommendations. These intelligent systems also serve as vital early-warning mechanisms, capable of detecting and responding rapidly to critical health conditions. The wearable health device market, already numbering hundreds of millions of units globally, is expected to experience substantial growth in the near future (Hafiz et al., 2024).

Implementing IoT in healthcare poses significant challenges, notably stringent regulatory requirements, including HIPAA compliance, as well as data security and privacy concerns. IoT medical solutions must maintain hospital-grade reliability and security, providing seamless, uninterrupted communication between remote patient-monitoring locations and healthcare facilities. Moreover, these systems must reliably function across varied environments, including patient homes, emergency response vehicles, and hospitals. Key IoT healthcare applications include (Lea, 2020):

- Home-based patient care and remote monitoring.
- Predictive analytics and preventative healthcare solutions.
- Monitoring and tracking systems for dementia patients and elderly care.
- Asset management for hospital equipment and medical supplies.
- Secure tracking of pharmaceuticals.
- Remote medical support and telemedicine services.
- Advanced research in pharmaceutical drug development.
- Fall detection and emergency response systems for patient safety.

Transportation and Logistics

The transportation and logistics sector is poised to become a major driver of IoT innovation, significantly influencing both operational efficiency and safety standards (Balfaqih et al., 2021). IoT technologies in this field primarily focus on real-time tracking and management of assets across various transportation modes, including trucks, trains, aircraft, and maritime vessels. Furthermore, IoT applications encompass connected vehicle solutions designed to enhance driver assistance, improve vehicle safety, and facilitate proactive maintenance (Khayyat et al., 2024).

Modern vehicles typically incorporate numerous sensors, a figure expected to substantially increase with the widespread adoption of vehicle-to-vehicle (V2V), vehicle-to-infrastructure (V2I), and autonomous driving technologies (Balfaqih et al., 2017). Beyond passenger vehicles, IoT solutions extend into critical sectors such as rail transport and commercial shipping, where operational downtime has significant financial implications. Additionally, IoT technologies enable specialized service vehicles to effectively monitor tools, construction machinery, and other high-value assets (Shepelev et al., 2023).

Transportation and logistics IoT solutions address essential operational needs, such as optimizing fleet deployment based on demand, automating vehicle routing, and enhancing asset management. Central to these IoT applications is precise geolocation tracking, often enabled through GPS technology, which provides crucial data points including asset location, transit duration, and geographical positioning (Khayyat et al., 2024) . Key IoT applications in transportation and logistics include:

- Real-time fleet tracking and geolocation services.
- Municipal fleet management, including optimized routing for snow removal and waste management.
- Safety monitoring for refrigerated transport and perishable goods delivery.
- Rail transportation asset identification and tracking systems.
- Package and asset tracking within transportation fleets.
- Proactive vehicle maintenance to minimize roadside breakdowns and enhance fleet reliability.

Agricultural and Environment

The agricultural and environmental sector leverages IoT solutions to enhance productivity, livestock health management, land and soil assessments, microclimate forecasting, optimal water resource management, and disaster prediction related to geological and meteorological events. Despite a slowdown in global population growth, rising economic prosperity continues to drive increased food demand, expected to nearly double by 2035. IoT technologies provide substantial opportunities to achieve efficiencies and higher productivity in agriculture (Balfaqih, 2023).

For instance, smart lighting systems capable of adjusting spectrum frequencies based on the age and growth stage of poultry can significantly improve growth rates and reduce mortality by minimizing stress. Transitioning from traditional incandescent lighting to smart lighting systems can also yield substantial energy savings, potentially reducing costs by approximately $1 billion annually. IoT sensors and devices further contribute to livestock management by monitoring animal movements and behaviors to detect early signs of illness, thereby preventing the spread of disease within herds through real-time analytics and predictive models.

A notable challenge in agricultural and environmental IoT applications arises from the remote or sparsely populated locations, such as volcanic regions or expansive agricultural fields, affecting the choice and implementation of suitable data communication solutions. Effective communication technologies tailored to these remote settings are essential and discussed further in subsequent chapters addressing non-IP-based wireless personal area networks (WPAN) and long-range communication systems. Prominent agricultural and environmental IoT use cases include:

- Precision irrigation and fertilization methods to enhance crop yields.
- Smart lighting systems in poultry and livestock environments to optimize production.
- Real-time tracking and health monitoring of livestock assets.
- Predictive maintenance of remotely located agricultural equipment.
- Drone-enabled surveys for precise land and crop assessment.
- Efficient farm-to-market supply chain management through asset tracking.
- Implementation of robotic solutions for farming automation.
- Monitoring volcanic activities and geological fault lines to predict and mitigate natural disasters.

Energy

The energy sector extensively leverages IoT technologies for comprehensive monitoring, spanning from the point of production to final consumer consumption. Significant research and innovation have been directed toward developing advanced energy-monitoring devices, including intelligent electricity meters capable of transmitting real-time data through efficient, long-range communication protocols (Abbas et al., 2023).

Energy production sites frequently exist in isolated or challenging locations, such as desert solar installations, wind turbines on mountainous terrains, or nuclear reactors in hazardous environments. The requirement for timely or instantaneous feedback in critical energy infrastructure control systems necessitates specific considerations when implementing IoT solutions in this field. Key IoT applications within the energy sector encompass:

- Comprehensive monitoring and analysis of sensor data from oil drilling platforms to enhance operational efficiency.
- Remote maintenance and performance monitoring of solar installations.
- Safety and hazard assessments in nuclear power plants.
- Deployment of smart electricity, gas, and water meters across urban areas to efficiently track consumption patterns and demand.

- Implementation of dynamic, usage-based tariff structures.
- Adaptive real-time blade positioning for wind turbines based on weather conditions to maximize energy production and equipment safety.

Military and Government

IoT technologies are increasingly essential for city, state, federal governments, and military organizations due to their potential to enhance operational efficiency, environmental sustainability, and public safety. For instance, the State of California's ambitious environmental targets, such as executive order B-30-15 aiming for a 40% reduction in greenhouse gas emissions below 1990 levels by 2030, rely heavily on IoT-enabled environmental monitoring, intelligent energy management, and adaptive technologies to meet sustainability goals without compromising economic vitality (Lea, 2020).

Military and defense sectors similarly capitalize on IoT capabilities through initiatives like the "Internet Battlefield of Things," which enhances strategic capabilities by deploying networked sensors and devices to improve situational awareness and countermeasure efficiency. Additionally, IoT solutions significantly contribute to smart city developments by facilitating real-time monitoring and maintenance of critical public infrastructure such as roads and bridges (Lea, 2020).

Governmental involvement in IoT also extends to regulatory responsibilities, including standard-setting, managing frequency spectrum allocations, and establishing compliance frameworks. The effective partitioning and safeguarding of frequency spectrums for diverse IoT applications reflect these regulatory roles, significantly influencing technology adoption and innovation (Lea, 2020). Prominent government and military IoT applications include:

- Analysis of potential terrorist threats using IoT-generated data patterns.
- Deployment of drone swarms equipped with sensor arrays for reconnaissance.
- Utilization of networked sensor devices strategically positioned in combat scenarios to detect and monitor enemy activities.
- Asset tracking systems employed by governmental bodies.
- Real-time location tracking and monitoring of military personnel.
- Synthetic sensing technologies to assess hazardous environments.
- Continuous monitoring of water levels in dams and flood-prone areas for proactive disaster management.

CONCLUSION

This chapter provided a comprehensive exploration of foundational technologies underpinning the development of smart cities. It discussed the pivotal role of IoT as a fundamental component that enabled intelligent urban infrastructure. Key IoT architectures, including both functional and sensing-to-cloud frameworks, were analyzed to highlight their significance in enhancing urban management, sustainability, and quality of life. The chapter further examined practical IoT applications across various sectors, including industrial and manufacturing, consumer technology, retail, healthcare, transportation, agriculture, energy, and government operations. Additionally, critical technological challenges and security considerations associated with large-scale IoT deployments were reviewed, emphasizing the importance of robust standards and protocols to ensure data integrity, reliability, and overall system efficiency.

REFERENCES

Abadía, J. J. P., Walther, C., Osman, A., & Smarsly, K. (2022). A systematic survey of Internet of Things frameworks for smart city applications. *Sustainable Cities and Society*, *83*, 103949.

Abbas, A. B., Almohammedi, A. A., Balfaqih, M., & Darshi, S. (2023, September). Conceptual Design of Wireless Smart Grid for the Optimization of Electric Transmission in Iraq. In *2023 3rd International Conference on Computing and Information Technology (ICCIT)* (pp. 577-580). IEEE.

Al-Ali, A. R., Gupta, R., Zualkernan, I., & Das, S. K. (2024). Role of IoT technologies in big data management systems: A review and Smart Grid case study. *Pervasive and Mobile Computing*, 101905.

Al-Fuqaha, A., Guizani, M., Mohammadi, M., Aledhari, M., & Ayyash, M. (2015). Internet of things: A survey on enabling technologies, protocols, and applications. *IEEE Communications Surveys and Tutorials*, *17*(4), 2347–2376.

Almohammedi, A. A., Balfaqih, M., Nahas, S., Bokhari, A., & Alqudsi, A. (2023b). Design and Implementation of IoT-Enabled Intelligent Fire Detection System Using Neural Networks. In *International Conference on AI and Mobile Services* (pp. 63-70). Cham: Springer Nature Switzerland.

Almohammedi, A. A., Raad, A., Darshi, S., Shepelev, V., Yusof, M. H. M., & Balfaqih, M. (2023a). High Gain and Efficiency Triple U-shaped Slots Microstrip Patch Antenna for 5G Applications. In *2023 3rd International Conference on Computing and Information Technology (ICCIT)* (pp. 573-576). IEEE.

Alzhrani, A. A., Balfaqih, M., Alsenani, F., Alharthi, M., Alshehri, A., & Balfagih, Z. (2024). Design and Implementation of an IoT-Integrated Smart Locker System utilizing Facial Recognition Technology. *Engineering, Technology &. Applied Scientific Research*, *14*(4), 16000–16010.

Astill, J., Dara, R. A., Fraser, E. D., Roberts, B., & Sharif, S. (2020). Smart poultry management: Smart sensors, big data, and the internet of things. *Computers and Electronics in Agriculture*, *170*, 105291.

Atzori, L., Iera, A., Morabito, G., & Nitti, M. (2012). The social internet of things (siot)–when social networks meet the internet of things: Concept, architecture and network characterization. *Computer Networks*, *56*(16), 3594–3608.

Balfaqih, M. (2023, November). Design and Implementation of IoT-enabled Intelligent Irrigation System Using Machine Learning. In *2023 28th Asia Pacific Conference on Communications (APCC)* (pp. 436-441). IEEE.

Balfaqih, M., & Alharbi, S. A. (2022). Associated information and communication technologies challenges of smart city development. *Sustainability, 14*(23), 16240.

Balfaqih, M., Alharbi, S. A., Alzain, M., Alqurashi, F., & Almilad, S. (2021). An accident detection and classification system using internet of things and machine learning towards smart city. *Sustainability, 14*(1), 210.

Balfaqih, M., Balfagih, Z., Lytras, M. D., Alfawaz, K. M., Alshdadi, A. A., & Al-solami, E. (2023). A blockchain-enabled IoT logistics system for efficient tracking and management of high-price shipments: A resilient, scalable and sustainable approach to smart cities. *Sustainability, 15*(18), 13971.

Balfaqih, M., Ismail, M., Nordin, R., Rahem, A. A., & Balfaqih, Z. (2017). Fast handover solution for network-based distributed mobility management in intelligent transportation systems. *Telecommunication Systems, 64*, 325–346.

Beard, C., & Stallings, W. (2002). *Wireless communication networks and systems* (2nd ed.). Pearson.

Caragliu, A., Del Bo, C., & Nijkamp, P. (2011). Smart cities in Europe. *Journal of Urban Technology, 18*(2), 65–82.

Catarinucci, L., De Donno, D., Mainetti, L., Palano, L., Patrono, L., Stefanizzi, M. L., & Tarricone, L. (2015). An IoT-aware architecture for smart healthcare systems. *IEEE Internet of Things Journal, 2*(6), 515–526.

Chen, T., Barbarossa, S., Wang, X., Giannakis, G. B., & Zhang, Z. L. (2019). Learning and management for Internet of Things: Accounting for adaptivity and scalability. *Proceedings of the IEEE, 107*(4), 778–796.

Esposito, M., Palma, L., Belli, A., Sabbatini, L., & Pierleoni, P. (2022). Recent advances in internet of things solutions for early warning systems: A review. *Sensors (Basel), 22*(6), 2124.

Fortino, G., Savaglio, C., Spezzano, G., & Zhou, M. (2020). Internet of things as system of systems: A review of methodologies, frameworks, platforms, and tools. *IEEE Transactions on Systems, Man, and Cybernetics. Systems, 51*(1), 223–236.

Gibson, D. V., Kozmetsky, G., & Smilor, R. W. (Eds.). (1992). *The technopolis phenomenon: Smart cities, fast systems, global networks*. Rowman & Littlefield.

Goyal, P., Sahoo, A. K., & Sharma, T. K. (2021). Internet of things: Architecture and enabling technologies. *Materials Today: Proceedings*, *34*, 719–735.

Hafiz, M. Y. M., Balfaqih, M., Khan, S., Almohammedi, A., & Shepelev, V. (2024). Review on Comparative Analysis of Univariate and Multivariate Input Performance Over Time Series Ids Prediction Models in Healthcare Infrastructure. Available at *SSRN*: https://ssrn.com/abstract=4931271

Hassan, M. M., Song, B., & Huh, E. N. (2009, February). A framework of sensor-cloud integration opportunities and challenges. In *Proceedings of the 3rd international conference on Ubiquitous information management and communication* (pp. 618-626).

Hong, J. H., & Shi, Y. T. (2023). Integration of heterogeneous sensor systems for disaster responses in smart cities: Flooding as an example. *ISPRS International Journal of Geo-Information*, *12*(7), 279.

Houssein, E. H., Othman, M. A., Mohamed, W. M., & Younan, M. (2024). Internet of Things in Smart Cities: Comprehensive Review, Open Issues and Challenges. *IEEE Internet of Things Journal.*

Huseien, G. F., & Shah, K. W. (2021). Potential applications of 5G network technology for climate change control: A scoping review of singapore. *Sustainability*, *13*(17), 9720.

Islam, M. M., Hasan, M. K., Islam, S., Balfaqih, M., Alzahrani, A. I., Alalwan, N., & Ghazal, T. M. (2024). Enabling pandemic-resilient healthcare: Narrowband Internet of Things and edge intelligence for real-time monitoring. *CAAI Transactions on Intelligence Technology*.

Jabbar, W. A., Kian, T. K., Ramli, R. M., Zubir, S. N., Zamrizaman, N. S., Balfaqih, M., & Alharbi, S. (2019). Design and fabrication of smart home with internet of things enabled automation system. *IEEE Access : Practical Innovations, Open Solutions*, *7*, 144059–144074.

Karunarathne, G. G. K. W. M. S. I. R., Kulawansa, K. A. D. T., & Firdhous, M. F. M. (2018, December). Wireless communication technologies in internet of things: A critical evaluation. In *2018 International conference on intelligent and innovative computing applications (ICONIC)* (pp. 1-5). IEEE.

Khalifeh, A., Rajendiran, K., Darabkh, K. A., Khasawneh, A. M., AlMomani, O., & Zinonos, Z. (2019). On the potential of fuzzy logic for solving the challenges of cooperative multi-robotic wireless sensor networks. *Electronics (Basel)*, *8*(12), 1513.

Khayyat, M., Balfaqih, M., Balfaqih, H., & Ismail, M. (2024). Challenges and Factors Influencing the Implementation of Green Logistics: A Case Study of Saudi Arabia. *Sustainability, 16*(13), 5617.

Law, K. H., & Lynch, J. P. (2019). Smart city: Technologies and challenges. *IT Professional, 21*(6), 46–51.

Lea, P. (2020). *IoT and Edge Computing for Architects: Implementing edge and IoT systems from sensors to clouds with communication systems, analytics, and security*. Packt Publishing Ltd.

Lee, S. K., Bae, M., & Kim, H. (2017). Future of IoT networks: A survey. *Applied Sciences (Basel, Switzerland), 7*(10), 1072.

Lin, J., Yu, W., Zhang, N., Yang, X., Zhang, H., & Zhao, W. (2017). A survey on internet of things: Architecture, enabling technologies, security and privacy, and applications. *IEEE Internet of Things Journal, 4*(5), 1125–1142.

Monzon, A. (2015). Smart cities concept and challenges: Bases for the assessment of smart city projects. In *2015 international conference on smart cities and green ICT systems (SMARTGREENS)* (pp. 1-11). IEEE.

Munawar, H. S., Mojtahedi, M., Hammad, A. W., Kouzani, A., & Mahmud, M. P. (2022). Disruptive technologies as a solution for disaster risk management: A review. *The Science of the Total Environment, 806*, 151351.

Naphade, M., Banavar, G., Harrison, C., Paraszczak, J., & Morris, R. (2011). Smarter cities and their innovation challenges. *Computer, 44*(6), 32–39.

Nathali Silva, B., Khan, M., & Han, K. (2017). Big data analytics embedded smart city architecture for performance enhancement through real-time data processing and decision-making. *Wireless Communications and Mobile Computing, 2017*(1), 9429676.

Noaman, M., Khan, M. S., Abrar, M. F., Ali, S., Alvi, A., & Saleem, M. A. (2022). Challenges in integration of heterogeneous internet of things. *Scientific Programming, 2022*(1), 8626882.

Ochoa, S. F., & Santos, R. (2015). Human-centric wireless sensor networks to improve information availability during urban search and rescue activities. *Information Fusion, 22*, 71–84.

Palmisano, S. J. (2008). A smarter planet: the next leadership agenda. *IBM. November, 6*, 1-8.

Petrolo, R., Loscri, V., & Mitton, N. (2017). Towards a smart city based on cloud of things, a survey on the smart city vision and paradigms. *Transactions on Emerging Telecommunications Technologies*, *28*(1), e2931.

Phuttharak, J., & Loke, S. W. (2023). An event-driven architectural model for integrating heterogeneous data and developing smart city applications. *Journal of Sensor and Actuator Networks*, *12*(1), 12.

Salman, M. Y., & Hasar, H. (2023). Review on environmental aspects in smart city concept: Water, waste, air pollution and transportation smart applications using IoT techniques. *Sustainable Cities and Society*, *94*, 104567.

Shepelev, V., Glushkov, A., Slobodin, I., & Balfaqih, M. (2023). Studying the relationship between the traffic flow structure, the traffic capacity of intersections, and vehicle-related emissions. *Mathematics*, *11*(16), 3591.

Sinha, A., Kumar, P., Rana, N. P., Islam, R., & Dwivedi, Y. K. (2019). Impact of internet of things (IoT) in disaster management: A task-technology fit perspective. *Annals of Operations Research*, *283*, 759–794.

Su, K., Li, J., & Fu, H. (2011). Smart city and the applications. In *2011 international conference on electronics, communications and control (ICECC)* (pp. 1028-1031). IEEE.

Sumi, L., & Ranga, V. (2016). Sensor enabled Internet of Things for smart cities. In *2016 fourth international conference on parallel, distributed and grid computing (PDGC)* (pp. 295-300). IEEE.

Syed, A. S., Sierra-Sosa, D., Kumar, A., & Elmaghraby, A. (2021). IoT in smart cities: A survey of technologies, practices and challenges. *Smart Cities*, *4*(2), 429–475.

Tan, L., & Wang, N. (2010, August). Future internet: The internet of things. In *2010 3rd international conference on advanced computer theory and engineering (ICACTE)* (Vol. 5, pp. V5-376). IEEE.

Vinodhini, V., Sathiyabhama, B., Sankar, S., Ramasubbareddy, S., Sundarakumar, M. R., & Balusamy, B. (2023). Enabling Technologies for Internet of Things (IoT)-Based Smart Cities. In *Convergence of IoT, Blockchain, and Computational Intelligence in Smart Cities* (pp. 41-57). CRC Press.

Vodák, J., Šulyová, D., & Kubina, M. (2021). Advanced technologies and their use in smart city management. *Sustainability*, *13*(10), 5746.

Wang, X. X., Cao, W. Q., Cao, M. S., & Yuan, J. (2020). Assembling nano–microarchitecture for electromagnetic absorbers and smart devices. *Advanced Materials*, *32*(36), 2002112.

Wang, Y. P. E., Lin, X., Adhikary, A., Grovlen, A., Sui, Y., Blankenship, Y., & Razaghi, H. S. (2017). A primer on 3GPP narrowband Internet of Things. *IEEE Communications Magazine*, *55*(3), 117–123.

Wu, M., Lu, T. J., Ling, F. Y., Sun, J., & Du, H. Y. (2010, August). Research on the architecture of Internet of Things. In *2010 3rd international conference on advanced computer theory and engineering (ICACTE)* (Vol. 5, pp. V5-484). IEEE.

Yaqoob, I., Ahmed, E., Hashem, I. A. T., Ahmed, A. I. A., Gani, A., Imran, M., & Guizani, M. (2017). Internet of things architecture: Recent advances, taxonomy, requirements, and open challenges. *IEEE Wireless Communications*, *24*(3), 10–16.

Younan, M., Elhoseny, M., Ali, A. A., & Houssein, E. H. (2021, December). Quantum chain of things (qcot): A new paradigm for integrating quantum computing, blockchain, and internet of things. In *2021 17th International computer engineering conference (ICENCO)* (pp. 101-106). IEEE.

Zeng, F., Pang, C., & Tang, H. (2023). Sensors on the internet of things systems for urban disaster management: A systematic literature review. *Sensors (Basel)*, *23*(17), 7475.

Chapter 2
Cloud Computing as a Catalyst for Technological Innovation and Industry Transformation

Hemesh Bhardwaj
https://orcid.org/0009-0008-7366-7304
NIELIT Shimla, India

Shobhana Kashyap
https://orcid.org/0000-0001-7351-2769
National Institute of Technology, Jalandhar, India

ABSTRACT

The convergence of cloud computing with emerging technologies is reshaping industries by enhancing efficiency, scalability, and innovation. This chapter examines how cloud platforms integrate with blockchain, serverless systems, artificial intelligence, machine learning, the Internet of Things, edge computing, and cybersecurity to deliver real-world benefits across sectors such as healthcare, manufacturing, retail, and supply chain management. It highlights the transformative potential of cloud-enabled innovations in reducing costs, fostering collaboration, and accelerating digital transformation, while also addressing key challenges including legacy system migration, regulatory compliance, and data privacy. By combining technological foundations with practical applications, the chapter offers valuable

DOI: 10.4018/979-8-3373-0154-9.ch002

insights for researchers and practitioners, demonstrating how the fusion of cloud computing and emerging technologies is driving industrial transformation and shaping the future of business.

INTRODUCTION

Cloud computing has become an essential component of the modern enterprise landscape, transforming how organizations and individuals use technology. By providing on-demand access to computing resources such as databases, storage, applications, and software over the internet, cloud computing eliminates the need for extensive on-premises infrastructure. Services are hosted and managed by providers such as Amazon Web Services, Microsoft Azure, and Google Cloud, while users benefit from flexible, scalable, and cost-efficient solutions. This model can be compared to renting resources as needed, with the ability to adjust usage and payment dynamically.

The scalability and adaptability of cloud platforms have driven innovation across industries by enabling seamless collaboration, efficient data management, and rapid deployment of new solutions. By offering Infrastructure as a Service (IaaS), Platform as a Service (PaaS), and Software as a Service (SaaS), cloud computing provides versatile frameworks for organizations to design, develop, and implement new business and research models. This flexibility has lowered barriers to entry, allowing both startups and established enterprises to evaluate, optimize, and quickly bring novel concepts to market.

The cloud has also accelerated the advancement of emerging technologies such as blockchain, machine learning (ML), artificial intelligence (AI), the Internet of Things (IoT), serverless computing, edge computing, and cybersecurity. Its vast storage capacity, processing power, and ability to scale resources on demand support the training of complex algorithms, processing of large data sets, and operation of intelligent systems. The adaptability of cloud platforms has enhanced the development of these fields, using their accessibility and computational capacity to enable breakthroughs in data modeling and intelligent systems. Additionally, robust security measures safeguard sensitive data in an era of heightened privacy concerns.

This chapter offers an in-depth examination of the pivotal role cloud computing plays in advancing modern technological innovation. It explains how cloud platforms act as catalysts for disruptive technologies such as machine learning, artificial intelligence, the Internet of Things, large-scale data analytics, edge computing, serverless architectures, and cryptocurrency. Particular emphasis is placed on the scalability and security features of cloud infrastructures, which enable the processing of vast volumes of data and the deployment of advanced applications. The chapter also highlights how cloud services are reshaping business operations by driving flexibility, efficiency, and agility. Through a discussion of industry transformations enabled by cloud adoption, it underscores the broader implications of this technology for future advancements.

Figure 1 illustrates several emerging technologies that are increasingly integrated with cloud computing, including blockchain, serverless systems, AI, IoT, edge computing, and advanced privacy and security mechanisms. These technologies are transforming industries by enabling scalability, efficiency, and innovation, while simultaneously introducing challenges related to data integrity, latency, and security. To frame the scope of this study and clarify its objectives, the following research questions are posed:

- **RQ1:** How can industries leverage cloud-based platforms to become more adaptable, scalable, and user-friendly in order to develop novel solutions?
- **RQ2:** In what ways can cloud-powered technologies transform traditional business practices and influence the direction of technological progress?
- **RQ3:** What recent technological advancements contribute most significantly to the field of cloud computing?

Figure 1. Several emerging technologies integrated with cloud computing

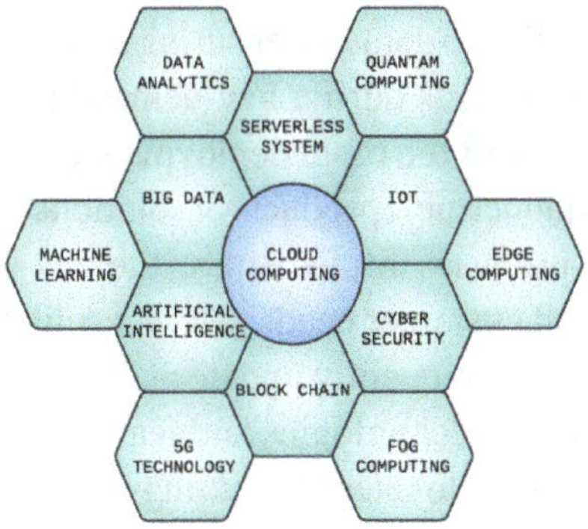

CLOUD-BASED PLATFORMS HELP INDUSTRIES TO DEVELOP NOVEL SOLUTIONS (RQ1)

Industries are increasingly adopting cloud-based platforms to enhance flexibility, scalability, and user experience, thereby enabling the development of innovative solutions. A key advantage of cloud computing lies in its dynamic scalability, allowing organizations to adjust resource usage according to demand and reduce operational costs. Cloud services also promote seamless collaboration across geographically distributed teams and provide a secure environment with reliable backup and recovery mechanisms. These capabilities allow businesses to accelerate the adoption of new ideas and technologies while maintaining data security and operational resilience.

This section examines the adaptability, scalability, and user-friendliness of cloud-based platforms, highlighting how these factors facilitate technological advancement. By leveraging these capabilities, industries can not only optimize existing processes but also explore new opportunities for innovation. Cloud platforms empower organizations to experiment, iterate, and develop novel solutions that address evolving market demands and customer expectations.

Adaptability

Cloud platforms provide a high degree of adaptability by offering resource flexibility, rapid deployment, and continuous innovation. They allow businesses to scale resources up or down according to demand, ensuring uninterrupted operations during peak periods while reducing costs during low-demand times (Dittakavi, 2021). In addition, cloud services support the rapid deployment of applications and services, enabling organizations to launch new products or solutions without the delays associated with establishing on-premises infrastructure, thereby allowing them to respond swiftly to market trends and customer needs (Mervat Adib Bamiah & Brohi, 2011). Furthermore, cloud providers frequently update and enhance their services, giving businesses access to the latest technologies and features. This cycle of continuous innovation empowers industries to integrate cutting-edge solutions without the need for extensive internal research and development investments (Atieh, 2021).

Scalability

Cloud platforms offer robust scalability through elasticity and global reach. Elastic computing capabilities allow resources to scale automatically in response to workload fluctuations, ensuring that applications maintain performance and reliability even under varying levels of traffic and data processing (Ghandour, El Kafhali, & Hanini, 2023). Beyond elasticity, cloud services provide access to globally distributed data centers, enabling organizations to expand their operations internationally without the need to establish physical infrastructure in multiple regions. This global presence minimizes latency and delivers seamless access for users across diverse geographic locations (Thakkalapelli, 2023).

User-Friendliness

Cloud platforms enhance user-friendliness through intuitive interfaces, collaborative tools, and personalized experiences. Many providers offer user-friendly dashboards that simplify resource management, performance monitoring, and data analysis, allowing employees to focus on their tasks without the burden of complex technical configurations (Chanthaweethip et al., 2013). In addition, cloud-based collaboration tools—such as real-time document editing, video conferencing, and instant messaging—facilitate effective teamwork and communication, even among geographically dispersed teams (Su, Wang, & Choo, 2022). Furthermore, cloud-powered applications can employ machine learning algorithms to analyze user behavior and preferences, enabling businesses to deliver tailored content, personalized product recommendations, and customized services that improve user engagement and satisfaction.

Development of Novel Solutions

Cloud platforms play a pivotal role in enabling the development of novel solutions by supporting advanced data analytics, integration of emerging technologies, and collaborative innovation. With powerful analytics capabilities, organizations can derive valuable insights from consumer behavior, market trends, and operational data stored in large-scale cloud databases, facilitating data-driven decision-making and fostering innovation (Kunduru, 2023). Cloud infrastructures also serve as a foundation for integrating disruptive technologies such as artificial intelligence (AI), the Internet of Things (IoT), and blockchain, allowing businesses to experiment and design innovative applications that address industry-specific challenges. Moreover, cloud-based services and tools enhance collaboration among businesses, startups, and developers by providing shared platforms, APIs, and development environments. This collaborative ecosystem reduces barriers to joint innovation and accelerates the creation of novel applications and services (Chiu et al., 2021).

TRANSFORM THE TRADITIONAL BUSINESS TO TECHNOLOGICAL PROGRESS (RQ2)

Cloud-powered technologies have transformed traditional business practices by enhancing efficiency, scalability, and innovation. Their adoption has reshaped organizational behavior, shifting enterprises toward technology-driven models across multiple dimensions. One major impact is the shift in resource accessibility: cloud platforms reduce upfront capital expenditures on infrastructure while providing scalable resources that accelerate experimentation and innovation. This allows organizations to evaluate new concepts rapidly and deploy advanced solutions with minimal financial risk.

In the past, access to sophisticated software and processing power was limited to large corporations; today, cloud services democratize these capabilities by making advanced technologies such as AI, ML, and big data analytics available to small and medium-sized enterprises. This democratization promotes technological progress across sectors and ensures broader competitiveness.

Cloud platforms also enhance collaboration and adaptability. Teams distributed across different time zones and locations can collaborate seamlessly through cloud-based tools, boosting productivity and innovation. Support for remote and flexible work arrangements further enables organizations to attract and retain talent in an increasingly globalized workforce. The dynamic nature of cloud systems empowers businesses to respond quickly to market demands and consumer preferences by deploying updates, features, and services at unprecedented speed.

Equally important, robust security mechanisms and reliable backup systems provide a trusted environment for data storage and operations. This fosters organizational confidence in adopting cloud solutions, reducing concerns about system failures or data breaches, and creating an environment conducive to continuous innovation. Table 1 illustrates how business practices have evolved over time, with real-world examples highlighting changes in production methods and customer engagement strategies.

Table 1. Some examples of traditional business transformation

Factors	Traditional Practice	Transformation
Data Storage and Accessibility	Businesses used to rely on physical storage devices, leading to limited storage capacity and potential data loss risks.	Cloud storage solutions like Amazon S3, Google Cloud Storage, and Dropbox offer virtually unlimited storage, ensuring data accessibility from anywhere, at any time. This accessibility raises collaboration and accelerates decision-making processes.
Collaboration and Communication	Email and on-premises communication systems were common, often causing delays and limited collaboration features.	Cloud-based collaboration tools include Google Workspace and Microsoft 365, which offer an array of features such as real-time document editing, video conferencing capabilities, and instant messaging functionalities. With these innovative tools, teamwork is greatly enhanced; employees can work together and communicate effectively and effortlessly without any geographical location or physical proximity to one another.
Software Development and Deployment	Developing and deploying software applications required significant infrastructure and time, often resulting in lengthy development cycles.	Tools for building, testing, and deploying applications are provided by cloud platforms like AWS, Azure, and Google Cloud Platform. Serverless computing platforms such as AWS Lambda and Azure Functions enable developers to focus solely on code and accelerate development while reducing operational overhead.
E-commerce and Retail	Brick-and-mortar stores were the primary retail channels, limiting geographical reach and requiring substantial investments in physical infrastructure.	Cloud-powered e-commerce platforms enable businesses to create online stores quickly. Cloud-based inventory management, payment processing, and analytics tools optimize operations and provide a seamless shopping experience. Additionally, AI-driven recommendation engines enhance customer engagement and increase sales.
Healthcare and Telemedicine	Healthcare services were predominantly delivered in person, limiting access, especially in remote areas.	Cloud-based electronic health record systems play a critical role in the healthcare industry, as they centralize patient data in a secure manner that allows healthcare professionals to easily access important information in real-time whenever it is needed. In addition, telemedicine platforms use the power of the cloud to allow remote consultations, thereby greatly improving the accessibility of a wide range of healthcare services for patients while working to reduce the overall costs of healthcare for both patients and providers.

continued on following page

Table 1. Continued

Factors	Traditional Practice	Transformation
Manufacturing and Supply Chain Management	Supply chain management often relied on manual processes and fragmented information, leading to inefficiencies and delays.	Cloud-based supply chain platforms make it possible to have real-time visibility of their inventory, production, and distribution processes. IoT and cloud solutions are integrated so that predictive maintenance is used, production schedules are optimized, and overall efficiency is improved. Blockchain technology helps ensure product integrity and traceability all along the supply chain.
Education	Education primarily occurred in physical classrooms, limiting access to learning resources and opportunities.	The prime importance in facilitating and improving online education experience is called a cloud-based learning management system. Educational content, interactive assessments, and collaborative projects can be accessed and completed online. Cloud-powered AI tools personalize learning experiences, providing tailored content to students' needs and preferences.

Ultimately, the transformative potential of cloud computing lies in its ability to democratize access to technology, foster collaboration, enhance flexibility, and provide a secure foundation for growth. These attributes drive traditional enterprises toward rapid technological advancement and enable them to adapt effectively within an ever-changing digital landscape.

Data Storage and Accessibility

The data storage and accessibility features of cloud platforms are critical to the growth and competitiveness of traditional businesses. With virtually unlimited capacity, companies can scale their storage needs without concerns about physical infrastructure, while the pay-as-you-go model minimizes capital expenses, making it particularly attractive for small and medium-sized enterprises. Cloud storage further ensures secure, convenient access from any location, enabling remote work and seamless collaboration across geographically dispersed teams. By allowing simultaneous access to shared data, productivity and efficiency are significantly enhanced.

In addition to accessibility, cloud platforms provide robust backup and disaster recovery solutions that protect data against hardware failures, cyber threats, or natural disasters. These safeguards ensure business continuity by minimizing downtime and enabling rapid data recovery in the event of disruptions. Moreover, the integration of cloud storage with advanced analytics tools empowers organizations to extract valuable insights, better understand customer behavior, and identify new opportunities for growth (Namasudra, 2021). The inherent scalability of these platforms supports business expansion by allowing storage capacity to increase seamlessly in line with organizational needs, without requiring costly infrastructure upgrades.

By combining scalability, cost-effectiveness, security, and accessibility, cloud platforms create a flexible and collaborative environment. This enables businesses to leverage data more effectively, make informed decisions, and adapt swiftly to evolving market demands, ultimately driving growth and long-term sustainability.

Collaboration and Communication

The collaboration and communication features of cloud technology play a vital role in the growth and competitiveness of traditional businesses. Cloud-based tools enable seamless teamwork among employees working across different locations, supporting real-time collaboration regardless of geographical boundaries. This capability is especially important in today's globalized workforce, as it fosters innovation, creativity, and cohesion within teams (Wang, 2020). Platforms offering video conferencing, instant messaging, and collaborative document editing enhance communication efficiency, accelerate decision-making, and shorten project timelines by reducing delays associated with conventional methods.

By streamlining workflows and simplifying document sharing and editing, cloud collaboration tools reduce time spent on administrative coordination, allowing employees to focus on core business activities and improving overall productivity. These platforms are accessible from multiple devices with an internet connection, providing employees with the flexibility to work from the office, home, or while traveling. This flexibility contributes to improved work-life balance, higher employee satisfaction, and stronger retention rates (Kai et al., 2021). Beyond internal operations, cloud collaboration also extends to external stakeholders by connecting businesses with partners, clients, and suppliers globally. Such expanded networks foster partnerships, open new markets, and drive business growth.

Importantly, cloud-based collaboration platforms incorporate robust security and compliance measures that safeguard sensitive information and maintain regulatory standards, instilling trust among stakeholders (Wu et al., 2020). Together, these features empower traditional businesses to operate more efficiently, adapt to evolving market dynamics, and achieve sustained growth through enhanced communication and collaboration.

Software Development and Deployment

The software development and deployment capabilities of cloud computing create significant growth opportunities for traditional businesses by enabling faster innovation, cost efficiency, and broader market reach. Cloud platforms provide tools and infrastructure that support agile methodologies, allowing organizations to iterate and deploy applications rapidly in response to changing market demands. With scalable infrastructure available on demand, businesses can handle fluctuating workloads and accommodate growth without investing heavily in physical hardware. The pay-as-you-go model further reduces capital expenditure, making development more cost-effective, particularly for small and medium-sized enterprises (M. Sharma, Gupta, & Acharya, 2021).

Cloud environments also enhance collaboration among distributed development teams by offering shared platforms and simplified integration with diverse tools and services, streamlining the software development lifecycle. Features such as continuous integration and continuous deployment (CI/CD) automate critical processes, reduce errors, and improve software quality. Faster deployment of applications enables businesses to introduce new features, products, and services more quickly, securing a competitive advantage in dynamic markets. In addition, leading cloud providers incorporate robust security protocols and compliance certifications, ensuring that applications are developed and deployed in secure environments that meet regulatory standards. By combining agility, scalability, cost-effectiveness, and security, cloud-based software development and deployment empower businesses to innovate efficiently, expand their market presence, and adapt to evolving industry needs, thereby driving sustainable growth.

E-commerce and Retail

The e-commerce and retail capabilities of cloud technology play a critical role in advancing the growth and competitiveness of traditional businesses. Cloud-based platforms provide scalable infrastructure that enables e-commerce systems to handle fluctuations in user traffic, particularly during peak seasons or promotional events, without concerns about server capacity. This scalability allows businesses to grow seamlessly while ensuring reliable customer experiences. By operating beyond geographical boundaries, cloud-enabled retail solutions also allow businesses to reach global markets without investing in extensive physical infrastructure, thereby expanding their customer base and driving growth.

Cloud platforms further enhance customer engagement through analytics and personalization tools that help businesses better understand consumer preferences. These tools support tailored shopping experiences, personalized recommendations, and improved satisfaction, which foster repeat business and strengthen brand loyalty. The pay-as-you-go model reduces upfront infrastructure costs, enabling businesses to allocate resources more efficiently and reinvest in innovation and growth initiatives (Xue et al., 2021). Additionally, cloud platforms support rapid deployment of features and updates, allowing retailers to respond quickly to shifting market trends, test new ideas, and innovate at greater speed for competitive advantage.

Robust security protocols and compliance certifications offered by leading providers ensure the protection of customer data, fostering trust and confidence in online transactions. Cloud-based solutions also integrate powerful analytics for examining customer behavior, sales data, and market trends, which inform decision-making and improve marketing strategies, product offerings, and operational efficiency. Moreover, operational tools for inventory management, order processing, and logistics streamline workflows, enabling businesses to manage higher transaction volumes while maintaining service quality (R. Yu et al., 2021). By combining scalability, cost-efficiency, security, analytics, and operational agility, cloud-based e-commerce and retail solutions empower businesses to expand their reach, improve customer experiences, and achieve sustainable growth in today's dynamic retail environment.

Healthcare and Telemedicine

The healthcare and telemedicine features of cloud technology offer significant opportunities for growth, particularly within the healthcare sector, by combining scalability, cost-efficiency, collaboration, and innovation. Cloud-based healthcare platforms provide scalable infrastructure that allows providers to expand services and accommodate rising patient volumes without extensive physical upgrades, ensuring adaptability to changing demands. Telemedicine solutions, powered by the cloud, remove geographical barriers by enabling remote consultations and access to medical specialists, thereby extending healthcare services to broader patient populations and improving accessibility (Chang et al., 2021).

Operating largely on a pay-as-you-go model, cloud solutions reduce upfront infrastructure investments, making them cost-effective and allowing providers to allocate resources toward advanced technologies or improved patient care. Cloud platforms also promote seamless collaboration among healthcare professionals through efficient data sharing and real-time consultations, enhancing diagnostics and treatment outcomes. Robust security protocols and compliance certifications safeguard sensitive patient information, instilling trust among patients and providers and encouraging wider adoption of digital healthcare solutions.

Additionally, cloud-based systems support advanced data analytics, enabling providers to extract insights from patient records, treatment outcomes, and population health trends. These insights inform evidence-based decisions, optimize treatments, and improve overall healthcare delivery. Administrative processes such as patient scheduling, billing, and electronic health record management are also streamlined, reducing operational burdens and allowing healthcare professionals to focus more on patient care (N. Sharma, Anand, & Husain, 2020). By combining scalability, security, analytics, and operational efficiency, cloud-enabled healthcare and telemedicine solutions empower organizations to expand services, improve patient outcomes, and remain competitive in an evolving digital healthcare landscape.

Manufacturing and Supply Chain Management

The integration of manufacturing and supply chain management (SCM) with cloud technology provides substantial benefits that enhance efficiency, collaboration, and innovation. Cloud-based manufacturing and SCM solutions deliver scalable infrastructure that allows businesses to respond quickly to fluctuations in production demand, ensuring more agile and resilient operations. By enabling real-time visibility across the supply chain, cloud platforms connect manufacturers with suppliers, distributors, and customers globally, fostering collaboration, streamlining logistics, and opening access to new markets (Lim, Xiong, & Lei, 2020).

Operating on a pay-as-you-go model, these solutions reduce upfront investments in hardware and software, lowering operational costs and allowing resources to be redirected toward innovation and growth initiatives. Real-time data analytics further enhance supply chain efficiency by optimizing inventory management, improving demand forecasting, and accelerating logistics decision-making. This reduces lead times and mitigates bottlenecks, ultimately improving productivity and operational performance.

Cloud platforms also facilitate seamless collaboration among diverse stakeholders—including suppliers, partners, manufacturers, and customers—by improving communication and coordination across the value chain. In addition, the integration of advanced technologies such as IoT, AI, and predictive analytics within cloud-enabled systems drives innovation by optimizing production processes, improving product quality, and enabling predictive maintenance. Cloud environments also strengthen operational resilience by providing backup and recovery mechanisms, safeguarding continuity in the face of disruptions (Tiwari, 2021). By combining scalability, analytics, collaboration, and resilience, cloud-enabled manufacturing and SCM solutions empower traditional businesses to optimize operations, foster innovation, and expand competitively in dynamic global markets.

Education

Cloud-based education provides traditional businesses with significant opportunities for growth by supporting skill development, fostering innovation, and creating a more adaptable workforce. Through accessible and flexible training programs, organizations can upskill or reskill employees in areas such as technology, leadership, and industry-specific knowledge, ensuring a competent and future-ready workforce (M. Sharma, Gupta, & Acharya, 2021). By equipping employees with the skills needed to adopt new technologies, cloud-based learning drives efficiency, innovation, and competitiveness.

The flexibility of cloud platforms allows employees to access training materials remotely, which is particularly valuable for distributed teams and businesses operating across multiple locations. Many platforms also incorporate collaborative features that encourage interaction and knowledge-sharing, promoting teamwork and a culture of continuous learning. Practical, problem-solving approaches common in cloud-based training further enhance critical thinking, enabling employees to address business challenges with innovative solutions.

Offering cloud-based education demonstrates an organization's commitment to employee development, which improves job satisfaction, retention, and talent acquisition. It also fosters a culture that embraces change and adaptability, preparing employees to respond effectively to evolving market dynamics and technological trends (Asadi, Abdekhoda, & Nadrian, 2020). Cloud-based education empowers businesses by nurturing skilled, adaptable employees, encouraging continuous improvement, and strengthening organizational resilience. These capabilities are essential for sustaining growth and maintaining competitiveness in rapidly changing business environments.

MOST RECENT TECHNOLOGICAL ADVANCEMENTS IN CLOUD COMPUTING (RQ3)

Cloud computing plays a pivotal role in advancing and supporting innovative technologies in the modern digital era. As a continuously evolving paradigm, it drives revolutionary developments that reshape how enterprises operate, manage data, and utilize computing resources. This section examines some of the most recent technologies at the forefront of this transformation.

Artificial Intelligence and Machine Learning

AI is a broad field of computer science focused on developing systems capable of performing tasks traditionally associated with human intelligence. A key subfield, ML, enables machines to learn from large datasets and make predictions or decisions autonomously, without explicit programming. Both AI and ML demand significant computational power, storage, and scalable infrastructure—capabilities effectively provided by cloud computing.

Cloud platforms act as critical enablers by offering cost-effective, elastic, and on-demand resources for training, deploying, and operating complex algorithms.

This democratizes access to advanced technologies, allowing organizations of all sizes to integrate AI and ML into their operations. Applications such as predictive analytics, natural language processing, and computer vision are increasingly used to enhance efficiency, foster innovation, and strengthen competitiveness (Balfaqih et al., 2021).

The synergy between AI, ML, and cloud computing has reshaped the technological landscape. Cloud services not only deliver the capacity to process vast volumes of data but also provide scalability and flexibility for model refinement and faster time-to-market of AI-driven solutions. Concerns around security and compliance are mitigated by providers' robust safeguards and certifications, reinforcing trust in sensitive data handling.

Additionally, cloud platforms offer specialized tools, frameworks, and pre-trained models that streamline development, enable rapid prototyping, and reduce barriers to adoption. These capabilities support iterative improvement, cross-team collaboration, and continuous innovation. As AI and ML continue to evolve, their integration with cloud computing will remain central to digital transformation—driving technology-enabled solutions and redefining how enterprises operate in the data-driven era.

Internet of Things

IoT is a vast network of interconnected devices equipped with sensors, embedded systems, advanced software, and reliable connectivity features. These devices—ranging from household appliances and wearable technologies to industrial machinery and connected vehicles—collect, transmit, and exchange data with centralized systems via the internet (Alzhrani et al., 2024). The continuous operation of such devices generates massive volumes of data that require efficient processing, secure storage, and advanced analytics.

Cloud computing provides the scalable infrastructure and services necessary to meet these demands. Acting as a central platform, the cloud enables real-time data processing, scalable storage, and powerful analytics that transform raw IoT data into actionable insights (Balfaqih, 2024). Once captured, data from IoT devices is transmitted to cloud-based platforms, where it is securely stored, analyzed, and integrated into diverse applications. These include monitoring activities, automating processes, optimizing resource use, and generating valuable business intelligence.

The seamless integration of IoT with cloud computing empowers organizations to unlock the full potential of connected devices and advanced analytics. This synergy creates opportunities for innovation, improves efficiency across industries, and drives the emergence of new business models that were previously unattainable (Almohammedi et al., 2023) .

Big Data Analytics

Cloud services will provide both the storage and processing scalability highly required to handle large amounts of data that organizations come to face in today's digital world. This capability will enable businesses to pursue complex analytics and extract deep meanings while allowing organizations to make decisions on data without significant investment in their infrastructure in-house. Big data analytics refers to a complex process of carrying out an in-depth analysis of large and complex data whereby the aim is to seek diverse patterns, correlations, trends, and insights. The discoveries will play a key role in helping businesses make better, more data-driven decisions crucial for their success and prosperity. This process essentially employs highly advanced and specialized software tools combined with complex algorithms for the analysis of data sets that, in general, are more than the capabilities and boundaries of usual data processing applications.

Generally, big data analytics heavily relies on cloud computing for most of the required functions such as data storage, processing, and analysis. These cloud platforms are designed to give organizations all the scalability and resources required to effectively manage and process large volumes of data. This enables organizations to conduct complex analytics tasks that may otherwise be difficult to achieve. The whole process entails gathering, storing, processing, and conducting thorough analyses of large amounts of structured and unstructured data that originates from a wide range of sources. In other words, Hadoop, Spark, and specialty databases all come into play here to ensure the right handling and processing of these enormous amounts of complex data sets. Machine learning algorithms and advanced data mining techniques will analyze this data in extracting valuable insights for purposes guiding decisions and innovations.

Scalability and on-demand resources provided by cloud computing play an incredibly vital and important role in enabling the advanced processes involved in big data analytics. The required infrastructure, along with a comprehensive set of tools, is offered by cloud computing, making it possible to process and analyze large volumes of data efficiently and effectively. This unique interaction between cloud computing and big data analytics allows businesses to fully harness the immense power of big data, which ultimately aids them in making more informed strategic decisions and achieving significant operational improvements.

Edge Computing

Cloud computing and edge computing function as complementary technologies that together optimize data processing and resource utilization. Cloud computing provides a centralized, scalable platform capable of handling the massive volumes of data generated at the network edge, while edge computing strategically places processing capabilities closer to the data source, such as IoT devices or local sensors. This distributed paradigm reduces latency, enables real-time analysis, and alleviates the strain on centralized servers by performing preliminary processing locally before transmitting only relevant or aggregated data to the cloud (Balfaqih et al., 2021).

Unlike traditional cloud models, where data is processed in remote data centers located far from its origin, edge computing extends computing resources to the periphery of the network. This proximity to the data source enhances responsiveness, reliability, and system performance, making it particularly effective for applications requiring immediate processing. Industries benefit from faster decision-making, improved privacy through localized processing, and greater efficiency in handling time-sensitive operations.

The integration of edge and cloud computing thus ensures a seamless flow of data between edge devices and central cloud platforms, combining the strengths of both paradigms. This synergy supports real-time applications, enhances system scalability, and provides businesses with a flexible, reliable, and efficient computing environment suitable for diverse use cases.

Serverless Computing

Cloud providers provide innovative serverless computing models that allow developers to build and run applications without the burden of having to manage the underlying infrastructure on their own. This significantly fosters faster development cycles, increases agility in responses to changing requirements, and generally encourages overall cost-effectiveness for businesses and developers alike. Serverless computing is also known as Function as a Service (FaaS), which depicts an advanced cloud computing model in which the necessary infrastructure and servers are managed by cloud providers. This allows the developer to focus all energies entirely on the important matters of writing and deploying the code, particularly in terms of modular functions that could easily be executed. In this particular model, developers are relieved of the provisioning or management burden of physical servers; they upload their code to the cloud. The cloud provider takes the task of dynamically allocating the required resources to execute the uploaded code whenever it is requested. This system can automatically scale the resources up or down based on the specific demands of the workload at any given time.

Serverless computing forms a part of the whole cloud computing service. Here, it abstracts an entire infrastructure layer for complete execution of code from its developers, freeing them up from concerns related to any underlying servers, OS's, or infra management. Developers write their codes in the form of specific event-driven or trigger-bound functions. They are uploaded into the platform provided by a cloud provider. At every point where an event happens, that can be one of various events such as receiving an HTTP request or a modification happening in the database. The cloud provider acts very promptly by automatically executing that particular function that is linked to that event. Along with this, it also allocates the required resources according to the needs of the function and dynamically regulates those resources by scaling them up or down based on fluctuating demand. Serverless computing greatly simplifies the processes involved in deploying as well as managing the applications. This streamlined process lets businesses focus their efforts much more effectively on developing solutions that are not only efficient but also scalable, without the complexities and challenges traditionally associated with server management.

Blockchain Technology

Blockchain technology represents one of the most advanced approaches to record-keeping, functioning as a decentralized and distributed ledger system that eliminates the need for central authorities or intermediaries. Transactions are recorded across a network of computers, where each block contains a cryptographic hash of the previous block, a timestamp, and a batch of validated transactions. This cryptographic linking creates an immutable chain of records that ensures transparency, security, and tamper resistance (Balfaqih et al., 2023). Consensus mechanisms further guarantee agreement on the validity of transactions before they are permanently added to the ledger, enhancing trust and reliability.

The integration of blockchain with cloud computing offers significant benefits by strengthening security, transparency, and data integrity in cloud environments. Blockchain can support secure identity verification systems, improve data consistency, and enable transparent, tamper-resistant transactions within cloud-based applications. By providing decentralized trust mechanisms, it mitigates risks associated with centralized control and unauthorized data manipulation (Balfaqih et al., 2023).

Cloud platforms also play a vital role in supporting blockchain-based applications by offering the computational resources necessary to deploy blockchain networks and run decentralized applications (dApps). Through this integration, organizations can leverage blockchain to achieve secure and verifiable data management, ensure trusted digital transactions, and create innovative business models that benefit from both the scalability of cloud computing and the resilience of decentralized ledger technology.

Cybersecurity and Privacy Solutions

Cybersecurity and privacy solutions encompass various technologies, practices, and measures designed to protect systems, networks, data, and users from unauthorized access, cyber threats, data breaches, and privacy violations. These solutions aim to ensure the confidentiality, integrity, and availability of sensitive information while maintaining user privacy. When it comes to cloud computing, cybersecurity and privacy solutions play a crucial role due to the nature of cloud services. Cloud providers offer robust security services and compliance measures to protect data and applications. These services include encryption, identity management, threat detection, and compliance tools, helping organizations enhance their security posture (Tajuddin et al., 2025).

In essence, cloud computing acts as a foundational layer that empowers the development and deployment of various innovative technologies. Its scalability, flexibility, and accessibility facilitate the adoption and advancement of these innovations, driving digital transformation across industries.

Cloud computing includes data storage processes as well as access to various applications over the internet, not dependent on local servers that can be in close proximity. In this modern landscape, cloud environment-specific cybersecurity solutions protect the integrity of cloud infrastructure, sensitive data housed in these clouds, as well as diverse applications running on remote servers distributed throughout geographical locations. A host of sophisticated technologies, such as but not limited to encryption, multi-factor authentication, secure access controls, and identity management techniques, are implemented within cloud environments so that data cannot be compromised in any way through unauthorized access. Encryption ensures that the intercepted data will be useless without the decryption keys. Cloud-based security solutions always monitor network traffic, system logs, and user activity to detect potential threats or anomalies. Incident response plans ensure that security breaches are immediately addressed and mitigated in the shortest time possible. Cloud services, by handling and processing sensitive and personal information, fundamentally have an important role to play in observing several regulations such as GDPR and HIPAA. In regard to security and privacy for the users, cybersecurity solutions are of prime importance for helping cloud service providers comply with and meet these regulations that add another layer of security to the user's data and personal information.

Cybersecurity and privacy solutions in cloud computing are essential for businesses to protect their assets, maintain customer trust, and ensure data privacy in an increasingly digital and interconnected world. To demonstrate the benefits of these technologies Table 2 highlights the significant capabilities of several technologies in this field. The study offers a thorough examination, emphasizing the distinct advantages of each technology. The table below summarizes the specific benefits offered by different technologies.

Table 2. Highlighting the strengths of different technologies

No.	Technology name	Advantages
1	**AI and Machine Learning**	• AI automates tasks, reducing manual effort and increasing operational efficiency. • ML algorithms analyze vast amounts of data to provide valuable insights and predictions for better decision-making. • AI enables personalized experiences, recommendations, and services tailored to individual preferences. • ML models continuously learn and improve accuracy over time, making them more effective in various applications.
2	**IoT (Internet of Things)**	• IoT enables automation of tasks, optimizing processes and resource utilization, leading to increased efficiency. • Real-time data collection and analysis from IoT devices provide insights for better decision-making. • IoT enables personalized experiences, predictive maintenance, and tailored customer services. • Through predictive maintenance and efficient resource usage, IoT can reduce operational costs and maintenance expenses.
3	**Big Data Analytics**	• Insights derived from big data analytics empower businesses to make data-driven decisions, improving efficiency and competitiveness. • Personalized recommendations, targeted marketing, and better customer service are possible by understanding customer behaviour through data analysis. • Optimizing operations and processes based on analytics findings can lead to cost savings and resource optimization. • Big data analytics fosters innovation by uncovering new insights and trends, giving businesses a competitive advantage.
4	**Edge Computing**	• Processing data close to the source where it is generated makes it possible to reduce the time the data takes to travel from centralized servers and then come back. This, therefore, brings about an improvement in the response times. • Edge computing reduces the data that has to be transmitted to the cloud, and further, it minimizes bandwidth usage, thereby aiding in network congestion. • By keeping the sensitive data closer to the source, a significant risk is reduced as far as the potential data breaches are concerned when data is being sent to the remote servers. • Edge computing can continue functioning uninterrupted even if connectivity to the cloud is completely lost. This implies that some core functions can be kept uninterrupted in certain scenarios thus sustaining ongoing operations.

continued on following page

Table 2. Continued

No.	Technology name	Advantages
5	**Serverless Computing**	• It scales automatically according to the demand, meaning that it doesn't require manual intervention and allows for cost optimization. • Businesses pay for actual compute time used; it does not maintain the necessity of constantly running servers, as in serverless. • By giving developers the luxury of focusing only on writing code, it takes away the responsibilities associated with managing infrastructure. The focus shift leads to quicker development and deployment cycles, making them more productive and efficient. • Supports various programming languages and can easily be integrated with many services that increase flexibility in the development process.
6	**Blockchain Technology**	• It is highly resistant to tampering and fraud because it has a decentralized nature with strong cryptographic security. • It makes the transactions on the blockchain visible to everyone involved in the network. This goes a long way in enhancing both transparency and trust among the users. • Added to the blockchain, this data becomes very hard to change after its inclusion, which proves beneficial for the overall integrity of the formed records. • By streamlining multiple processes, cutting down on intermediaries, and employing smart contracts in implementing automation, companies will ultimately realize massive cost-cutting benefits.
7	**Cybersecurity and Privacy**	• These solutions provide advanced security measures, encryption, and access controls, ensuring data confidentiality and integrity. • Cloud-based security solutions often offer a cost-effective alternative to maintaining in-house security infrastructure. • Cloud-based solutions can scale with the business's needs and can be accessed from anywhere, providing flexibility without compromising security. • They help businesses adhere to various industry regulations and data protection laws, reducing the risk of legal and financial penalties.

Table 3 presents a comprehensive mapping of technologies and their applications across various business domains. It offers a detailed overview of how each technology aligns with different industries and sectors, highlighting practical examples and real-world implementations. By illustrating these connections, the table provides a clear perspective on how diverse technologies are leveraged to address specific business needs and challenges.

Table 3. Mapping business applications across diverse technologies

No.	Technology name	Applications in Business
1	**AI and ML**	• Businesses use AI and ML for predictive modeling, forecasting trends, and making informed decisions based on historical data. • AI helps analyze customer behaviour, preferences, and sentiments to personalize marketing strategies and improve customer experiences. • It automates repetitive tasks, streamlines workflows, and enhances operational efficiency across various industries. • AI and ML algorithms analyze patterns in data to identify potential risks, detect anomalies, and enhance cyber security.
2	**IoT**	• IoT devices in factories and supply chains monitor equipment, optimize processes, and enable predictive maintenance, improving efficiency and reducing downtime. • IoT sensors in urban infrastructure monitor traffic flow, manage energy consumption and enhance public safety by collecting and analyzing data in real-time. • IoT-enabled devices in healthcare monitor patient health, track medical assets, and facilitate remote patient monitoring, improving healthcare delivery. • IoT devices enable inventory management, personalized marketing, and enhanced customer experiences through smart devices and systems.
3	**Big Data Analytics**	• Analyzing customer data helps understand behaviours, preferences, and trends, enabling personalized marketing and improving customer experiences. • Big data analytics optimizes operations by analyzing supply chain data, improving inventory management, and enhancing production processes. • Businesses use analytics to identify and mitigate risks by analyzing patterns in financial data, fraud detection, and cyber security. • Analyzing patient data, medical records, and genomic data leads to advancements in treatments, personalized medicine, and disease prevention.
4	**Edge Computing**	• This is very crucial for IoT because, with edge computing, there is the capability to process data in real-time right from the sensors within an application - be it a smart home, industrial IoT, smart city, and many more. • The acceleration of content delivery with the use of edge computing in the form of content caching and serving from the edge servers closer to end-users minimizes latency for streaming services and web applications. • Edge computing is one of the factors that enable real-time decision-making in autonomous vehicles. For instance, processing high volumes of data received from multiple sensors makes real-time decision-making possible in autonomous vehicles. • It enables a personalized in-store experience, inventory management, and real-time analytics in retail.
5	**Serverless Computing**	• Serverless computing is suitable for building parts of web applications that require rapid scaling based on varying user demands. • It's used for processing real-time data streams, IoT data, or event-driven applications. • Businesses utilize serverless for automation tasks, like file processing, data manipulation, and scheduling tasks. • Developers use serverless architecture to build and deploy microservices that perform specific functions within larger applications.

continued on following page

Table 3. Continued

No.	Technology name	Applications in Business
6	**Blockchain Technology**	• Healthcare: It is able to securely store and manage the records of the patient by using blockchain technology so that the integrity and privacy of the data will be maintained. • Supply Chain Management: This technology will be able to track and authenticate products through every stage of the supply chain, hence ensuring that there is no fraud involved. • Smart Contracts: These are advanced, self-executing contracts that automatically execute through code and programmed instructions. They automatically start and perform certain functions when certain conditions are met or fulfilled. This new system has greatly compressed out most business processes and activities while also reducing the number of mediators needed, making it more efficient while cutting costs. • Financial Services: The technology is heavily used in financial services for secure and transparent transactions, cross-border payments, and digital currencies. • Identity Verification: This technology can help build secure, decentralized systems for identity verification, hence reducing the likelihood and risks associated with identity theft.
7	**Cybersecurity and Privacy**	• Secure Data Storage: Businesses use cloud services to store and access their data. Cybersecurity solutions ensure the security of this data against unauthorized access or data breaches. • Secure Collaboration: Cloud-based collaboration tools and platforms require robust security measures to protect sensitive information shared among employees, partners, or clients. • Scalable Security Solutions: Cloud-based cybersecurity solutions provide scalability, allowing businesses to adapt their security measures according to their needs as they grow.

CONCLUSION

Cloud computing has become a transformative force for industries by delivering cost efficiencies, scalability, flexibility, and improved accessibility. Through its pay-as-you-go model, organizations can minimize upfront infrastructure investments and scale resources dynamically to meet changing demands. Cloud-based platforms further enhance collaboration and innovation by enabling remote access, seamless teamwork, faster deployment cycles, and integration with advanced technologies such as blockchain, artificial intelligence (AI), serverless systems, and the Internet of Things (IoT). These capabilities provide industries with opportunities to optimize operations, drive growth, and remain competitive in the digital era. Despite these advantages, cloud adoption also presents significant challenges. Migrating legacy systems and applications to cloud environments can be complex, often requiring reengineering and additional investment. Concerns surrounding data privacy, regulatory compliance, and cybersecurity remain paramount, particularly in sensitive sectors such as healthcare and finance. Moreover, successful adoption requires cultural and organizational adjustments, including training employees to embrace new tools and workflows. These challenges highlight the importance of careful planning, strategic investment, and expert guidance throughout the transition process. The convergence of cloud computing with emerging technologies is reshaping industries and fostering new opportunities for research and innovation. This chapter underscores the pivotal role of cloud-enabled innovations in transforming business practices, streamlining operations, and unlocking creative solutions to address evolving technological and societal challenges. By offering a flexible and secure foundation, cloud platforms not only illuminate the current landscape but also chart a path for future exploration. Researchers and practitioners alike are encouraged to refine these technologies, develop novel applications, and advance cybersecurity measures. The evolving nature of cloud-enabled innovations thus serves as both a catalyst for industrial transformation and a beacon for future research, ensuring that enterprises can thrive in an increasingly dynamic digital environment.

REFERENCES

Almohammedi, A. A., Balfaqih, M., Nahas, S., Bokhari, A., & Alqudsi, A. (2023). Design and Implementation of IoT-Enabled Intelligent Fire Detection System Using Neural Networks. In *International Conference on AI and Mobile Services* (pp. 63-70). Cham: Springer Nature Switzerland.

Alzhrani, A. A., Balfaqih, M., Alsenani, F., Alharthi, M., Alshehri, A., & Balfagih, Z. (2024). Design and Implementation of an IoT-Integrated Smart Locker System utilizing Facial Recognition Technology. *Engineering, Technology &. Applied Scientific Research*, *14*(4), 16000–16010.

Asadi, Z., Abdekhoda, M., & Nadrian, H. (2020). Cloud computing services adoption among higher education faculties: Development of a standardized questionnaire. *Education and Information Technologies*, *25*(1), 175–191. DOI: 10.1007/s10639-019-09932-0

Atieh, A. T. (2021). The next generation cloud technologies: A review on distributed cloud, fog and edge computing and their opportunities and challenges. *ResearchBerg Review of Science and Technology, 1*(1), 1–15. https://researchberg.com/

Balfaqih, M. (2024). Enhancing security and flexibility in smart locker systems: A multi-authentication approach with IoT integration. In *2024 21st Learning and Technology Conference (L&T)* (pp. 325–329). IEEE.

Balfaqih, M., Alharbi, S. A., Alzain, M., Alqurashi, F., & Almilad, S. (2021). An accident detection and classification system using internet of things and machine learning towards smart city. *Sustainability*, *14*(1), 210. DOI: 10.3390/su14010210

Balfaqih, M., Balfagih, Z., Lytras, M. D., Alfawaz, K. M., Alshdadi, A. A., & Al-solami, E. (2023). A blockchain-enabled IoT logistics system for efficient tracking and management of high-price shipments: A resilient, scalable and sustainable approach to smart cities. *Sustainability*, *15*(18), 13971. DOI: 10.3390/su151813971

Balfaqih, M., Jabbar, W., Khayyat, M., & Hassan, R. (2021). Design and development of smart parking system based on fog computing and internet of things. *Electronics (Basel)*, *10*(24), 3184. DOI: 10.3390/electronics10243184

Bamiah, M. A., & Brohi, S. N. (2011). Cloud deployment and service delivery models cloud service models. *International Journal of Research and Reviews in Information Sciences*, *1*(3), 77–80.

Chang, S. C., Lu, M. T., Pan, T. H., & Chen, C. S. (2021). Evaluating the e-health cloud computing systems adoption in Taiwan's healthcare industry. *Life (Chicago, Ill.), 11*(4), 310. DOI: 10.3390/life11040310 PMID: 33918246

Chanthaweethip, W., Han, X., Crespi, N., Chen, Y., Farahbakhsh, R., & Cuevas, A. (2013). "Current city" prediction for coarse location based applications on Facebook. In *GLOBECOM - IEEE Global Telecommunications Conference* (pp. 3188–3193). IEEE. https://doi.org/DOI: 10.1109/GLOCOM.2013.6831562

Chiu, M. C., Huang, J. H., Gupta, S., & Akman, G. (2021). Developing a personalized recommendation system in a smart product service system based on unsupervised learning model. *Computers in Industry, 128*, 103421. DOI: 10.1016/j.compind.2021.103421

Dittakavi, R. S. S. (2021). Deep learning-based prediction of CPU and memory consumption for cost-efficient cloud resource allocation. *Sage Science Review of Applied Machine Learning, 4*(1), 45–58.

Ghandour, O., El Kafhali, S., & Hanini, M. (2023). Computing resources scalability performance analysis in cloud computing data center. *Journal of Grid Computing, 21*(4), 96. Advance online publication. DOI: 10.1007/s10723-023-09696-5

Kai, C., Zhou, H., Yi, Y., & Huang, W. (2021). Collaborative cloud-edge-end task offloading in mobile-edge computing networks with limited communication capability. *IEEE Transactions on Cognitive Communications and Networking, 7*(2), 624–634. DOI: 10.1109/TCCN.2020.3018159

Kunduru, A. R. (2023). Cloud BPM application (Appian) robotic process automation capabilities. *Asian Journal of Research in Computer Science, 16*(3), 267–280. DOI: 10.9734/ajrcos/2023/v16i3361

Lim, M. K., Xiong, W., & Lei, Z. (2020). Theory, supporting technology and application analysis of cloud manufacturing: A systematic and comprehensive literature review. *Industrial Management & Data Systems, 120*(8), 1585–1614. DOI: 10.1108/IMDS-10-2019-0570

Namasudra, S. (2021). Data access control in the cloud computing environment for bioinformatics. *International Journal of Applied Research in Bioinformatics, 11*(1), 40–50. DOI: 10.4018/ijarb.2021010105

Sharma, M., Gupta, R., & Acharya, P. (2021). Analysing the adoption of cloud computing service: A systematic literature review. *Global Knowledge, Memory and Communication, 70*(1–2), 114–153. DOI: 10.1108/GKMC-10-2019-0126

Sharma, N., Anand, A., & Husain, A. (2020). Cloud based healthcare services for telemedicine practices using internet of things. *Journal of Critical Reviews*, *7*(14), 2605–2611. DOI: 10.31838/jcr.07.14.510

Su, M., Wang, G., & Choo, K. K. R. (2022). Prediction-based resource deployment and task scheduling in edge-cloud collaborative computing. *Wireless Communications and Mobile Computing*, *2022*, 2568503. Advance online publication. DOI: 10.1155/2022/2568503

Tajuddin, A. F. B., Yusof, M. H. M., Balfaqih, M., Almohammedi, A. A., Darshi, S., & Qaid, N. (2025). Comparative analysis of AES, RC and Twofish algorithms for data encryption and decryption. In *2025 4th International Conference on Computing and Information Technology (ICCIT)* (pp. 14–19). IEEE.

Thakkalapelli, D. (2023). Cost analysis of cloud migration for small businesses. *Journal Name*, *44*(4), 4778–4784.

Tiwari, S. (2021). Supply chain integration and Industry 4.0: A systematic literature review. *Benchmarking*, *28*(3), 990–1030. DOI: 10.1108/BIJ-08-2020-0428

Wang, K. (2020). Migration strategy of cloud collaborative computing for delay-sensitive industrial IoT applications in the context of intelligent manufacturing. *Computer Communications*, *150*, 413–420. DOI: 10.1016/j.comcom.2019.12.014

Wu, H., Zhang, Z., Guan, C., Wolter, K., & Xu, M. (2020). Collaborate edge and cloud computing with distributed deep learning for smart city internet of things. *IEEE Internet of Things Journal*, *7*(9), 8099–8110. DOI: 10.1109/JIOT.2020.2996784

Xue, M., Xiu, G., Saravanan, V., & Montenegro-Marin, C. E. (2021). Cloud computing with AI for banking and e-commerce applications. *The Electronic Library*, *39*(4), 539–552. DOI: 10.1108/EL-07-2020-0207

Yu, R., Wu, C., Yan, B., Yu, B., Zhou, X., Yu, Y., & Chen, N. (2021). Analysis of the impact of big data on e-commerce in cloud computing environment. *Complexity*, *2021*, 5613599. Advance online publication. DOI: 10.1155/2021/5613599

Chapter 3
Blockchain for Smart Cities:
Enabling Sustainable and Inclusive Urban Development

Krishna Khanal
https://orcid.org/0009-0003-1369-0952
Westcliff University, USA

Neha Arora
Geeta University, India

ABSTRACT

This chapter explores the role of blockchain technology in enabling sustainable and inclusive smart cities amid rapid urbanization and resource constraints. It examines how blockchain's core features—decentralization, transparency, and immutability—integrate with smart city frameworks to enhance governance, energy management, urban planning, and transportation. Case studies from Dubai, Zug, and Telangana demonstrate practical applications in healthcare, public services, and land registries, showcasing blockchain's transformative potential. The chapter also discusses key challenges such as scalability and energy consumption, while highlighting emerging trends like decentralized finance, DAOs, and integration with AI, IoT, and 5G. By combining technological foundations with real-world implementations, the chapter underscores blockchain's capacity to strengthen transparency, citizen engagement, and resilience in smart cities, paving the way for more efficient, secure, and sustainable urban futures.

DOI: 10.4018/979-8-3373-0154-9.ch003

INTRODUCTION

Rapid urbanization, demographic shifts, climate change, and limited resources present significant challenges for modern societies. Currently, more than half of the global population resides in cities, and this number is expected to increase by 2.5 billion by 2050 (United Nations, 2018). Urban areas attract large-scale migration due to socio-economic opportunities, yet this growth creates immense pressure on infrastructure, governance, and resource management. Addressing these issues requires innovative approaches that enhance sustainability, efficiency, and quality of life. The concept of smart cities has emerged as a response, integrating information and communication technologies (ICT) into urban systems to improve healthcare, energy and water management, public transportation, education, and governance (Silva et al., 2018).

At the forefront of this transformation is blockchain technology, originally introduced in 2008 as the foundation of Bitcoin but now recognized as a versatile tool far beyond cryptocurrencies (Zachariadis et al., 2019). Blockchain is a decentralized digital ledger that records transactions securely and transparently across a distributed network. Each block contains a cryptographic hash of the previous one, a timestamp, and transaction data, all validated through consensus mechanisms such as Proof of Work (PoW) or Proof of Stake (PoS). This structure ensures immutability, decentralization, and trustworthiness, offering a paradigm shift in data management and digital security (Hameed et al., 2022).

Smart cities, meanwhile, can be viewed both as physical infrastructure—buildings, transportation, energy, and water systems—and as governance and service networks that sustain urban life. By integrating technologies such as the Internet of Things (IoT), artificial intelligence (AI), and big data analytics, smart cities aim to improve efficiency, reduce resource consumption, enhance transparency, and foster citizen engagement (Bhushan et al., 2021; Alnahari & Ariaratnam, 2022). A city's "smartness" is often evaluated across four interdependent pillars: social infrastructure (healthcare, education, community services), physical infrastructure (transport, utilities, waste management), institutional infrastructure (governance, law enforcement), and economic infrastructure (business and employment) (Haque et al., 2021).

The intersection of blockchain and smart city development represents a natural progression in urban innovation. Blockchain's features—decentralization, transparency, immutability, and efficiency—align closely with the objectives of smart cities. For instance, blockchain can secure land and property records, streamline e-governance services such as voting or tax collection, enable decentralized energy grids, and enhance trust in public transportation and logistics (Amadi-Echendu, 2021; Hammouti et al., 2023; Rejeb et al., 2021). Smart contracts further expand these applications by automating agreements between parties, ensuring tamper-proof execution without intermediaries (Margret & Julie, 2024). Beyond operational efficiency, blockchain can address societal challenges such as corruption and disinformation by improving accountability, securing content integrity, and ensuring transparency in government operations.

Real-world implementations already demonstrate its potential: the United Nations World Food Program successfully used blockchain in Jordan to facilitate cash transfers to Syrian refugees without third-party intermediaries, showcasing how blockchain can enhance aid distribution and public trust (Connolly et al., 2022). Such cases illustrate blockchain's role not only in strengthening technical systems but also in fostering inclusivity, resilience, and citizen engagement within smart cities.

This chapter explores the synergy between blockchain and smart cities by discussing the fundamentals of blockchain, its diverse applications in urban contexts, the challenges it faces—including scalability, energy use, and regulation—and the future directions that may redefine how cities operate. The goal is not merely to understand blockchain as a technology but to envision how it can enable sustainable, transparent, and citizen-centric smart cities in the decades ahead.

APPLICATION AREAS IN SMART CITIES

The integration of blockchain technology into smart cities extends across multiple domains, offering innovative solutions to enhance governance, resource management, and urban living. Its decentralized, transparent, and tamper-proof nature addresses challenges such as inefficiency, fraud, and lack of trust in traditional systems. The following subsections highlight key application areas where blockchain demonstrates significant potential: strengthening government and governance processes, advancing energy management through smart grids and renewable distribution, improving urban planning and infrastructure, and transforming transportation and mobility systems. Together, these applications illustrate how blockchain contributes to building more sustainable, efficient, and citizen-centric smart cities.

Government and Governance: Voting Systems, Public Records

Blockchain technology introduces a transformative approach to governance in smart cities (Zygiaris et al., 2023). In the process of voting systems, blockchain can help in secure and transparent electronic voting, mitigating concerns of fraud and tampering (Mostafa & El-Masry, 2013). By creating an immutable ledger of votes, blockchain ensures the integrity of each vote, building trust in the democratic process. The COVID-19 pandemic brought unique challenges to traditional voting systems. With social distancing measures and health risks related to large gatherings, traditional in-person voting posed significant public health risks (Kamil et al., 2021). This situation demanded the need for secure e-voting procedures.

Furthermore, in managing public records, blockchain provides a decentralized platform where records such as property deeds, licenses, and civil registrations can be stored securely and accessed transparently. Blockchain and smart contracts can automate public services like tax collection, benefits distribution, property registries, identity management, and record-keeping (Li et al., 2019). Using Blockchain technology in the real estate industry guarantees trustworthiness and dependability in maintaining land records and helps in the generation of national wealth (Veujer, 2018). This technology improves public administration by offering secure information access to citizens, building stronger engagement in public management. This not only streamlines administrative processes but also significantly reduces the chances of corruption and bureaucratic inefficiency.

Blockchain technology significantly benefits government operations, enhancing efficiency, transparency, and security (Ning et al., 2021). It streamlines bureaucratic processes, improves public benefit management, and transforms e-business models for more integrated services. Additionally, blockchain helps in building resilient societies by enabling real-time data tracking and ensuring task authenticity and execution. This technology advances transparent and accountable governments, reducing corruption and increasing public trust. In essence, blockchain's application in government leads to more efficient, transparent, and effective governance (Joia et al., 2021). US, the UK, Estonia, Singapoand re, and Sweden are adopting e-government services, and few other countries are following suit (Park et al., 2018; Lapointe & Fishbane, 2019).

Energy Management: Smart Grids, Renewable Energy Distribution

In energy management, blockchain technology is important in the making and operation of smart grids. These grids employ blockchain to decentralize energy distribution, allowing for more efficient and sustainable energy management systems. By enabling peer-to-peer energy trading, residents can buy, sell, or exchange renewable energy (like solar or wind power) directly with each other, optimizing energy usage and costs. Blockchain technology can serve as a basis for creating a peer-to-peer (P2P) platform for energy transactions, linking energy resources with home appliances to ensure the availability of affordable energy anytime and anywhere (Park et al., 2018). A peer-to-peer network removes the need for middlemen and trusted third parties in trading, allowing buyers and sellers to have more control over their preferences, options, and pricing (Alskaif et al., 2022; Otjacques et al., 2018). This decentralization not only reduces dependency on traditional energy sources but also encourages the use of renewable energy. Moreover, blockchain's transparent nature allows for accurate tracking and billing of energy usage, fostering a more conscientious and efficient approach to energy consumption in urban areas (Mauladdawilah et al., 2025).

Blockchain technology is enhancing the transparency and trustworthiness of energy transactions in microgrids. It makes sure that the energy traded is genuinely sourced from renewable resources, as verified by green certificates (Imbault et al., 2017). Blockchain technology helps in tracking the carbon footprint of energy appliances, ensuring data security and unalterability for reliable information (Miglani et al., 2020). This helps both manufacturers and consumers to assess and comprehend the environmental impact of a product's carbon footprint. Microgrids offer various advantages, including the potential to reduce costs related to the main grid, ensure access to electricity during blackouts due to weather or maintenance, empowering individuals to have greater control over their energy usage within a community, and enable the utilization of local resources that may be too small or unreliable for the primary utility grid to incorporate (Karandikar et al., 2019). This technology helps in accurately tracing the origin of energy, guaranteeing its eco-friendly nature. It also helps in protection from cyber-attacks and stops fraud and other malicious activities (Dong et al., 2018). Additionally, blockchain incorporates a reputation system, which helps prevent fraudulent activities by assigning trust scores to participants, fostering a reliable and honest energy trading environment (Khaqqia et al., 2018).

Urban Planning and Management: Smart Home, Land Registry, Infrastructure Development

Blockchain technology improves urban infrastructure management through key features like decentralization, reducing central failure risks, immutability, ensuring data integrity; and a democratic consensus process for decision-making (Sun et al., 2016). Additionally, it offers pseudonymity and transparency, increasing security while maintaining system transparency (Hughes et al., 2019) This technology has the power to revolutionize various elements of urban living, from enhancing governance through transparent and secure voting systems to enabling smart grids for more efficient energy distribution. Blockchain's capability to provide secure, tamper-proof systems facilitates the way for more accountable and efficient urban governance, thereby fostering trust among residents and stakeholders.

Blockchain's application in urban planning and management is significantly impactful. The technology can revolutionize land registry systems, providing a secure, transparent, and immutable record of property ownership and transactions (Shuaib et al., 2022). This reduces the risk of fraud and land disputes, ensuring smoother property transfer processes. Additionally, blockchain can play an important role in infrastructure development. By maintaining transparent records of contracts, project progress, and resource allocation, blockchain can enhance the efficiency and accountability of infrastructure projects. This results in more effective use of resources and improved project delivery times, contributing to the overall development and sustainability of smart cities.

The other most impactful application of blockchain technology in the water infrastructure would likely be in enhancing the efficiency and transparency of water supply and quality management. By integrating blockchain, cities could develop a decentralized and tamper-proof ledger system to monitor and record water usage, distribution patterns, and water quality data across the city (Naqash et al., 2023). This system would enable real-time tracking of water flow, detect leaks or contaminations swiftly, and ensure equitable distribution of water resources. Additionally, the technology could facilitate smart contracts for water trading and allocation, allowing for more dynamic and responsive management of water resources, especially critical in regions facing water scarcity or drought conditions. This application not only promises to improve operational efficiencies and water conservation efforts but also helps in community engagement and awareness regarding water sustainability in urban cities.

Smart cities, leveraging IoT technologies, have been pivotal in the evolution of smart homes. Smart home appliances utilize an IoT-enabled framework, allowing these smart devices to communicate and share information (Ali et al., 2023). These homes represent a synergy of various advanced technologies interconnected through home networks, aiming to enhance the quality of life. IoT devices and Internet appliances in smart homes offer unprecedented control and automation of home functions, from energy management to security systems (Sripan et al., 2012). This integration allows for real-time monitoring and efficient management of resources, leading to energy savings and increased convenience. Smart homes also adapt to the needs and preferences of residents, learning from their behaviors to optimize settings like lighting, temperature, and even entertainment systems. The connectivity offered by IoT in smart homes not only simplifies daily tasks but also opens up possibilities for remote control and monitoring, providing peace of mind for homeowners (Siswipraptini et al., 2021). Thus, smart homes stand as a testimony to how smart city initiatives are directly enhancing individual living spaces through technology.

Transportation and Mobility: Supply Chain Transparency, Traffic Management

In the transportation sector, blockchain technology improves supply chain transparency and traffic management. Blockchain technology is identified as an effective solution for addressing transportation challenges such as congestion, emissions, and accidents. It improves the security of connected automated vehicles against various threats, including hacking, traffic congestion, and a range of cyber-attacks (Gupta et al., 2021). In traffic management, blockchain can be employed to securely store and analyze traffic data, helping in the optimization of traffic flows and reduction of congestion. This data-driven approach aids in developing more effective public transportation systems and urban mobility solutions, making travel within the city more efficient and environmentally friendly. Smart mobility is another important feature of smart cities. Smart mobility is the development of safe, clean, and efficient public transportation infrastructure. The application of blockchain technology can help in tracking passengers and vehicles (Ferdous et al.; 2019) and also store data regarding drivers and vehicle maintaining a detailed, immutable record of accidents, maintenance, and other events (Rahman et al., 2019). Autonomous Transport as a Service (ATaaS), involving driverless public transport vehicles like buses and shuttles, is revolutionizing urban design by employing advanced sensors, blockchain, and algorithms for navigation. This innovation paves the way for the development of "mobility hubs" in cities, integrating various transportation modes such as autonomous shuttles, bicycles, and scooters, thereby enhancing urban mobility and accessibility (Khayyat et al., 2024).

For supply chains, blockchain offers an unalterable ledger that tracks the movement of goods from origin to destination (Balfaqih et al., 2023). This level of transparency ensures the authenticity of products, reduces delays, and minimizes fraud, leading to more efficient supply chains. For example: Blockchain technology can significantly enhance the pharmaceutical supply chain by tracking medicine flow and reducing counterfeit risks, minimizing waste and data manipulation, and efficiently tracing drug origins through transaction data logged by IoT devices (Rejeb et al., 2021).

CASE STUDIES AND CURRENT IMPLEMENTATIONS

The practical adoption of blockchain in smart cities is best illustrated through real-world case studies. Several pioneering regions have already integrated blockchain into their urban ecosystems, showcasing its potential to enhance governance, transparency, efficiency, and citizen engagement. The following subsections present three notable examples: Dubai, often recognized as the first blockchain-powered government; Zug, Switzerland, known as "Crypto Valley" and a hub for blockchain innovation; and Telangana, India, which has applied blockchain to land registries, education, and financial inclusion. Together, these cases demonstrate the diverse ways blockchain can transform urban services and provide valuable insights for cities seeking to adopt similar innovations.

City Of Dubai

One notable case study is the city of Dubai, which aims to become the world's first blockchain-powered smart city. Dubai's blockchain strategy includes various sectors, including healthcare, transportation, and energy (Alam, 2021; Khan et al., 2022). In healthcare, blockchain is used to securely store patient records, allowing for seamless sharing of information between authorized healthcare providers, improving efficiency and patient care quality. In transportation, the city has employed a blockchain-based system for vehicle registration and tracking, enhancing the management and security of vehicle-related data. The Dubai Blockchain Strategy has successfully made the city the world's first blockchain-powered government, streamlined processes and enhanced transparency. The local blockchain startup scene has flourished, attracting international talent and investments, while blockchain research and development centers have proliferated. As a result, Dubai remains a trailblazer in blockchain technology, setting an example for other cities worldwide and playing a pivotal role in shaping the future of decentralized technologies.

Zug, Switzerland

Another exemplary project is in Zug, Switzerland, often referred to as "Crypto Valley." It is home to a thriving blockchain and crypto start-up ecosystem (Gambazzi et al., 2021). Zug has implemented blockchain in its e-governance initiatives. The city conducted a blockchain-based e-voting pilot project, which demonstrated the feasibility and security of using blockchain for public voting. This initiative highlighted blockchain's potential to enhance citizen participation in governance and increase the transparency of electoral processes. The region's authorities have shown a progressive approach to blockchain, facilitating the use of cryptocurrencies and creating a regulatory framework that balances innovation with investor protection. As a result, Zug has become synonymous with blockchain excellence, hosting blockchain conferences, incubators, and a vibrant community of blockchain enthusiasts, proving its status as a premier destination for blockchain entrepreneurs and investors worldwide (Wüst & Gervais., 2018).

Telangana, India

Telangana has initiated a blockchain-based land registry project in some parts of Hyderabad to improve the transparency and efficiency of land record management (Perera et al., 2020). This project was lunched in the year 2018 and is part of the state's broader efforts to modernize land administration and reduce land disputes. In this project, blockchain technology is used to secure and streamline land records.

Anyone interested can see who holds the land since the records of the blockchain are made available on an open platform. At the same time, the Telangana state government has introduced the idea of establishing a "blockchain district" with the aim of accommodating prominent blockchain technology firms (Pandey & Sen, 2022). The district is set to establish an incubation center and a state-of-the art facility dedicated to fostering research, innovation, and collaboration within the blockchain industry. The Telangana state educational department has also started hosting SSC certificates on blockchain for verifying students' certificates leading the blockchain innovation in India. In the state, there is another blockchain project named T-chits that monitors chit fund transactions (Government of Telangana, 2024). Additionally, to support women's empowerment and financial inclusion, the state employs a blockchain-based solution called B-Post. This system utilizes an unmodifiable record of transactions on a decentralized blockchain platform to establish credit ratings for StreeNidhi Self Help Groups (SHGs). These ratings can also be shared with other groups.

TECHNICAL CHALLENGES

While blockchain offers significant promise for enhancing smart city ecosystems, its implementation is not without obstacles. Technical challenges, particularly related to scalability and energy consumption, pose critical barriers to widespread adoption. The vast amount of data generated by IoT devices in urban environments strains blockchain's transaction throughput, while energy-intensive consensus mechanisms raise concerns about environmental sustainability. Addressing these issues is essential to ensure that blockchain can support the complex, data-driven infrastructure of smart cities without compromising efficiency or sustainability. The following subsections explore these key challenges and potential solutions in greater detail.

Scalability Issues and Potential Solutions

One of the primary technical challenges facing blockchain in the context of smart cities is scalability (Zhao, 2022). As blockchain networks grow and the number of transactions increases, the ability of the blockchain to process and record these transactions efficiently becomes a concern. This is particularly challenging in smart cities, where the volume of data generated by various IoT devices and sensors can be immense. The conventional blockchain models, which require each transaction to be recorded on every node of the network, can lead to congestion and slow processing times.

Several potential solutions are being examined to address scalability. One approach is the development of Layer 2 solutions, such as off-chain transactions or sidechains, which can handle transactions off the main blockchain, thereby reducing the load and speeding up processing (Anupama & Sunitha., 2023). Layer Two solutions, also called off-chain solutions, represent protocols made to improve the scalability and efficiency of the Bitcoin blockchain (Hasan et al., 2023). They aim to solve issues related to transaction throughput and time latency by handling specific transactions off the primary blockchain. This approach helps alleviate the strain on the main blockchain while still upholding security and reliability. Notable examples of such solutions currently in use include Lightning Networks (LN), Side Chains, and various others. Another solution is the use of sharding, a method that partitions the blockchain into smaller, more manageable pieces, allowing multiple transactions to be processed simultaneously. Additionally, the advancement of more efficient consensus mechanisms, such as Proof of Stake (PoS), can also enhance scalability by reducing the computational power required for transaction validation (Balfaqih, M., & Alharbi, 2022).

Energy Consumption Concerns, Especially with POW Blockchains

Another significant challenge is the high energy consumption associated with blockchain, particularly those that use Proof of Work (PoW) consensus mechanisms, like Bitcoin (Tomatsu & Han, 2023). To successfully implement blockchain in smart cities, it will be essential to tackle the challenges of scalability and energy consumption. These systems must be capable of handling the demands of fast-expanding urban areas while contributing to the broader objectives of sustainability and resilience (Dritsas & Trigka, 2024). PoW requires a substantial amount of computational power, as nodes on the network compete to solve complex mathematical problems to validate transactions and create new blocks. Bitcoin uses this algorithm to mine the blocks (Wang et al., 2019; Aljabr et al., 2019). This process, known as mining, consumes a large amount of electricity, raising concerns about the environmental sustainability of blockchain technologies, especially in the context of smart cities aiming for eco-friendliness and sustainability.

There is a growing shift towards more energy-efficient consensus mechanisms to mitigate these concerns. Proof of Stake (PoS) has emerged as a popular alternative, where the capacity to validate transactions is based on a node's stake or ownership of the currency, rather than computational power (Asif & Hassan, 2023). This significantly reduces the amount of energy required. Another approach is the use of hybrid models that combine PoW and PoS, attempting to balance security and energy efficiency. Additionally, the exploration of renewable energy sources for blockchain operations and the optimization of blockchain network designs for energy efficiency are also critical steps towards creating sustainable blockchain solutions in smart cities.

EMERGING TRENDS IN BLOCKCHAIN

As blockchain continues to mature, a range of emerging trends is shaping its potential role in smart cities. Advances in scalability and energy-efficient architectures, along with the rise of decentralized finance (DeFi) and decentralized autonomous organizations (DAOs), are expanding blockchain's applications beyond conventional use cases. At the same time, its integration with other cutting-edge technologies such as AI, the IoT, and 5G is creating powerful synergies that can enhance urban governance, service delivery, and citizen engagement. The following subsections explore these developments, highlighting how technological evolution and cross-disciplinary integration could transform blockchain into a cornerstone of next-generation smart city ecosystems.

Developments in Blockchain Technology That Could Impact Smart Cities

Blockchain technology is continuously evolving, with new developments poised to significantly impact smart cities. One such advancement is the emergence of more scalable and energy-efficient blockchain architectures. These advancements promise to handle the vast amounts of data generated in smart cities more effectively, addressing the current limitations of scalability and energy consumption (Al Ghamdi, 2022). Another significant trend is the advent of DeFi and DAOs. These concepts extend blockchain's application beyond traditional boundaries, enabling more democratic and transparent management of resources in urban settings. For instance, DeFi could lead to innovative models of funding urban projects, while DAOs could facilitate community-led governance and decision-making processes (Khanna et al., 2021).

By leveraging DeFi, smart cities can create more efficient and transparent financial ecosystems, enabling seamless transactions and innovative funding models for urban projects. This could facilitate improved access to capital for infrastructure development and enable residents to directly invest in their city's growth. DAOs, on the other hand, could revolutionize urban governance. They provide a platform for decentralized decision-making, where residents can directly participate in the management and evolution of their cities. This participatory approach could lead to more responsive, inclusive, and adaptive urban management, aligning with the dynamic needs of a smart city's inhabitants and infrastructure. Thus, the integration of DeFi and DAOs into smart city frameworks could guide an era of enhanced financial fluidity and democratic governance, fundamentally transforming urban living.

Integration of Blockchain With Other Cutting-Edge Technologies

Blockchain technology is being applied as a decentralized and widely distributed method to fulfil security needs in the realm of other technologies like IoT and others as it serves as an unalterable storage solution (Gangwani et al., 2023). The integration of blockchain with other technologies like AI, IoT, and 5G is set to create synergies that could transform smart cities (Dwivedi et al., 2021). Blockchain can provide a secure and transparent framework for the vast amounts of data generated by IoT devices, while AI can analyze this data to optimize city services and infrastructure. Similarly, the incorporation of 5G technology can enhance the speed and efficiency of blockchain transactions, supporting real-time data processing and decision-making in smart urban environments.

The integration of blockchain with other technologies in smart cities extends further to enhance security, efficiency, and innovation. AI-driven analytics can use the data stored on the blockchain to improve predictive maintenance of critical infrastructure, optimize traffic management, and even enhance public safety using predictive crime analysis. The seamless synergy between blockchain, IoT, and 5G can lead to the creation of autonomous, self-regulating systems that respond to real-time data and events, making smart cities more adaptive and resilient. Furthermore, as blockchain fosters trust and transparency in supply chains, people can have greater confidence in the quality and origins of goods and services, promoting sustainable consumption and fostering a thriving urban environment (Bhushan et al., 2020). This convergence of cutting-edge technologies promises to redefine urban living and pave the way for more connected, efficient, and sustainable smart cities in the future. Interoperability, privacy concerns, and ethical challenges pose significant hurdles to the integration of IoT devices, blockchain platforms, and machine-learning frameworks in smart cities, necessitating the development of universal standards, privacy-preserving technologies, and robust governance structures to ensure compatibility, transparency, and responsible deployment while addressing the risks of algorithmic bias, data security, and personal privacy (Dritsas & Trigka, 2024; Biswas et al., 2023; Ziosi et al., 2022; Yang et al., 2024).

CONCLUSIONS

Increasing global population and rapid urbanization have placed significant pressure on the sustainability of cities. Considering this change, urban cities are going through a period of technological change with the help of blockchain technologies. Interoperability, efficiency, democratization, and intractability are the features of blockchain that make this technology effective in solving these problems in emerging markets. This technology has the power to revolutionize urban cities, making them more inclusive, efficient, and livable. Incorporating blockchains into several domains of smart cities like transportation, public administration, supply chains, and healthcare represents a promising paradigm shift in urban development. As blockchain technology gains wider public acceptance, it is important to collaborate and engage various stakeholders to help adapt to the changing technological landscape. The future and the success of smart cities depend on the implementation of blockchain technology along with urban planning, and together, they promise more resilient, connected, and sustainable cities for future generations.

REFERENCES

Al Ghamdi, M. A. (2022). An Optimized and Secure Energy-Efficient Blockchain-Based Framework in IoT. *IEEE Access : Practical Innovations, Open Solutions.*

Alam, T. (2021). Blockchain cities: The futuristic cities driven by Blockchain, big data and the internet of things. *GeoJournal.*

Ali, Y., Shah, S. W., & Khan, W. A. (2023). Security at the Internet of Things. In Machine Tools: An Industry 4.0 Perspective (pp. 31–48). CRC Press.

Aljabr, A. A., Sharma, A., & Kumar, K. (2019). Mining Process in Cryptocurrency Using Blockchain Technology: Bitcoin as a Case Study. *Journal of Computational and Theoretical Nanoscience, 16*(10), 4293–4298.

Alnahari, M. S., & Ariaratnam, S. T. (2022). The Application of Blockchain Technology to Smart City Infrastructure. *Smart Cities, 5*(3), 979–993. DOI: 10.3390/smartcities5030049

Alskaif, T., Crespo-Vazquez, J. L., Sekuloski, M., Leeuwen, G. V., & Catalao, J. P. S. (2022). Blockchain-based fully peer-to-peer energy trading strategies for residential energy systems. *IEEE Transactions on Industrial Informatics, 18*, 231–241. DOI: 10.1109/TII.2021.3050018

Amadi-Echendu, A. P. (2021). Using blockchain technology to facilitate property transactions. *South African Journal of Information Management, 23*(1), a1421.

Anupama, B. S., & Sunitha, N. R. (2023). Survey on blockchain scalability addressing techniques. In S. J. Patel, N. K. Chaudhary, B. N. Gohil, & S. S. Iyengar (Eds.), *Information Security, Privacy and Digital Forensics. ICISPD 2022.* (Lecture Notes in Electrical Engineering, Vol. 1075). Springer. https://doi.org/DOI: 10.1007/978-981-99-5091-1_28

Asif, R., & Hassan, S. R. (2023). Shaping the future of Ethereum: Exploring energy consumption in Proof-of-Work and Proof-of-Stake consensus. *Frontiers in Blockchain, 6*, 1151724. Advance online publication. DOI: 10.3389/fbloc.2023.1151724

Balfaqih, M., & Alharbi, S. A. (2022). Associated information and communication technologies challenges of smart city development. *Sustainability, 14*(23), 16240.

Balfaqih, M., Balfagih, Z., Lytras, M. D., Alfawaz, K. M., Alshdadi, A. A., & Alsolami, E. (2023). A blockchain-enabled n nIoT logistics system for efficient tracking and management of high-price shipments: A resilient, scalable and sustainable approach to smart cities. *Sustainability, 15*(18), 13971.

Bhushan, B., Khamparia, A., Martin Sagayam, K., Sharma, S. K., Ahad, M. A., & Debnath, N. C. (2020). Blockchain for smart cities: A review of architectures, integration trends and future research directions. *Sustainable Cities and Society*.

Bhushan, B., Sinha, P., Sagayam, K. M., & Andrew, J. (2021). Untangling blockchain technology: A survey on state of the art, security threats, privacy services, applications and future research directions. *Computers & Electrical Engineering*, *90*, 106897.

Biswas, S., Yao, Z., Yan, L., Alqhatani, A., Bairagi, A. K., Asiri, F., & Masud, M. (2023). Interoperability benefits and challenges in smart city services: Blockchain as a solution. *Electronics (Basel)*, *12*(1036). Advance online publication. DOI: 10.3390/electronics12051036

Connolly, D., Nam, S., & Goodman, K. (2022). Solving old problems or making new ones? Blockchain technology for the protection of refugees and migrants. *Journal of Human Rights*, *22*(02), 1–26. DOI: 10.1080/14754835.2022.2100984

Dong, Z., Luo, F., & Liang, G. (2018). Blockchain: A secure, decentralized, trusted cyber infrastructure solution for future energy systems. *Journal of Modern Power Systems and Clean Energy*, *6*, 958–967.

Dritsas, E., & Trigka, M. (2024). Machine learning for blockchain and IoT systems in smart cities: A survey. *Future Internet*, *16*(9), 324. DOI: 10.3390/fi16090324

Dwivedi, A. D., Singh, R., Kaushik, K., Mukkamala, R. R., & Alnumay, W. S. (2021). Blockchain and AI for 5G-enabled IoT: Challenges, opportunities and solutions. *Transactions on Emerging Telecommunications Technologies (Online)*, Article e4329. https://doi.org/DOI: 10.1002/ett.4329

Ferdous, M. S., Biswas, K., Chowdhury, M. J. M., Chowdhury, N., & Muthukkumarasamy, V. (2019). Chapter Two—Integrated platforms for blockchain enablement. In Kim, S., Deka, G. C., & Zhang, P. (Eds.), *Role of Blockchain Technology in IoT Applications* (Vol. 115, pp. 41–72). Academic Press.

Gambazzi, L., Schaller, P., Mermoud, A., & Lenders, V. (2021). Blockchain in Cyberdefence: A Technology Review from a Swiss Perspective. *The Swiss Technology Observatory*. https://doi.org//arXiv.2103.02606DOI: 10.48550

Gangwani, P., Perez-Pons, A., Joshi, S., Upadhyay, H., & Lagos, L. (2023). *Integration of Data Science and IoT with Blockchain for Industry 4.0. In S. Namasudra & K. Akkaya (Eds.), Blockchain and its Applications in Industry 4.0* (Studies in Big Data, vol 119, pp. 6). Springer. https://doi.org/DOI: 10.1007/978-981-19-8730-4_6

Gupta, R., Kumari, A., & Tanwar, S. (2021). A taxonomy of blockchain envisioned edge-as-a-connected autonomous vehicles. *Transactions on Emerging Telecommunications Technologies*, *32*(6), e4009. DOI: 10.1002/ett.4009

Government of Telangana. (2024). *T-chits*. Retrieved February 01, 2024 from Registration and Stamps Department: https://t-chits.telangana.gov.in

Hameed, K., Barika, M., Garg, S., Amin, M. B., & Kang, B. (2022). A taxonomy study on securing Blockchain-based Industrial applications: An overview, application perspectives, requirements, attacks, countermeasures, and open issues. *Journal of Industrial Information Integration*, *26*, 100312. DOI: 10.1016/j.jii.2021.100312

Hammouti, I., Addaim, A., & Guennoun, Z. (2023). Decentralized Energy Trading in Smart Grids: A Blockchain-based Framework for Efficiency and Transparency. *IEEE Access : Practical Innovations, Open Solutions*.

Haque, A. K. M. B., Bhushan, B., & Dhiman, G. (2021). Conceptualizing smart city applications: Requirements, architecture, security issues, and emerging trends. *Expert Systems: International Journal of Knowledge Engineering and Neural Networks*.

Hasan, H. F., Najwadi, M., Yusoff, B., & Ali, S. M. A. (2023). Bitcoin Layer Two Scaling Solutions: Lightening Payment Channels Network Comprehensive Review, Mechanisms, Challenges, Open Issues and Future Research Directions. *Iraqi Journal for Computer Science and Mathematics*, *5*(1), 25–59.

Hughes, L., Dwivedi, Y. K., Misra, S. K., Rana, N. P., Raghavan, V., & Akella, V. (2019). Blockchain research, practice and policy: Applications, benefits, limitations, emerging research themes and research agenda. *International Journal of Information Management*, *49*, 114–129. DOI: 10.1016/j.ijinfomgt.2019.02.00

Imbault, F., Swiatek, M., de Beaufort, R., & Plana, R. (2017). The green blockchain: Managing decentralized energy production and consumption. In *IEEE International Conference on Environment and Electrical Engineering and 2017 IEEE Industrial and Commercial Power Systems Europe (EEEIC / I&CPS Europe).*

Joia, L. A., & Vieira, J. A. P. (2021). Going beyond cryptocurrencies: The social representation of blockchain. *Sustainability*, *13*, 14054.

Karandikar, N., Chakravorty, A., & Rong, C. (2019). Transactive Energy on Hyperledger Fabric. In 2019 *Sixth international Conference on Internet of Things Systems, Management and Security (*IOTSMS), pp.539-546.

Khan, S., Shael, M., Majdalawieh, M., Nizamuddin, N., & Nicho, M. (2022). Blockchain for governments: The case of the Dubai Government. *Sustainability*, *14*(11), 6576. DOI: 10.3390/su14116576

Khanna, A., Sah, A., Bolshev, V., Jasinski, M., Vinogradov, A., Leonowicz, Z., & Jasiński, M. (2021). Blockchain: Future of e-Governance in Smart Cities. *Sustainability, 13*(21), 11840. DOI: 10.3390/su132111840

Khaqqia, K. N., Sikorskib, J. J., Hadinotoa, K., & Kraft, M. (2018). Incorporating seller/buyer reputation-based system in blockchain-enabled emission trading application. *Applied Energy, 209*, 8–19.

Khayyat, M., Balfaqih, M., Balfaqih, H., & Ismail, M. (2024). Challenges and factors influencing the implementation of green logistics: A case study of Saudi Arabia. *Sustainability, 16*(13), 5617.

Lapointe, C., & Fishbane, L. (2019). The Blockchain Ethical Design Framework. *Innovations: Technology, Governance, Globalization, 12*, 50–71.

Li, J., Greenwood, D., & Kassem, M. (2019). Blockchain in the built environment and construction industry: A systematic review, conceptual models and practical use cases. *Automation in Construction, 102*, 288–307.

Margret, M. K., & Julie, E. G. (2024). Smarter and resilient smart contracts applications for smart cities environment using blockchain technology. *Automatika (Zagreb), 65*(2), 572–583. DOI: 10.1080/00051144.2024.2307228

Mauladdawilah, H., Balfaqih, M., Balfagih, Z., Pegalajar, M. D. C., & Gago, E. J. (2025). Deep Feature Selection of Meteorological Variables for LSTM-Based PV Power Forecasting in High-Dimensional Time-Series Data. *Algorithms, 18*(8), 496.

Miglani, A., Kumar, N., Chamola, V., & Zeadally, S. (2020). Blockchain for Internet of Energy management: Review, solutions, and challenges. *Computer Communications, 151*, 395–418. DOI: 10.1016/j.comcom.2020.01.006

Mostafa, M. M., & El-Masry, A. A. (2013). Citizens as consumers: Profiling e-government services' users in Egypt via data mining techniques. *International Journal of Information Management, 33*, 627–641.

Naqash, M. T., Syed, T. A., Alqahtani, S. S., Siddiqui, M. S., Alzahrani, A., & Nauman, M. (2023). A Blockchain Based Framework for Efficient Water Management and Leakage Detection in Urban Areas. *Urban Science (Basel, Switzerland), 7*(4), 99. DOI: 10.3390/urbansci7040099

Ning, X., Ramirez, R., & Khuntia, J. (2021). Blockchain-enabled government efficiency and impartiality: Using blockchain for targeted poverty alleviation in a city in China. *Information Technology for Development, 27*(3), 599–616. DOI: 10.1080/02681102.2021.1925619

Otjacques, B., Hitzelberger, P., Naumann, S., & Wohlgemuth, V. (2018). *From Science to Society: New Trends in Environmental Informatics*. Springer.

Pandey, S., & Sen, C. (2022). Blockchain Technology in Real-Time Governance: An Indian Scenario. *The Indian Journal of Public Administration*, *68*(3), 397–413. DOI: 10.1177/00195561221105241

Park, J. S., Kim, Y. S., Choi, C. H., & Shim, J. (2018). How to define value on data under blockchain driven open data system for e-government. In *Proceedings of the 11th International Conference on Theory and Practice of Electronic Governance* (pp. 670–672). Galway, Ireland, April 4–6, 2018.

Park, L. W., Lee, S., & Chang, H. (2018). A sustainable home energy prosumer-chain methodology with energy tags over the blockchain. *Sustainability*, *10*(3), 658. DOI: 10.3390/su10030658

Perera, S., Nanayakkara, S., Rodrigo, M. N. N., Senaratne, S., & Weinand, R. (2020). Blockchain technology: Is it hype or real in the construction industry? *Journal of Industrial Information Integration*, *17*, 100125.

Rahman, M. A., Rashid, M. M., Shamim Hossain, M., Hassanain, E., Alhamid, M. F., & Guizani, M. (2019). Blockchain and IoT-based cognitive edge framework for sharing economy services in a smart city. *IEEE Access : Practical Innovations, Open Solutions*, *7*, 18611–18621.

Rejeb, A., Rejeb, K., Simske, S., & Treiblmaier, H. (2021). Blockchain Technologies in Logistics and Supply Chain Management: A Bibliometric Review. *Logistics (Basel)*, *5*(72). Advance online publication. DOI: 10.3390/logistics5040072

Shuaib, M., Hassan, N. H., Usman, S., Alam, S., Bhatia, S., Koundal, D., Mashat, A., & Belay, A. (2022). Identity model for blockchain-based land registry system: A comparison. *Wireless Communications and Mobile Computing*, *2022*. Advance online publication. DOI: 10.1155/2022/5670714

Silva, B. N., Khan, M., & Han, K. (2018). Towards sustainable smart cities: A review of trends, architectures, components, and open challenges in smart cities. *Sustainable Cities and Society*, *38*, 697–713. DOI: 10.1016/j.scs.2018.01.053

Siswipraptini, P. C., Aziza, R. N., & Sangadji, I., Indrianto, Siregar, R. R. A., & Sondakh, G. (2021). IoT for smart home system. *Indonesian Journal of Electrical Engineering and Computer Science*, *23*(2), 733–739. DOI: 10.11591/ijeecs.v23.i2.pp733-739

Sripan, M., Lin, X., Petchlorlean, P., & Ketcham, M. (2012). Research and thinking of smart home technology. In *Proceedings of the International Conference on Systems and Electronic Engineering (ICSEE'2012)* (pp. 61–63). Phuket, Thailand.

Sun, J., Yan, J., & Zhang, K. Z. K. (2016). Blockchain-based sharing services: What blockchain technology can contribute to smart cities. *Financial Innovation*, *2*(26), 1–9. DOI: 10.1186/s40854-016-0040-y

Tomatsu, Y., & Han, W. (2023). Bitcoin and Renewable Energy Mining: A Survey. *Blockchains*, *1*(2), 90–110. DOI: 10.3390/blockchains1020007

United Nations, Department of Economic and Social Affairs, Population Division. (2018). World Urbanization Prospects (WUP): The 2018 Revision, Methodology (Working Paper No. ESA/P/WP.252). New York, NY: United Nations.

Veuger, J. (2018). Trust in a viable real estate economy with disruption and blockchain. *Facilities*, *36*, 103–120.

Wang, W., Hoang, D. T., Hu, P., Xiong, Z., Niyato, D., & Wang, P.. (2019). A survey on consensus mechanisms and mining strategy management in blockchain networks. *IEEE Access : Practical Innovations, Open Solutions*, *7*, 22328–22370.

Wüst, K., & Gervais, A. (2018). Do you need a blockchain? In *Proceedings of the 2018 Crypto Valley Conference on Blockchain Technology (CVCBT)* (pp. 45–54). Zug, Switzerland.

Yang, R., Zhao, T., Yu, F. R., Li, M., Zhang, D., & Zhao, X. (2024). Blockchain-based federated learning with enhanced privacy and security using homomorphic encryption and reputation. *IEEE Internet of Things Journal, 11*, 21674–21688.

Zachariadis, M., Hileman, G., & Scott, S. V. (2019). Governance and control in distributed ledgers: Understanding the challenges facing blockchain technology in financial services. *Information and Organization*, *29*(2), 105–117. DOI: 10.1016/j.infoandorg.2019.03.001

Zhao, W. (2022). On Blockchain: Design Principle, Building Blocks, Core Innovations, and Misconceptions. *IEEE Systems, Man, and Cybernetics Magazine*, *8*(4), 6–14.

Ziosi, M., Hewitt, B., Juneja, P., Taddeo, M., & Floridi, L. (2022). Smart cities: Mapping their ethical implications. SSRN *Electronic Journal, 10*. https://doi.org/ DOI: 10.2139/ssrn.1234567

Zygiaris, S., Saleh, M. F., & Al-Imamy, S. Y. (2023). The Smart City Blockchain Governance: A Literature Review. *Central European Management Journal,* 31(1).

Chapter 4
AI-Enabled Applications of Medical Internet of Things in Smart Hospitals

Arun Agrawal
https://orcid.org/0000-0001-7233-6660
Institute of Technology and Management, Gwalior, India

Prashant Laxmanrao Mandale
https://orcid.org/0009-0000-9329-4893
International Institute of Information Technology, Pune, India

Deepika Dattatraya Walanjkar
International Institute of Information Technology, Pune, India

Deepak Gupta
https://orcid.org/0000-0003-3929-1362
Institute of Technology and Management, Gwalior, India

Sharda Thete
https://orcid.org/0009-0002-5891-5804
International Institute of Information Technology, Pune, India

Atul Kumar
https://orcid.org/0000-0003-3859-0686
Dnyaan Prasad Global University, Pune, India

ABSTRACT

This chapter explores how Artificial Intelligence (AI) works together with the Medical Internet of Things (MIoT) in modern healthcare. It focuses on smart hospitals. AI helps MIoT by making it possible to predict health problems, track disease progress, and make better decisions using data. IoT tools like asset tracking, energy monitoring, and staff management help reduce hospital costs and save resources. A cost analysis shows that MIoT can reduce hospital readmissions by 30% and cut healthcare costs by 20%. It also offers a return on investment within 2–3 years. There are some challenges, such as device compatibility, system expansion, and data security. The chapter provides simple solutions like using cloud or edge computing, blockchain, and common international standards. Real-life examples from South

DOI: 10.4018/979-8-3373-0154-9.ch004

Korea and the UK show that MIoT improves patient care and hospital performance. The chapter also covers new trends like 5G, AI-based automation, and robots doing routine tasks. These trends point toward a future of smart, patient-focused, and technology-driven healthcare.

INTRODUCTION

The Medical Internet of Things (MIoT) has emerged as a major innovation in healthcare, linking medical devices, hospital systems, and patient data into a unified digital ecosystem (Akhtar, Rahman, Sadia, & Perwej, 2021). This integration enables the development of smart hospitals, where IoT devices enhance patient care, optimize resources, and minimize delays in operations (Kamal, 2024). Through wearable sensors, remote monitoring tools, and connected diagnostic equipment, health information can be collected and transmitted in real time. These devices continuously monitor patient conditions and send data to centralized systems, allowing clinicians to respond quickly and make informed medical decisions. For example, wearable devices that track heart rate and blood pressure can generate alerts when irregularities are detected, supporting proactive management of chronic illnesses (Dias & Paulo Silva Cunha, 2018).

Smart hospitals leverage IoT to automate routine tasks such as patient monitoring, inventory checks, and predictive maintenance of medical equipment. These systems improve safety, reduce errors, and enhance patient compliance. Examples include smart medication dispensers that ensure accurate dosing and infection-control systems that help maintain sanitary environments (Uslu, Okay, & Dursun, 2020; Yesmin, Carter, & Gladman, 2022). Beyond clinical applications, MIoT improves hospital operations by tracking medical tools, managing energy consumption, and reducing environmental impact. When integrated with artificial intelligence, MIoT can analyze medical histories to detect early signs of disease and deliver personalized treatments (Muthunagai, Girija, Kumar, & Anitha, 2024).

Despite these advantages, several challenges remain. Security and privacy must be ensured, devices must be interoperable, and systems must be capable of scaling with demand (Awotunde et al., 2021). Cloud and edge computing provide promising solutions by enabling large-scale data processing and real-time analytics. In contrast to traditional healthcare—where records are fragmented, monitoring is manual, and care is typically reactive—MIoT enables continuous monitoring, centralized data integration, and early intervention. Traditional systems often record vital signs only during hospital visits, while MIoT devices such as wearables track patients around the clock, immediately notifying clinicians of abnormalities (Codina Barberà, 2024; Reddy et al., 2025a).

The integration of artificial intelligence (AI) with MIoT is further transforming healthcare delivery by unlocking the full potential of data collected from connected devices. AI algorithms enable faster decision-making, more accurate predictions, and highly personalized treatment strategies. Predictive analytics, such as long short-term memory (LSTM) models, can analyze patient data over time to forecast disease progression, while convolutional neural networks (CNNs) interpret medical images with precision, often identifying abnormalities missed by human observation. Reinforcement learning adapts treatment plans based on past outcomes, and natural language processing (NLP) extracts key insights from unstructured clinical notes or laboratory reports. In addition, AI-driven real-time monitoring systems can detect anomalies such as irregular heart rhythms or oxygen desaturation and instantly alert clinicians. Together, AI and MIoT create a healthcare ecosystem that is predictive, preventive, and personalized, enabling hospitals to deliver safer, faster, and more efficient care.

Several key features distinguish MIoT from traditional healthcare technologies. Unlike conventional care, which typically measures vital signs only during hospital visits, MIoT enables real-time and continuous monitoring through wearable devices and sensors that track patient health around the clock (Zhao et al., 2022). It also supports remote patient care, allowing doctors to monitor and communicate with individuals without requiring physical hospital visits, enabling patients to remain at home while staying connected to their care teams (Tantray et al., 2024). By analyzing large volumes of health data, MIoT systems can provide early warnings of potential health issues, whereas traditional systems often detect problems too late due to limited data (Anikwe et al., 2022). Furthermore, MIoT facilitates personalized treatment by collecting detailed patient information that helps physicians tailor care plans to individual needs, moving beyond the one-size-fits-all approach common in conventional healthcare (Vuong, 2024).

Another advantage is improved data integration, as MIoT connects devices and systems to enable seamless data sharing, addressing the fragmentation often seen in traditional hospital environments (Awotunde et al., 2021). Importantly, MIoT also promotes patient engagement by granting individuals access to their health data via mobile applications and monitoring tools, encouraging them to actively participate in their own care—a contrast to traditional systems that restrict access mainly to doctors (Reddy et al., 2025b). Finally, MIoT provides automated support for clinicians by issuing alerts and recommending interventions when a patient's condition changes, reducing reliance on manual monitoring and decision-making (Shanmathi & Jagannath, 2018).

This chapter investigates the application of MIoT in smart hospitals, focusing on its role in improving healthcare delivery, operational efficiency, and sustainability. It reviews real-world implementations, highlights benefit and challenges and discusses future opportunities for combining MIoT with artificial intelligence, cloud computing, and edge technologies. The aim is to provide guidance for researchers, healthcare practitioners, and policymakers on harnessing MIoT to create more adaptive, efficient, and patient-centered healthcare systems.

HISTORICAL CONTEXT OF MIoT

The Internet of Things (IoT) in healthcare represents a key technological milestone in modern medicine, with its origins dating back to the 1960s when basic telemetry systems were first used to monitor vital signs (Anorboyevich, 2024). During the 1970s and 1980s, hospitals began adopting early remote monitoring tools, though these systems relied on extensive infrastructure and had limited wireless functionality. By the 1990s, improvements in sensor design and communication technologies made it possible to introduce the first home monitoring devices, which played a crucial role in tracking chronic conditions such as diabetes and cardiovascular health outside hospital settings (Manavi, Zafar & Sharif, 2024).

The early 2000s brought a major transformation with the spread of broadband internet and smartphones, enabling the development of wearable devices and advanced monitoring platforms for chronic disease management (Yalli, Hasan & Badawi, 2024). By 2010, cloud computing had revolutionized healthcare by providing secure storage and analysis of patient data from multiple sources, while new privacy regulations were introduced to strengthen data protection and system interoperability (Javaid et al., 2022). Between 2015 and 2020, the integration of artificial intelligence with IoT accelerated the creation of smart healthcare systems, allowing hospitals to monitor room conditions, optimize workflows, and track medical equipment more effectively (Maleki Varnosfaderani & Forouzanfar, 2024).

Today, MIoT encompasses devices ranging from implants and wearables to smart hospital infrastructure, enabling faster response times, better patient care, and enhanced data-driven decision-making (Kumar et al., 2023). Yet, challenges persist. Cybersecurity remains a critical concern, with studies highlighting the rising risks of cyberattacks on MIoT systems and advocating for robust safeguards such as blockchain-based tamper-proof records (Madanian et al., 2024; Sargiotis, 2024). Interoperability is also a key issue, as heterogeneous devices must seamlessly communicate with hospital information systems. To address this, standards such as HL7 FHIR have been recommended to ensure consistent and reliable data exchange (Saleem et al., 2018).

Economic considerations further influence MIoT adoption. While evidence suggests long-term benefits such as reduced readmission rates, improved resource utilization, and task automation (Haghani, 2023; Poyyamozhi et al., 2024), smaller hospitals often face barriers due to high initial costs associated with equipment, training, and system upgrades (Koppel, 2022). Recent studies have also emphasized resilience and adaptability. For example, narrowband IoT (NB-IoT) and edge computing have been applied to support patient monitoring in both urban and rural environments, proving especially useful during the COVID-19 pandemic (Osama et al., 2023). Other innovations include embedding patient data within Wi-Fi HaLow (IEEE 802.11ah) signals using steganography to enhance privacy without degrading performance (Ferre et al., 2021), and blockchain-based logistics systems that employ smart contracts to secure medical records and supply chains (Aloini et al., 2023).

APPLICATIONS IN PATIENT CARE AND SAFETY

MIoT has introduced transformative applications in patient care and hospital safety, enabling smarter, more proactive healthcare delivery. By integrating connected devices such as wearable sensors, smart beds, automated alert systems, and intelligent medication dispensers, MIoT improves monitoring, reduces errors, and enhances patient outcomes. These technologies not only support clinical decision-making through real-time data but also optimize hospital operations by reducing workloads, improving resource management, and strengthening infection control. The following subsections outline key applications—ranging from remote patient monitoring to infection prevention—that highlight how MIoT is reshaping modern healthcare environments.

Remote Patient Monitoring (RPM)

Remote patient monitoring (RPM) is one of the most important applications of MIoT in smart hospitals. It enables continuous and proactive care for patients. IoT devices such as wearables and implantable devices can track vital signs such as heart rate, blood pressure, blood oxygen levels, and blood sugar. These devices send real-time data to hospital systems or secure cloud platforms. This helps doctors monitor patients remotely and take quick action if any health issues are detected (Mahmmod et al., 2024; Abdulmalek et al., 2022) (see Figure 1).

Figure 1. Real time data driven using wearable devices (Babu, 2024)

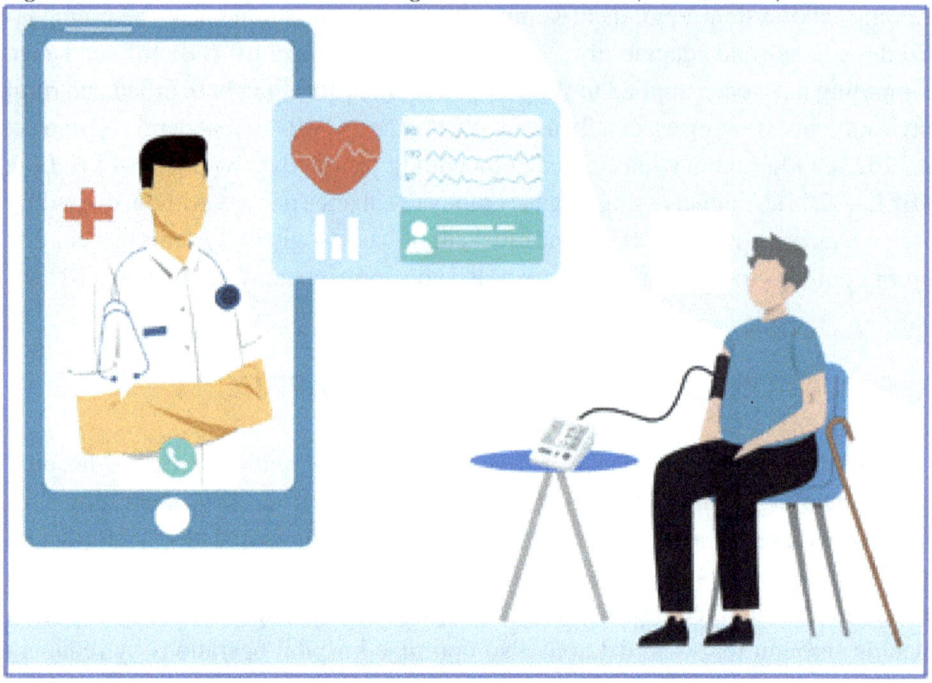

RPM is especially useful for patients with long-term conditions such as diabetes, high blood pressure, and respiratory problems. It helps reduce emergency department visits and readmissions. It also improves patient adherence to treatment plans and increases satisfaction with their care (Dogheim & Hussain, 2023; Haghani, 2023). For example, if a respiratory patient's blood oxygen levels drop, the system will sound an alarm and the medical team can respond quickly.

From a medical perspective, RPM helps detect problems early, reduces emergency cases, and improves control of chronic conditions. From an operational perspective, it reduces the burden on nurses, increases bed availability, and reduces healthcare costs. RPM provides a good return on investment and improves the overall quality of care. This is a clear example of how MIoT can help bridge gaps in traditional healthcare delivery.

Smart Beds

Smart beds equipped with IoT technology are bringing significant improvements to patient care. They can continuously monitor important factors such as patient activity, weight, posture, and vital signs (Karvounis et al., 2021). This continuous tracking can help reduce the risk of falls, especially for elderly or high-risk patients.

These beds can also automatically adjust their position based on patient needs, thereby supporting personalized care. If the system detects sudden movement or changes, it sends an alert to medical staff so that they can respond quickly. As shown in Figure 2, smart beds help reduce the workload of medical teams. They do this by reducing response time and completing daily adjustments without human intervention. This not only improves patient safety and comfort but also allows medical staff to focus more on critical tasks, thereby improving care efficiency.

Figure 2. Smart bed for personalized care (Kamal, 2025)

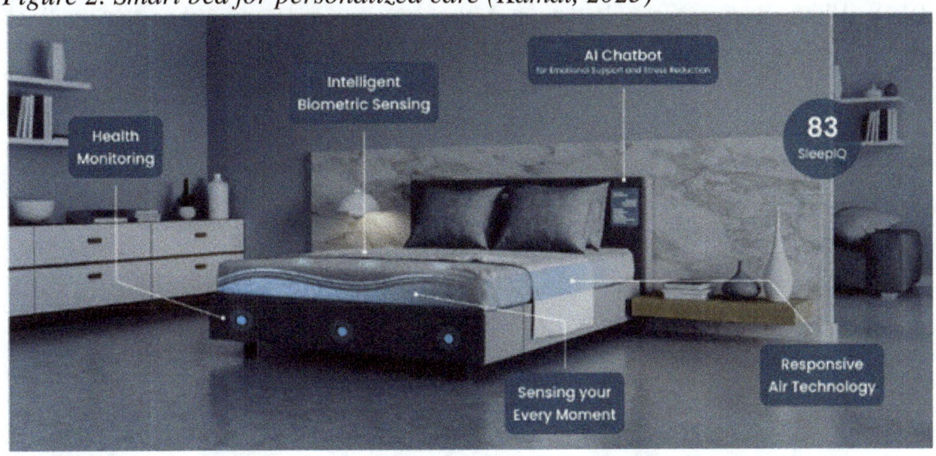

These smart bed systems connect to the hospital network. When they detect important changes in a patient's condition, they send alerts to caregivers. These changes may include unusual movements or sudden changes in health status (Guarascio-Howard, 2011). The system can also improve hospital operations, providing data that helps manage resources more efficiently. Clinical results are encouraging. Falls were reduced by 45%. Post-operative recovery time was reduced by 20%. Pressure ulcers were reduced by 60% through automated repositioning (Wen et al., 2024). Response time to patient distress was reduced by 30 minutes. Nurse workload was reduced by 35%. Sleep quality tracking was improved by 40%.

However, smart beds also face some challenges. A smart bed costs between $15,000 and $30,000. Annual maintenance costs are about 12% of the purchase price. Technical issues include a 5% annual sensor failure rate. Hospitals may face unstable network connections. Employees must be retrained to use the new system (Koppel, 2022). Other issues include the storage of massive amounts of data. Batteries need to be replaced regularly. Firmware must be updated frequently. Despite these challenges, many hospitals can still achieve a return on investment within two to three years. Smart beds are particularly useful for high-risk patients and can increase the early detection of health problems by up to 50%.

Automated Emergency Alerts

Automatic emergency alerts through MIoT devices are critical to improving patient safety and treatment outcomes in smart hospitals. Devices such as wearable electrocardiograms (ECGs) can continuously track vital signs to detect serious health issues such as irregular heartbeats or breathing problems (Lewandowski, 2014). When an emergency is detected, the system sends an alert to the medical team through a connected server. This allows medical staff to respond quickly and provide immediate care. As shown in Figure 3, this setup supports faster intervention and helps prevent complications. Even if there is no one at the bedside, it ensures that patients receive timely treatment when needed.

Figure 3. Illustration of wearable ECG telemonitoring system (Shao et al., 2020)

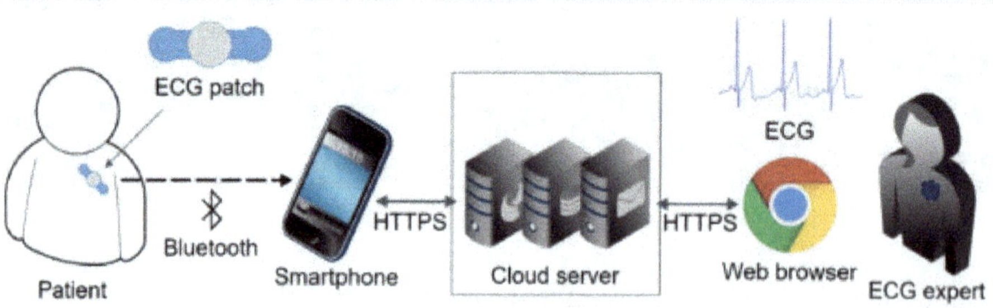

Wearable ECG monitors can detect early signs of arrhythmia or cardiac arrest. If a problem is detected, the system immediately alerts medical staff, allowing them to take immediate action. In life-threatening situations, a fast response is critical. These timely alerts help shorten response times and increase patients' chances of survival.

Medication Management

IoT-enabled smart medicine cabinets and pill dispensers help ensure that patients take the right medication at the right time. These systems automatically monitor medication schedules, tracking how and when medications are taken. They can also detect problems such as missed doses or possible overdoses (Raisa et al., 2023).

When problems arise, the system sends alerts to patients or healthcare providers so they can take quick action.

By automatically administering medications, these devices reduce the risk of errors. They help prevent adverse drug reactions and improve the overall safety of treatment. Smart pill boxes connected to mobile apps send reminders when it's time for patients to take their medications. Caregivers are notified if a dose is missed. This helps patients stick to their treatment plans and makes it easier for caregivers to stay involved. This technology creates a safer and more reliable approach to medication management in smart healthcare systems.

Infection Control

IoT sensors play an important role in improving infection control in hospitals. They monitor environmental conditions such as humidity, temperature, and air quality. These sensors check whether the environment is maintained at safe and sterile levels. The system immediately alerts if anything goes outside the normal range (Piaggio et al., 2023). This real-time monitoring helps staff to take quick action. It can also prevent hospital-acquired infections (HAIs), which are a serious risk in healthcare facilities (Blot et al., 2022).

IoT systems help hospitals adhere to strict hygiene and safety regulations by keeping the environment stable. This reduces infection rates and improves patient safety and quality of care (Thakur & Rao, 2024). In Japan, hospitals use these systems to maintain a sterile environment in operating rooms. Sensors continuously track vital indicators. When any value is out of range, staff are immediately alerted. This helps prevent contamination before it occurs, especially during surgery (Abdulmalek et al., 2022). These technologies support safer care and a cleaner hospital environment.

OPERATIONAL EFFICIENCY ENHANCEMENTS

Beyond improving patient care, MIoT also plays a pivotal role in strengthening hospital operations. Smart systems powered by IoT technologies streamline asset tracking, workforce management, and energy use, ensuring that hospitals can deliver high-quality care while reducing costs and waste. From RFID-enabled inventory systems to intelligent rostering tools and energy-efficient infrastructure, these innovations optimize resource utilization, enhance staff productivity, and support sustainability goals. The following subsections outline how MIoT applications drive greater efficiency in asset management, workforce allocation, and environmental control within smart hospitals.

Asset and Inventory Management

Asset and inventory management in smart hospitals becomes more efficient through the use of RFID tags and IoT-based tracking systems. These tools provide real-time updates on the location, condition, and usage of medical equipment, consumables, and medicines. By continuously monitoring inventory, IoT systems help avoid common problems such as stock-outs, overstocking, or lost equipment. Each item is tagged with an RFID tag for easy identification and tracking. This reduces the need for manual record keeping and routine checks. The system also uses data to predict when items will run out and alert staff to restock them in time. This helps prevent treatment delays due to lack of resources. Accurate tracking can reduce waste from expired or unused supplies. It can also help hospitals meet regulatory requirements and improve the way they order and store items. These improvements can reduce costs and make better use of existing resources.

Workforce Optimization

IoT-based rostering and location tracking can help improve staff management in smart hospitals. These systems can provide real-time information about the location of medical staff and the number of patients requiring care. This helps hospital managers adjust staff shifts based on current patient demand. This ensures that there are enough staff in busy areas, especially during emergencies or peak hours. The system also tracks changes in a patient's condition. For example, if a patient's vital signs change or an emergency alarm is triggered, the system notifies staff immediately. This helps caregivers focus on the most urgent tasks and provide timely care. By reducing the need for manual planning, IoT systems can assign tasks faster and more accurately. This reduces the workload of employees and improves the hospital's operational efficiency. It also helps to distribute tasks fairly, thereby reducing employee stress and increasing their job satisfaction.

Energy Management Systems

Smart sensors built into hospital infrastructure help manage energy use and support environmental goals. These IoT sensors monitor and control systems such as heating, cooling, lighting, and other high-energy-consuming devices. They work continuously and automatically adjust based on room occupancy, time of day, and current conditions. This helps reduce energy waste, lower electricity bills, and reduce the hospital's carbon footprint.

These systems not only save energy but also create a comfortable and safe indoor environment for patients. They regulate temperature, humidity and air quality in real time. This is especially important in areas such as operating rooms and intensive care units, where a stable environment is crucial for patients' recovery. Smart energy systems offer two major benefits: reduced operating costs and improved patient comfort. These systems can also support hospitals' efforts to be more sustainable and environmentally friendly. All in all, smart sensors make hospitals more efficient and better suited to providing high-quality, patient-centered care.

CHALLENGES IN MIoT IMPLEMENTATION

While MIoT offers significant opportunities to enhance patient care and hospital operations, its deployment faces several obstacles that limit widespread adoption. Key issues include ensuring the security and privacy of sensitive health data, overcoming interoperability barriers among heterogeneous devices, and managing scalability as hospitals expand and the volume of connected devices grows. Addressing these challenges requires robust cybersecurity frameworks, standardized communication protocols, and advanced computing solutions such as cloud and edge technologies. The following subsections examine these challenges in detail and outline potential strategies for overcoming them.

Security and Privacy

The integration of IoT into healthcare has raised serious concerns about data security and patient privacy. As sensitive health data is constantly transmitted over the network, the risk of cyberattacks and data breaches increases. To reduce these risks, hospitals must take strong cybersecurity measures. This includes using encryption technology to protect data, setting up firewalls, and controlling who can access patient information. In addition, regular security checks and system updates are required to address new threats and comply with data protection regulations (Balfaqih et al., 2023a, Balfaqih, 2024).

A comprehensive cybersecurity program helps protect patient records, build trust, and ensure the security of IoT medical systems. Blockchain technology can also improve security. Its decentralized and tamper-proof design prevents unauthorized access and ensures data accuracy. It also increases transparency in data processing. By using blockchain, hospitals can create a more secure system for storing and sharing medical information. This reduces the risk of data breaches and increases patient confidence in connected medical technology (Bin Tajuddin et al., 2025).

Interoperability

One of the major challenges in applying IoT in healthcare is the lack of standardization between devices. Many IoT devices run on different platforms and use different communication protocols. This leads to compatibility issues, making it difficult for them to work together (Saleem et al., 2018). As a result, these devices often fail to connect smoothly with hospital systems. This limits data sharing and makes it difficult for departments to coordinate care effectively.

To address this, it's important to adopt shared standards and ensure devices can communicate across platforms. Standardization can simplify integration and improve data flow between systems. When devices work together, hospitals can build a more connected environment. This helps improve care coordination, supports faster decision-making, and leads to better treatment outcomes for patients.

Scalability

As hospitals continue to grow and serve more patients, IoT systems must handle more devices and larger amounts of data (Farahani et al., 2018). This puts pressure on traditional IT infrastructure. When the system is overloaded, it may slow down or even crash.

To cope with this growth, cloud computing and edge computing offer effective solutions. Cloud computing enables hospitals to store and process data on remote servers. It provides virtually unlimited storage and computing power. This prevents local systems from being overwhelmed even as IoT devices continuously collect and send data (Sachdeva et al., 2024).

Edge computing processes data close to where it is generated, such as on IoT devices. This can reduce latency and speed up decision making. By combining cloud computing and edge computing, hospitals can spread workloads. This can improve system performance and help manage data more efficiently (Nain, Pattanaik, & Sharma, 2022).

Combined, these technologies ensure that IoT systems remain responsive and reliable even as the number of patients and device usage grow. They help hospitals with scale operations without sacrificing speed or quality of care.

Cost-Benefit and Return on Investment

Introducing MIoT in hospitals requires significant upfront investments. These upfront costs include building the necessary infrastructure, purchasing IoT-enabled medical devices, integrating the systems, training staff, and setting up IT support. Hospitals may also need to upgrade their network systems, improve network security, and adjust existing workflows to support the new technology. For smaller hospitals with limited budgets, these expenses can be challenging. However, MIoT has been shown to deliver significant economic benefits over time. Through real-time monitoring and faster medical response, MIoT reduces hospital readmission rates, especially for chronic conditions such as heart failure, diabetes, and COPD. It also reduces medication errors through automated alerts and helps manage resources more efficiently by tracking device usage and monitoring inventory (Balfaqih & Alharbi, 2022).

Hospitals that adopt targeted MIoT solutions first, such as remote patient monitoring or smart medication management, can typically recoup their costs within 18 to 24 months. These early wins can help fund MIoT expansion to other departments. Hospital-wide deployments may take three to five years to break even. But these initiatives can often pay back 1.5 to 3 times the original investment due to increased efficiency and improved patient care.

REAL-WORLD CASE STUDIES

The practical adoption of the MIoT in healthcare highlights its transformative potential beyond theory. Real-world deployments demonstrate how smart hospitals can leverage IoT devices, predictive analytics, and connected infrastructures to improve patient care, optimize resources, and enhance operational efficiency. Case studies from South Korea and the United Kingdom provide concrete examples of how MIoT technologies are reshaping healthcare delivery, from improving treatment outcomes and patient satisfaction to streamlining workflows and reducing costs. The following subsections present these implementations in detail.

Smart Hospital Implementation in South Korea

A leading hospital in South Korea has fully deployed a MIoT system. The system includes wearables, smart beds, and predictive analytics tools (Dimitrov, 2016). Together, these technologies have brought significant improvements to patient care and hospital operations. Wearables can track vital signs such as heart rate and blood oxygen levels in real time, which can help detect health problems early and help doctors respond quickly. Smart beds can automatically adjust to improve patient comfort and help prevent pressure sores. Predictive analytics studies patient data to identify health trends and suggest better treatment options. The results are impressive. By managing chronic conditions more effectively, the hospital has reduced its readmission rate by 15%. Personalized care has also increased patient satisfaction through real-time health data and customized treatments. Patients now feel more involved in their treatment, and hospitals can respond to their needs more quickly. This case shows how MIoT can lead to better treatment outcomes and a more efficient healthcare system.

IoT-Driven Operational Efficiency in the UK

A hospital in the UK has adopted RFID and IoT sensor technology to improve equipment tracking and overall workflow (Antunes et al., 2018). These tools allow staff to see the real-time location and status of medical equipment. This means that essential tools are always available when needed. As a result, emergency response times become faster because staff no longer waste time looking for equipment. The system also records the use of equipment. This helps the hospital plan maintenance before problems occur. By resolving issues early, they have avoided breakdowns and reduced equipment maintenance costs by 30%.

FUTURE TRENDS IN MIoT

The future of MIoT lies in its integration with advanced digital technologies that promise to reshape healthcare delivery. Innovations such as 5G and edge computing are enabling real-time data processing and ultra-low latency applications, while blockchain ensures secure, transparent, and compliant management of sensitive health information. At the same time, robotic process automation (RPA) is streamlining hospital operations and enhancing precision in both clinical and administrative tasks. Together, these emerging trends point toward a more connected, secure, and efficient healthcare ecosystem, advancing the vision of fully digital and patient-centered smart hospitals.

5G and Edge Computing

5G combined with edge computing is revolutionizing smart healthcare by increasing data transmission speeds and reducing latency. In hospitals, fast decision making is critical, and 5G allows large amounts of data to be transmitted between devices and systems with very low latency. Edge computing can process this data close to its source, such as at the patient's bedside or inside a device. This allows healthcare providers to act on information in real time. This setup is particularly useful for telemedicine. High-quality video calls and remote diagnosis require strong and stable connections, and 5G can meet these needs. In robotic surgery, even the slightest delay can pose risks, and 5G's ultra-low latency and edge computing can ensure safe and precise surgery (Balfaqih et al., 2021).

Together, these technologies support faster, more reliable health care. They improve patient monitoring, emergency response, and remote procedures. Whether in urban or rural areas, 5G and edge computing can make modern healthcare more accessible, more responsive, and more efficient.

Blockchain Integration

Blockchain technology provides a secure and transparent way to manage sensitive patient data in healthcare. It uses a decentralized ledger to store health records. This makes the data immutable and ensures that only authorized users can access it. Therefore, the data remains accurate and protected from unauthorized disclosure or alteration. Blockchain also supports smart contracts. This enables hospitals, labs, insurance companies, and patients to share data securely and automatically. The system controls who can access what without compromising patient privacy (Balfaqih et al., 2023b).

In addition, blockchain helps hospitals comply with strict data privacy laws, such as the Health Insurance Portability and Accountability Act (HIPAA) and the General Data Protection Regulation (GDPR). It clearly records every data exchange, making it easier to track and audit.

Robotic Process Automation

RPA combined with the Internet of Things (IoT) will bring about significant changes in healthcare. IoT robots can help in high-precision surgeries, thereby improving accuracy and reducing complication rates. Within hospitals, these robots can also deliver medicines, consumables, and meals without human assistance. This improves the daily operations of hospitals. RPA can also handle many routine tasks. It can register patients, schedule appointments, and update medical records. This reduces the workload of hospital staff. Doctors and nurses can therefore have more time to directly care for patients. By reducing human errors and speeding up daily processes, RPA makes healthcare more consistent and efficient. It plays a key role in supporting smart hospitals and improving the overall quality of patient care.

CONCLUSION

MIoT is transforming healthcare by enabling continuous monitoring, personalized treatment, and data-based decision making. Tools such as smart beds, wearables, emergency alerts, and AI analytics are increasing safety of care, reducing readmissions, and improving hospital workflows. MIoT is also driving efficiency through improved equipment tracking, energy utilization, and staff management. While costly to set up, its long-term benefits—such as reduced medication errors, improved patient outcomes, and cost savings—make it a worthwhile investment. Key challenges such as cybersecurity, data privacy, and device compatibility must be addressed through encryption, blockchain, and standard communication protocols. Case studies in South Korea and the UK have shown that medical care and operations have been significantly improved through MIoT. New technologies such as 5G, edge computing, blockchain, and robotic process automation are further expanding the capabilities of smart hospitals. These advances can help provide fast and reliable healthcare services in both urban and rural areas. In India, MIoT can play a major role in improving the accessibility and efficiency of healthcare, especially in resource-limited areas. All in all, MIoT is a solid and pragmatic step towards building future-proof hospitals that will be able to provide high-quality, patient-centric healthcare on a scale.

REFERENCES

Abdulmalek, S., Nasir, A., Jabbar, W. A., Almuhaya, M. A., Bairagi, A. K., Khan, M. A. M., & Kee, S. H. (2022, October). IoT-based healthcare-monitoring system towards improving quality of life: A review. In Healthcare (Vol. 10, No. 10, p. 1993). MDPI.

Akhtar, N., Rahman, S., Sadia, H., & Perwej, Y. (2021). A holistic analysis of Medical Internet of Things (MIoT). *Journal of Information and Computational Science*, *11*(4), 209–222.

Aloini, D., Benevento, E., Stefanini, A., & Zerbino, P. (2023). Transforming health-care ecosystems through blockchain: Opportunities and capabilities for business process innovation. *Technovation*, *119*, 102557.

Anikwe, C. V., Nweke, H. F., Ikegwu, A. C., Egwuonwu, C. A., Onu, F. U., Alo, U. R., & Teh, Y. W. (2022). Mobile and wearable sensors for data-driven health monitoring system: State-of-the-art and future prospect. *Expert Systems with Applications*, *202*, 117362.

ANORBOYEVICH. A. R. (2024). Exploring the evolutionary path: A historical analysis of Internet of Things (IoT) technologies. Global Book Publishing Services, 1-70.

Antunes, R. S., Seewald, L. A., Rodrigues, V. F., & Costa, C. A. D.Jr, L. G., Righi, R. R. & Campanatti, G. (2018). A survey of sensors in healthcare workflow monitoring. *ACM Computing Surveys*, *51*(2), 1–37.

Awotunde, J. B., Adeniyi, A. E., Ogundokun, R. O., Ajamu, G. J., & Adebayo, P. O. (2021). MIoT-based big data analytics architecture, opportunities and challenges for enhanced telemedicine systems. Enhanced telemedicine and e-health: advanced IoT enabled soft computing framework, 199-220.

Babu, P. (2024), Future Trends in Remote Patient Monitoring: Innovations and Possibilities, *Firminiq*, Online: https://firminiq.com/future-trends-in-remote-patient-monitoring-innovations-and-possibilities/ [28/09/2025]

Balfaqih, M. (2024). Enhancing security and flexibility in smart locker systems: A multi-authentication approach with IoT integration. In *2024 21st Learning and Technology Conference (L&T)* (pp. 325-329). IEEE.

Balfaqih, M., & Alharbi, S. A. (2022). Associated information and communication technologies challenges of smart city development. *Sustainability*, *14*(23), 16240.

Balfaqih, M., Balfagih, Z., Almohammedi, A. A., & Alfawaz, K. M. (2023a). A Smart and Privacy-Preserving Logistics System Based on IoT and Blockchain Technologies. In *2023 1st International Conference on Advanced Innovations in Smart Cities (ICAISC)* (pp. 1-5). IEEE.

Balfaqih, M., Balfagih, Z., Lytras, M. D., Alfawaz, K. M., Alshdadi, A. A., & Al-solami, E. (2023b). A blockchain-enabled IoT logistics system for efficient tracking and management of high-price shipments: A resilient, scalable and sustainable approach to smart cities. *Sustainability*, *15*(18), 13971.

Balfaqih, M., Jabbar, W., Khayyat, M., & Hassan, R. (2021). Design and development of smart parking system based on fog computing and internet of things. *Electronics (Basel)*, *10*(24), 3184.

Bin Tajuddin, A. F., Yusof, M. H. M., Almohammedi, A. A., Dael, F. A., Balfaqih, M., & Qaid, N. (2025, May). Developing Secure MFA through QR code, SHA-256 OTP & Device Verification. In *2025 International Conference on Innovation in Artificial Intelligence and Internet of Things (AIIT)* (pp. 1-6). IEEE.

Blot, S., Ruppé, E., Harbarth, S., Asehnoune, K., Poulakou, G., Luyt, C. E., & Zahar, J. R. (2022). Healthcare-associated infections in adult intensive care unit patients: Changes in epidemiology, diagnosis, prevention and contributions of new technologies. *Intensive & Critical Care Nursing*, *70*, 103227. PMID: 35249794

Codina Barberà, M. (2024). Gait-Analysis MIoT Platform for Health Assessment.

Dias, D., & Paulo Silva Cunha, J. (2018). Wearable health devices—Vital sign monitoring, systems and technologies. *Sensors (Basel)*, *18*(8), 2414. PMID: 30044415

Dimitrov, D. V. (2016). Medical internet of things and big data in healthcare. *Healthcare Informatics Research*, *22*(3), 156–163. PMID: 27525156

Dogheim, G. M., & Hussain, A. (2023). Patient care through AI-driven remote monitoring: Analyzing the role of predictive models and intelligent alerts in preventive medicine. *Journal of Contemporary Healthcare Analytics*, *7*(1), 94–110.

Farahani, B., Firouzi, F., Chang, V., Badaroglu, M., Constant, N., & Mankodiya, K. (2018). Towards fog-driven IoT eHealth: Promises and challenges of IoT in medicine and healthcare. *Future Generation Computer Systems*, *78*, 659–676.

Ferre, M., Batista, E., Solanas, A., & Martinez-Balleste, A. (2021). Smart health-enhanced early mobilisation in intensive care units. *Sensors (Basel)*, *21*(16), 5408. PMID: 34450850

Guarascio-Howard, L. (2011). Examination of wireless technology to improve nurse communication, response time to bed alarms, and patient safety. *HERD: Health Environments Research & Design Journal*, *4*(2), 109–120. PMID: 21465438

Haghani, N. (2023). Reduction of Blood Pressure Among Adult Hypertensive Patients Using Remote Patient Monitoring Followed by Telehealth (Doctoral dissertation, University of Massachusetts Global).

Javaid, M., Haleem, A., Singh, R. P., Rab, S., Suman, R., & Khan, I. H. (2022). Evolutionary trends in progressive cloud computing based healthcare: Ideas, enablers, and barriers. *International Journal of Cognitive Computing in Engineering*, *3*, 124–135.

Kamal, H. (2024). IoT-enabled Smart Hospital Infrastructure for Efficient Resource Management: Designing IoT-enabled systems to optimize resource allocation and management within hospitals. *Journal of AI in Healthcare and Medicine*, *4*(2), 45–56.

Kamal, R. (2025), Sleep Better with AI and IoT-Enabled Smart Beds, *Intuz*, Online: https://www.intuz.com/blog/sleep-smarter-with-multifunctional-smart-bed [28/09/2025]

Karvounis, E., Polymeni, S., Tsipouras, M., Koritsoglou, K., & Tzovaras, D. (2021, September). Smart beds and bedding surfaces for personalized patient care: a review. In 2021 6th South-East Europe Design Automation, Computer Engineering, Computer Networks and Social Media Conference (SEEDA-CECNSM) (pp. 1-8). IEEE.

Koppel, R. (2022). Estimating the United States' cost of healthcare information technology. In *Healthcare Information Management Systems: Cases, Strategies, and Solutions* (pp. 3–38). Springer International Publishing.

Kumar, A., Nanthaamornphong, A., Selvi, R., Venkatesh, J., Alsharif, M. H., Uthansakul, P., & Uthansakul, M. (2023). Evaluation of 5G techniques affecting the deployment of smart hospital infrastructure: Understanding 5G, AI and IoT role in smart hospital. *Alexandria Engineering Journal*, *83*, 335–354.

Kumar, M., Kumar, A., Verma, S., Bhattacharya, P., Ghimire, D., Kim, S. H., & Hosen, A. S. (2023). Healthcare Internet of Things (H-IoT): Current trends, future prospects, applications, challenges, and security issues. *Electronics (Basel)*, *12*(9), 2050.

Lewandowski, J. (2014). Mobile application of artificial intelligence to vital signs monitoring: multi parametric, user adaptable model for ubiquitous well-being monitoring (Doctoral dissertation, Coventry University).

Madanian, S., Chinbat, T., Subasinghage, M., Airehrour, D., Hassandoust, F., & Yongchareon, S. (2024). Health IoT Threats: Survey of Risks and Vulnerabilities. *Future Internet, 16*(11), 389.

Mahmmod, B. M., Naser, M. A., Al-Sudani, A. H. S., Alsabah, M., Mohammed, H. J., Alaskar, H., & Abdulhussain, S. H. (2024). Patient Monitoring System Based on Internet of Things: A Review and Related Challenges with Open Research Issues. *IEEE Access : Practical Innovations, Open Solutions.*

Maleki Varnosfaderani, S., & Forouzanfar, M. (2024). The role of AI in hospitals and clinics: Transforming healthcare in the 21st century. *Bioengineering (Basel, Switzerland), 11*(4), 337. PMID: 38671759

Manavi, T., Zafar, H., & Sharif, F. (2024). An era of digital healthcare—A comprehensive review of sensor technologies and telehealth advancements in chronic heart failure management. *Sensors (Basel), 24*(8), 2546. PMID: 38676163

Muthunagai, S. U., Girija, M. S., Kumar, B. P., & Anitha, R. (2024). Storage Management Techniques for Medical Internet of Things (MIoT). In Intelligent Systems and Sustainable Computational Models (pp. 34-47). Auerbach Publications.

Nain, G., Pattanaik, K. K., & Sharma, G. K. (2022). Towards edge computing in intelligent manufacturing: Past, present and future. *Journal of Manufacturing Systems, 62*, 588–611.

Osama, M., Ateya, A. A., Sayed, M. S., Hammad, M., Pławiak, P., Abd El-Latif, A. A., & Elsayed, R. A. (2023). Internet of medical things and healthcare 4.0: Trends, requirements, challenges, and research directions. *Sensors (Basel), 23*(17), 7435. PMID: 37687891

Piaggio, D., Zarro, M., Pagliara, S., Andellini, M., Almuhini, A., Maccaro, A., & Pecchia, L. (2023). The use of smart environments and robots for infection prevention control: A systematic literature review. *American Journal of Infection Control, 51*(10), 1175–1181. PMID: 36924997

Poyyamozhi, M., Murugesan, B., Rajamanickam, N., Shorfuzzaman, M., & Aboelmagd, Y. (2024). IoT—A Promising Solution to Energy Management in Smart Buildings: A Systematic Review, Applications, Barriers, and Future Scope. *Buildings (Basel, Switzerland), 14*(11), 3446.

Raisa, F. F., Orthy, N. T., Nibir, M. D., & Faysal, H. M. (2023). IoT based medicine dispenser for improving medication adherence.

Reddy, A., Ghantasala, G. P., Kurra, M., & Ayyappa, R. M. K. (2025a). PATIENT EMPOWERMENT THROUGH SECURE DATA MANAGEMENT. Using Blockchain Technology in Healthcare Settings: Empowering Patients with Trustworthy Data, 153.

Reddy, K. V., Saketh, G., Priyanshu, S. S., Nitesh, M., Hussain, S. K., & Sharma, K. V. (2025b). AI-Driven Healthcare Management Platform: Enhancing Accessibility, Efficiency, and Security in Digital Health Systems. Synthesis: A Multidisciplinary Research Journal, 3(1), 1-14.

Sachdeva, S., Bhatia, S., Al Harrasi, A., Shah, Y. A., Anwer, K., Philip, A. K., & Halim, S. A. (2024). Unraveling the role of cloud computing in health care system and biomedical sciences. *Heliyon*. PMID: 38601602

Saleem, J., Hammoudeh, M., Raza, U., Adebisi, B., & Ande, R. (2018, June). IoT standardisation: Challenges, perspectives and solution. In *Proceedings of the 2nd international conference on future networks and distributed systems* (pp. 1-9).

Sargiotis, D. (2024). Data Security and Privacy: Protecting Sensitive Information. In *Data Governance: A Guide* (pp. 217–245). Springer Nature Switzerland.

Shanmathi, N., & Jagannath, M. (2018). Computerised decision support system for remote health monitoring: A systematic review. *Ingénierie et Recherche Biomédicale : IRBM = Biomedical Engineering and Research*, *39*(5), 359–367.

Shao, M., Zhou, Z., Bin, G., Bai, Y., & Wu, S. (2020). A wearable electrocardiogram telemonitoring system for atrial fibrillation detection. *Sensors (Basel)*, *20*(3), 606. DOI: 10.3390/s20030606 PMID: 31979184

Tantray, J., Kosey, S., Prajapati, J. B., & Prajapati, B. (2024). The Future of the Healthcare System: A Meta-Analysis of Remote Patient Monitoring. In Handbook on Augmenting Telehealth Services (pp. 105-126). CRC Press.

Thakur, H., & Rao, R. (2024). Emphasis of infection prevention and control: A review. *Journal of Population Therapeutics and Clinical Pharmacology = Journal de la Thérapeutique des Populations et de la Pharmacologie Clinique*, *31*, 2238–2249.

Uslu, B. Ç., Okay, E., & Dursun, E. (2020). Analysis of factors affecting IoT-based smart hospital design. *Journal of Cloud Computing (Heidelberg, Germany)*, *9*(1), 67. PMID: 33532168

Vuong, Q. P. (2024). The potential for artificial intelligence and machine learning in healthcare: the future of healthcare through smart technologies.

Wen, M. H., Chen, P. Y., Lin, S., Lien, C. W., Tu, S. H., Chueh, C. Y., & Bai, D. (2024). Enhancing Patient Safety Through an Integrated Internet of Things Patient Care System: Large Quasi-Experimental Study on Fall Prevention. *Journal of Medical Internet Research*, *26*, e58380. PMID: 39361417

Yalli, J. S., Hasan, M. H., & Badawi, A. (2024). Internet of things (iot): Origin, embedded technologies, smart applications and its growth in the last decade. *IEEE Access : Practical Innovations, Open Solutions*.

Yesmin, T., Carter, M. W., & Gladman, A. S. (2022). Internet of things in healthcare for patient safety: An empirical study. *BMC Health Services Research*, *22*(1), 278. PMID: 35232433

Zhao, J., Zhang, S., Sun, Y., Zhou, N., Yu, H., Zhang, H., & Jia, D. (2022). Wearable optical sensing in the medical internet of things (MIoT) for pervasive medicine: Opportunities and challenges. *ACS Photonics*, *9*(8), 2579–2599.

Chapter 5
Digitalization and Automation of Logistics and Supply Chain Management in the Industry 4.0 Era

Hasan Balfaqih
https://orcid.org/0000-0003-2289-1142
Effat University, Saudi Arabia

ABSTRACT

This chapter examines the impact of Industry 4.0 on the digitalization and automation of logistics and supply chain management (SCM). It explores how technologies such as the Internet of Things (IoT), big data analytics, artificial intelligence (AI), blockchain, and automation are transforming supply chain processes by improving efficiency, transparency, and decision-making. Particular emphasis is given to data-driven models, predictive analytics, and digital twins, which enable real-time optimization, operational agility, and resilience in global supply networks. The findings demonstrate that digital transformation streamlines logistics operations, reduces inefficiencies, and supports proactive disruption management. By integrating these technologies, organizations can build smarter, adaptive, and sustainable supply chains. This study provides insights into how Industry 4.0 is reshaping logistics and SCM, offering practical guidance for academics, practitioners, and policymakers.

DOI: 10.4018/979-8-3373-0154-9.ch005

INTRODUCTION

The rapid advancement of digitalization has transformed numerous industries, including transportation, education, and healthcare. Supply Chain Management (SCM) and logistics are also experiencing a significant shift as digital technologies reshape traditional operations, enhancing efficiency, transparency, and overall performance (Alshibli et al., 2024). Digitalization plays a crucial role in integrating supply chain processes, ensuring operational continuity, and fostering long-term business prosperity. The emergence of cyber-physical systems, which naturally result from digital breakthroughs, provides access to diverse and abundant data from various sources. This wealth of data enables businesses and organizations to make informed decisions, optimize operations, and gain a competitive edge. As a result, the growing interest in digitalization has led to an increased focus on leveraging big data and analytics for supply chain improvements.

This chapter explores the key factors driving digital transformation in SCM and logistics, highlighting the role of data utilization, automation, and advanced technologies. The most widely cited research on digital supply chains in SCM and logistics is the work of Wang et al. (2016), which categorizes the impact of work analytics on supply chain processes. Further research by Wang et al. (2017) examines the evolution of big data, emphasizing the need for action-oriented mechanisms in logistics processes such as distribution, storage, and transportation. These studies underscore how leveraging big data can lead to significant benefits for supply chain stakeholders. Similarly, Zhong et al. (2017) analyzed the use of big data in SCM within both manufacturing and service industries, identifying current trends and future applications. Tiwari et al. (2018) also contributed to this field by studying the role of big data analytics in SCM between 2010 and 2016, providing a sectoral perspective on its impact.

Beyond big data, digital innovations in manufacturing, logistics, and supply chain operations have been revolutionized by automation, the Internet of Things (IoT), and Industry 4.0. Technological advancements such as smart sensors, robotics, automated warehouses, self-driving vehicles, intelligent transportation systems, and cloud computing are redefining logistics and SCM (Evtodieva, 2019). Additionally, advancements in information management—such as improved data quality, structured data availability, complete supply chain visibility, and enhanced data processing capabilities—further reinforce the benefits of digitalization in logistics. Other critical technologies, including IoT, blockchain, 3D printing, drones, augmented reality, automated guided vehicles, and wireless sensor networks, are integral to this transformation (Rajković et al., 2022).

Implementing digitalization within organizations is essential for enhancing operational efficiency and increasing automation. Digital transformation minimizes business errors and optimizes human resource allocation, reducing reliance on manual processes. One of the most significant benefits of supply chain digitalization is improved customer satisfaction, particularly in logistics and delivery services. Real-time tracking, seamless documentation processing, and enhanced visibility enable businesses to offer more reliable and timely deliveries. Additionally, digitalization leads to substantial cost and time savings, with automation streamlining repetitive tasks such as invoice processing, delivery documentation, and order fulfillment (Rajković et al., 2022).

Over the past decade, digital platforms and applications for logistics have steadily expanded, driven by companies' strategic efforts to enhance productivity and reduce costs. Automation and digitalization significantly impact supply chain networks by improving transparency, scheduling accuracy, and operational agility. Furthermore, digital transformation has reshaped communication and social interactions within organizations, altering how information is acquired and transmitted through advanced devices such as smartphones, tablets, drones, and intelligent wearables. As industries across all sectors undergo digital transformation, SCM is no exception. The digitalization of supply chains requires organizations to adopt new strategies and technologies to remain competitive in an increasingly digital landscape (Gritsenko et al., 2022).

This chapter delves into the critical aspects of digitalization and automation in supply chain and logistics processes, emphasizing their role in shaping Industry 4.0. The discussion will highlight key technologies, challenges, and opportunities, providing insights into how businesses can effectively leverage digital transformation to enhance supply chain efficiency and resilience.

INDUSTRY 4.0: CONCEPT AND EVOLUTION

The concept of Industry 4.0 was introduced in 2011 by Henning Kagermann, a former top executive of SAP, a leading software corporation in Germany (Paprocki, 2016). Industry 4.0 is often referred to as the "Fourth Industrial Revolution," "smart manufacturing," "industrial internet," or "integrated industry" (Hofmann & Rusch, 2017). Over the past decade, this concept has gained increasing attention from researchers, policymakers, and industry professionals worldwide (Rennung et al., 2016; Liao et al., 2017). The growing significance of Industry 4.0 is reflected in the rising number of Google searches containing terms such as "Industry 4.0" and "Fourth Industrial Revolution" since December 2018, indicating the expanding global interest in this transformative movement.

Industry 4.0 has presented new research opportunities while also providing business leaders with strategic insights into how they can integrate advanced digital technologies into their operations (Lopes de Sousa Jabbour et al., 2018). However, Industry 4.0 extends beyond mere digitization; it encompasses a broad spectrum of technologies such as Artificial Intelligence (AI), 3D printing, big data analytics, Augmented Reality (AR), and IoT as shown in Figure 1. The fundamental goal of Industry 4.0 is to achieve the highest levels of quality and operational autonomy by optimizing human-machine interaction and synchronizing data flow across interconnected systems (Balfaqih & Alharbi, 2022).

Figure 1. Technologies of Industry 4.0 (Öztuna, 2021)

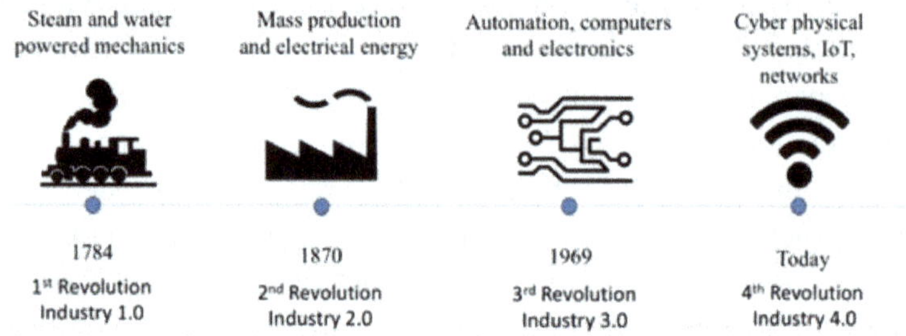

Figure 1 highlights the core technologies driving Industry 4.0, including IoT, AI, cloud computing, blockchain, big data analytics, 3D printing, augmented reality, automation, and wireless sensor networks (Öztuna, 2021). These technologies enhance decision-making, improve operational efficiency, and enable predictive analytics for optimizing logistics and supply chain processes.

In addition to technological advancements, Industry 4.0 also represents a paradigm shift in organizational culture and business models. It encourages companies to move away from traditional linear value chains toward integrated and dynamic value networks, where information flows seamlessly across stakeholders. This transformation enables greater transparency, responsiveness, and collaboration across supply chains, paving the way for resilient and adaptive enterprises in a volatile global market (Udeh et al., 2024).

Moreover, the evolution of Industry 4.0 has been closely associated with sustainability objectives. Smart technologies, when implemented effectively, can reduce waste, optimize resource utilization, and minimize environmental impact, thereby contributing to the broader goals of sustainable development (Balfaqih, 2022). For instance, predictive maintenance enabled by IoT sensors reduces unnecessary replacements, while additive manufacturing (3D printing) supports more sustainable production through on-demand and localized manufacturing.

Another important dimension of Industry 4.0 is its impact on the workforce. While automation and artificial intelligence raise concerns about job displacement, they also create new opportunities for highly skilled roles in areas such as data science, robotics engineering, and cyber-physical systems design. This shift highlights the urgent need for upskilling and reskilling initiatives to prepare the workforce for the digital future (Bécue et al., 2021).

Finally, Industry 4.0 is not a static concept but a continuously evolving phenomenon. The integration of emerging technologies such as 6G, quantum computing, and bio-digital convergence is likely to accelerate the transition toward what some scholars are already calling Industry 5.0, a stage that emphasizes the synergy between advanced technology and human creativity (Peters et al., 2022). This perspective suggests that Industry 4.0 serves not only as a technological revolution but also as a bridge toward a future where innovation, sustainability, and human-centric design converge.

Evolution of Industry 4.0 and Core Technologies

The emergence of Industry 4.0 has transformed the way supply chains and logistics operate, leading to more efficient, data-driven, and interconnected supply networks. This paradigm shift enhances communication among supply chain agents and extends structural connectivity across global supply networks, ultimately simplifying supply chain functionalities (Görçün, 2016). At the heart of this transformation is IoT, which facilitates seamless connectivity among smart devices to create value in supply chain management. Figure 2 illustrates the evolution of industrial revolutions, depicting the transition from mechanization (Industry 1.0) to mass production (Industry 2.0), automation (Industry 3.0), and finally, the integration of smart technologies in the Fourth Industrial Revolution (Industry 4.0) (Dorduncu & Oruç, 2021).

Figure 2. Evolution of Industry 4.0 (Dorduncu & Oruç, 2021)

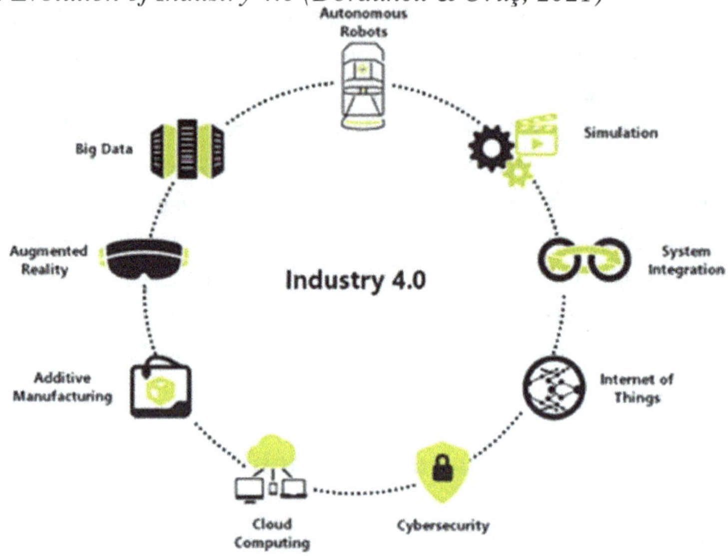

The evolution of Industry 4.0 has been accelerated by the convergence of multiple technologies working in synergy. Big Data analytics, for example, plays a central role in transforming raw data into actionable insights, enabling companies to make predictive and prescriptive decisions in real time. This capacity is particularly valuable in logistics, where demand forecasting, inventory optimization, and transportation planning can be enhanced through advanced data models (Sarker, 2021).

Another defining element of Industry 4.0 is the integration of Artificial Intelligence (AI) and Machine Learning (ML) into supply chain systems. These technologies empower organizations to detect patterns, automate decision-making, and respond proactively to market disruptions. In combination with cloud computing, AI ensures that vast amounts of data can be processed efficiently, shared across organizational boundaries, and used to support collaborative planning (Sundaramurthy et al., 2022).

Furthermore, the development of cyber-physical systems (CPS) has reinforced the digital-physical integration that characterizes Industry 4.0. CPS links sensors, machines, and control systems in real time, allowing for continuous monitoring and autonomous adjustments in production and logistics. When combined with blockchain technology, CPS also ensures transparency, security, and trust in supply chain transactions, thus addressing long-standing challenges in global trade such as fraud and counterfeit products (Vermesan & Friess, 2022).

In addition, additive manufacturing (3D printing) has emerged as a disruptive technology within Industry 4.0, enabling decentralized production and reducing dependency on large inventories. This technology enhances supply chain resilience by allowing organizations to manufacture spare parts and customized products on demand, thereby shortening lead times and minimizing waste (Choi et al., 2022).

The trajectory of Industry 4.0 highlights not only the technological advancements but also their profound implications for business models and strategic management. Organizations are increasingly compelled to embrace digital transformation as a means of maintaining competitiveness in dynamic global markets. As Industry 4.0 continues to evolve, it is expected that future supply chains will become even more autonomous, sustainable, and customer-centric, laying the foundation for the transition toward Industry 5.0.

Applications of Industry 4.0 in Manufacturing and Supply Chains

The central aim of Industry 4.0 is to establish intelligent, fully automated, and data-driven manufacturing ecosystems where machines, systems, and products interact autonomously with minimal human intervention (Sokolov & Ivanov, 2015). This paradigm shift has fostered the development of smart industrial systems that integrate multiple interconnected components. Smart products are embedded with sensors that enable real-time monitoring and communication, while smart logistics relies on automation and tracking technologies to optimize delivery networks. Smart engineering incorporates artificial intelligence, digital twin models, and advanced simulations to enhance design and operational efficiency. Supply chains are strengthened through smart suppliers, who use digital integration to improve coordination and transparency across networks. At the production level, smart manufacturing processes leverage IoT-enabled automation and adaptive systems to achieve higher flexibility and productivity. Complementing these are smart equipment and devices, including robotics, autonomous vehicles, and machine learning-powered predictive maintenance, which ensure operational continuity and reduced downtime. Together, these innovations are transforming traditional supply chain models into more resilient, data-driven, and responsive systems. For the logistics sector, adapting to these technologies is no longer optional but essential for sustaining competitiveness in an interconnected global economy (Kamble et al., 2018).

DIGITAL TRANSFORMATION AND LOGISTICS AUTOMATION

The digital transformation of logistics is expected to permeate all levels of supply chain operations, fundamentally reshaping how goods, information, and value move across global networks. Its promise lies in enhancing efficiency, transparency, and automation, with technologies enabling real-time decision-making and seamless coordination among stakeholders (Winkle & Zinsmeister, 2019). While digital technologies are often associated with productivity improvements and process automation, full automation in logistics remains constrained by structural and operational challenges. Flexibility, adaptability, and the necessity of human oversight remain central, as logistics environments are dynamic and demand context-specific responses to fluctuating conditions such as seasonal variations, geopolitical disruptions, and sudden demand surges (Grosse et al., 2017). Consequently, rather than focusing exclusively on complete physical automation, digitalization efforts often emphasize process optimization, data-driven decision-making, and enhanced integration across the supply chain.

Beyond process optimization, digital transformation offers a significant leap in achieving real-time visibility and traceability across global logistics operations. The deployment of IoT sensors, RFID tagging, and GPS-enabled tracking systems provides continuous monitoring of goods in transit, delivering end-to-end situational awareness that empowers companies to proactively respond to risks such as shipment delays, product theft, or environmental hazards. This transparency not only improves operational responsiveness but also fosters trust among customers and partners, as accurate and timely data sharing has become indispensable in today's demand-driven and customer-centric markets. Furthermore, visibility contributes to compliance with regulatory frameworks, particularly in industries such as pharmaceuticals and food, where traceability is critical to ensuring quality and safety standards.

Simultaneously, big data analytics and artificial intelligence (AI) are redefining logistics planning and forecasting, offering predictive and prescriptive insights that go far beyond traditional approaches. By analyzing large, heterogeneous datasets from sales channels, market trends, and multimodal transportation networks, firms can enhance demand forecasting accuracy, balance inventory levels, and minimize the risks of stockouts or overstocks. Predictive analytics enables logistics managers to anticipate potential disruptions—such as extreme weather events or supplier delays—while prescriptive analytics supports the identification of optimal routing, warehousing, or sourcing strategies in real time. These tools strengthen supply chain resilience by embedding agility and responsiveness into decision-making processes, positioning organizations to navigate uncertainty more effectively.

Another transformative dimension of digital logistics is the rise of platform-based ecosystems, which foster unprecedented levels of collaboration across the supply chain. Digital platforms and marketplaces are reshaping logistics models by integrating manufacturers, logistics service providers, and retailers into interconnected networks. Leveraging cloud computing, these ecosystems facilitate real-time data sharing and process integration, while blockchain technologies provide tamper-proof transaction records, enhancing security and accountability in cross-border operations. This evolution marks a shift away from fragmented, siloed operations toward highly integrated ecosystems, where freight and supply chain services are transacted within digital marketplaces that prioritize efficiency, transparency, and trust.

Despite these advances, organizational and workforce challenges remain critical hurdles to the effective adoption of digital logistics. The transition requires not only investment in digital infrastructure but also significant cultural and structural changes within organizations. Employees must continuously adapt to new digital workflows, necessitating reskilling in areas such as data analytics, digital platform management, and human–machine collaboration. While automation reduces reliance on manual labor for repetitive tasks, human oversight is indispensable for areas requiring judgment, creativity, and ethical decision-making. Organizations that succeed in striking the right balance—leveraging digital tools while empowering human expertise—are better positioned to achieve long-term competitive advantages and sustainability in digital supply chains.

Looking forward, the convergence of emerging technologies such as 5G, edge computing, and digital twins is expected to accelerate logistics transformation. 5G promises ultra-reliable, low-latency communication, enabling faster and more stable data transmission between logistics nodes. Edge computing complements this by decentralizing decision-making and processing data closer to its source, which is critical for real-time logistics operations such as fleet management and warehouse robotics. Digital twins—virtual replicas of supply chains—introduce new possibilities for simulation, predictive maintenance, and risk analysis, allowing firms to test strategies in digital environments before deploying them in reality. Together, these advancements pave the way for logistics systems that are not only efficient and transparent but also adaptive, resilient, and aligned with global sustainability objectives.

Industry 4.0 and Logistics Automation

Industry 4.0 is transforming logistics and SCM by introducing advanced digitalization and automation capabilities that reshape business operations. Concepts such as IoT, Digital Supply Networks (DSN), SCM 4.0, e-logistics, and Logistics 4.0 are frequently used to describe its role in creating highly interconnected and intelligent supply chains (Kayikci, 2018). At its core, Industry 4.0 facilitates real-time communication and seamless data exchange across organizational components, improving supply chain visibility, operational agility, and decision-making accuracy.

The impact of Industry 4.0 extends across four critical dimensions of supply chain management: integration, purchasing, distribution, and operations. Integration focuses on interoperability and seamless data sharing among supply chain partners, enabling transparency and collaboration across networks. Purchasing processes are increasingly automated, allowing real-time interaction and closer collaboration with suppliers. In distribution, advanced technologies optimize transportation networks, enhance last-mile delivery, and streamline logistics workflows. Meanwhile, operations benefit from AI-driven decision-making, predictive analytics, and process automation, which collectively improve efficiency and adaptability (Kayikci, 2018).

By embedding these capabilities, Industry 4.0 enables supply chains to become more resilient and responsive. Real-time data analytics enhance the ability to anticipate disruptions, reduce lead times, and support proactive, data-driven strategies. Ultimately, these advancements contribute to greater efficiency, improved decision-making, and stronger competitive positioning for organizations navigating the complexities of modern logistics. One of the most critical areas where these digital innovations are applied is warehousing and order fulfillment, which represents the backbone of logistics operations.

Smart Warehousing and Order Fulfillment

Warehouse order-picking is one of the most labor-intensive and cost-critical processes in logistics, making it a primary target for digital transformation under Industry 4.0. To improve efficiency, accuracy, and worker productivity, several smart picking technologies have been introduced, each leveraging advanced tools such as augmented reality, voice interaction, and automation. Among the most widely adopted are Pick-by-Vision, Pick-by-Voice, and Part-to-Picker systems, which provide innovative alternatives to traditional manual or paper-based methods. These technologies streamline warehouse operations, reduce errors, and enable faster fulfillment in increasingly complex supply chains.

Pick-by-Vision systems leverage AR and wearable devices to enhance warehouse picking operations. These systems use smart glasses or tablets connected via Wi-Fi to a warehouse management system (WMS), providing real-time visual instructions to workers. This technology enables warehouse employees to efficiently locate products, reducing errors and increasing speed. Some Pick-by-Vision models also incorporate built-in scanners (e.g., cameras) to eliminate the need for separate handheld devices, further optimizing the picking process (Lager et al., 2021).

Pick-by-Voice systems utilize voice-guided picking technology to replace traditional paper-based or handheld scanner-based picking methods. In this system, workers wear a headset connected to a terminal that communicates directly with the WMS. The system verbally instructs pickers on which items to collect, eliminating the need for manual data entry. Once an item is picked, the worker confirms the action using voice commands, ensuring hands-free and efficient operations. This technology improves worker productivity, reduces errors, and enhances overall warehouse efficiency (Lager et al., 2021).

Unlike traditional picker-to-part systems, which require workers to move throughout the warehouse, part-to-picker systems bring items directly to the worker. One example is monorail conveyor systems, which transport products to designated picking stations. While these systems increase picking efficiency (measured in picks per hour), they may lack flexibility in warehouse planning and require significant infrastructure investments (Lager et al., 2021).

While these technologies demonstrate tangible progress in warehouse efficiency, research perspectives provide a broader view of how Industry 4.0 frameworks are shaping logistics transformation.

Evolving Research Perspectives in Industry 4.0

Research on Industry 4.0 has increasingly emphasized its transformative role in logistics and supply chain management, offering a range of frameworks and models for implementation. Barata et al. (2018) highlighted the importance of mobile supply chain management and system integration, demonstrating how digital connectivity strengthens coordination across logistics networks. Similarly, Balfaqih et al. (2022) examined the role of IoT in smart manufacturing, illustrating its capacity to enhance automation and improve efficiency. Dallasega et al. (2018) contributed a framework that identifies the significance of Industry 4.0 in fostering collaborative supply chains, emphasizing the value of interconnected systems and cooperative strategies. Collectively, these studies underscore how Industry 4.0 serves as both a managerial and technological enabler of digital logistics.

As supply chains continue to evolve, experts argue that traditional models are becoming increasingly obsolete, giving way to digitally integrated and automated networks (Witkowski, 2017). Core technologies such as Cyber-Physical Systems (CPS), the Industrial Internet of Things (IIoT), smart factories, big data analytics, and blockchain are reshaping the landscape of logistics (Balfaqih et al., 2023; Ivanov et al., 2018; Douaioui et al., 2018; Kache & Seuring, 2017). CPS integrates physical assets with digital control, enabling seamless automation and responsiveness. IIoT builds on this foundation by deploying smart sensors and interconnected devices for real-time monitoring and tracking. Smart factories leverage artificial intelligence, robotics, and machine learning to create adaptive and intelligent production environments, while big data analytics provides predictive insights that enhance decision-making and resilience. Complementing these innovations, blockchain introduces secure and transparent transaction systems, fostering trust and traceability across supply chains.

Looking ahead, Industry 4.0 research in logistics points toward several promising directions. The integration of artificial intelligence and machine learning is expected to optimize logistics operations further, enabling predictive and autonomous decision-making. Blockchain continues to gain traction as a tool for enhancing transparency and traceability, particularly in complex global supply networks. Emerging technologies such as autonomous vehicles and drones are anticipated to redefine last-mile delivery, improving both efficiency and accessibility (Khayyat et al., 2024). At the same time, the growing digitalization of logistics highlights the need to address cybersecurity and data privacy challenges, which remain critical barriers to adoption. Finally, sustainability has become a central theme, with increasing focus on how Industry 4.0 can enable green logistics through energy-efficient technologies, optimized resource usage, and environmentally conscious supply chain practices.

AUTOMATION IN SUPPLY CHAIN MANAGEMENT (LSCM)

Automation in LSCM has evolved from mechanized material handling to the sophisticated, data-driven ecosystems enabled by Industry 4.0. While advances in AI, IoT, robotics, and predictive analytics have accelerated the shift toward smart, self-optimizing supply chains, the path to automation is not without barriers. High implementation costs, system integration challenges, workforce adaptation, and cybersecurity risks continue to complicate adoption, particularly for small and medium-sized enterprises. At the same time, research and practice are moving toward future-oriented solutions that integrate intelligence, autonomy, and sustainability, including AI-driven predictive logistics, blockchain-enabled transparency, hyper automation in warehousing, and green logistics innovations. Together, these perspectives frame automation as both a transformative opportunity and a complex challenge, highlighting the need for balanced strategies that ensure competitiveness, resilience, and long-term efficiency in global supply networks.

In addition to improving efficiency, automation in supply chains has a profound impact on decision-making and risk management. By leveraging real-time data streams from IoT sensors, automated systems can detect disruptions such as equipment failures, traffic delays, or demand fluctuations and respond proactively with corrective measures. This predictive and adaptive capability allows firms to minimize downtime, reduce losses, and maintain service levels even in volatile environments. For instance, digital twins combined with AI analytics create virtual models of supply chains that can simulate disruptions, enabling organizations to evaluate alternative responses before committing resources.

Moreover, the integration of collaborative robotics (cobots) and autonomous vehicles has redefined logistics operations, particularly in warehousing and last-mile delivery. Cobots support human workers by automating repetitive or hazardous tasks, thereby improving safety and productivity while allowing employees to focus on higher-value activities. Similarly, drones and autonomous delivery vehicles are being piloted to enhance speed and efficiency in urban logistics, especially under the growing demand for e-commerce fulfillment. These innovations not only improve responsiveness but also expand the scope of automation beyond traditional industrial environments.

Sustainability has also become a central consideration in automated supply chain strategies. Automated routing systems, for example, can optimize transportation schedules to reduce fuel consumption and carbon emissions. Smart energy management systems in automated warehouses further enhance eco-efficiency by minimizing energy waste and integrating renewable energy sources. As consumers and regulators push for greener operations, automation offers companies the ability to align operational efficiency with environmental responsibility.

Looking ahead, the future of automation in supply chain management is expected to move toward hyperconnected ecosystems where machine learning, blockchain, and edge computing converge. This integration will enable seamless interoperability between systems and stakeholders, ensuring greater transparency and coordination across global networks. However, the successful realization of this vision depends on organizations' ability to address ethical and governance issues, such as data privacy, cybersecurity, and equitable workforce transformation. Building trust among stakeholders and investing in digital literacy will be critical for ensuring that automation delivers sustainable value rather than short-term efficiency gains alone.

Evolution and Dimensions of Automation in LSCM

Logistics and supply chain management (LSCM) have experienced a steady but profound evolution in automation, shaped by the pursuit of efficiency, cost reduction, and operational resilience (Junge et al., 2019). Early efforts centered on mechanization, such as forklifts and conveyor belts, followed by the computerization of inventory tracking through Enterprise Resource Planning (ERP) systems and the introduction of automated order processing with barcoding and early warehouse technologies. These developments marked the initial transition from manual operations toward digital integration. With the advent of Industry 3.0, computer-assisted logistics planning became the foundation of modern supply chain digitalization. However, Industry 4.0 has dramatically accelerated this transformation by introducing smart automation, real-time analytics, and self-optimizing supply chains powered by AI, IoT, and robotics (Nitsche et al., 2021).

The drivers of this transformation are both technological and structural. Growing demand for speed and accuracy in supply chain operations, rising complexity in global trade networks, and labor shortages coupled with escalating wages have made automation not only desirable but necessary. At the same time, advances in AI, robotics, and predictive analytics have expanded the possibilities of logistics automation. Additionally, the rising expectations for sustainable and energy-efficient supply chains have reinforced the adoption of digital tools that optimize resource utilization and minimize environmental impact (Khayyat et al., 2025).

Automation in LSCM manifests across two complementary dimensions: physical and informational. Physical automation encompasses machines and autonomous systems that execute material handling, order fulfillment, and transportation tasks. This includes Autonomous Guided Vehicles (AGVs) in warehouses, Robotic Process Automation (RPA) for sorting and packaging, Automated Storage and Retrieval Systems (AS/RS) for optimized inventory management, and the deployment of self-driving trucks and drones to redefine last-mile delivery. Smart warehouses, integrating IoT sensors, AI-driven analytics, and cloud-based control systems, exemplify how physical automation can enhance efficiency, precision, and safety while reducing reliance on manual labor.

Informational automation, while less visible, plays an equally critical role in reshaping supply chain performance. AI-driven demand forecasting supports more accurate inventory planning, minimizing shortages and overstock. Blockchain technologies strengthen transparency and trust across supply networks by providing secure and immutable transaction records. Digital twins offer virtual replicas of supply chains that allow simulation, scenario testing, and predictive maintenance. Big data analytics identifies inefficiencies and informs strategic decision-making, while cloud-based logistics platforms enable real-time collaboration across suppliers, manufacturers, and distributors. Collectively, these innovations ensure that supply chains remain adaptive, data-driven, and capable of responding dynamically to disruptions.

The COVID-19 pandemic further accelerated the pace of automation by exposing vulnerabilities in global supply chains (Belhadi et al., 2021; Govindan et al., 2020). Widespread factory shutdowns disrupted traditional flows, while surging e-commerce demand underscored the need for automated warehouses and robotics to meet customer expectations. Workforce shortages intensified reliance on automated systems, and organizations recognized the importance of digital resilience as a safeguard against future crises. As a result, businesses worldwide have significantly expanded investments in automation to build more agile and crisis-ready supply chains.

Challenges in Implementing Automation in LSCM

Despite the numerous benefits of automation in supply chains, several challenges hinder its widespread adoption. One of the most significant barriers is the high initial cost associated with implementing AI-driven robotics, IoT devices, and smart logistics systems. The substantial capital investment required for these technologies makes it difficult for many businesses, particularly small and medium-sized enterprises (SMEs), to afford cutting-edge automation solutions. As a result, while larger corporations may successfully integrate automation into their supply chain operations, smaller businesses often struggle to keep pace, widening the gap between technologically advanced enterprises and those that lack the financial resources to adopt automation.

Another major challenge is integration and compatibility issues between legacy logistics systems and modern automation technologies. Many existing supply chain management systems were not designed to support AI, IoT, or cloud-based logistics platforms, creating significant bottlenecks integration. Interoperability challenges also arise when companies attempt to automate processes across multiple supply chain partners. This issue is particularly prevalent in global supply chains, where different companies may use various technological infrastructures that are not easily compatible. The lack of standardized automation frameworks makes it difficult to ensure seamless communication between different logistics systems, slowing down the efficiency gains that automation promises.

In addition to technological and financial barriers, workforce adaptation and resistance to automation remain a critical concern. Many logistics professionals fear job displacement as automated systems and AI-driven processes replace traditional manual tasks. This resistance to change can hinder the successful implementation of automation initiatives. To address these concerns, businesses must focus on retraining employees to work alongside AI-driven and robotic systems rather than replacing them entirely. By upskilling workers and fostering a culture of collaboration between humans and machines, organizations can ease the transition toward automated logistics operations while maintaining workforce engagement and productivity (Klumpp, 2018).

Finally, cybersecurity and data privacy risks pose a significant challenge in the automation of supply chains. As businesses become increasingly reliant on automated digital networks, they also become more vulnerable to cyber threats, including data breaches, ransomware attacks, and unauthorized access to sensitive supply chain information. Protecting customer and supplier data from security breaches is a critical priority, as compromised data can lead to financial losses, reputational damage, and regulatory penalties. Companies must implement robust cybersecurity measures, such as encrypted communication channels, multi-layered authentication protocols, and real-time threat detection systems, to safeguard their automated logistics networks.

While automation offers unparalleled opportunities for improving efficiency, reducing costs, and enhancing decision-making in supply chain management, addressing these challenges is essential for successful implementation. Businesses must carefully balance technological advancements with cost considerations, workforce engagement, system integration, and cybersecurity to ensure a smooth transition toward fully automated supply chains in the Industry 4.0 era.

Evolving Research Perspectives in LSCM

The future of automated supply chains will be shaped by emerging technologies that integrate intelligence, autonomy, and sustainability into logistics. One of the most transformative advancements is the integration of AI-powered autonomous supply chains, which will enable logistics systems to self-optimize by predicting demand, adjusting inventory levels, and optimizing transportation routes in real-time. By leveraging machine learning algorithms, supply chains will continuously refine their processes through predictive analytics, enhancing efficiency, reducing waste, and minimizing disruptions. These AI-driven systems will allow businesses to proactively respond to market fluctuations, ensuring greater resilience and adaptability in global supply networks (Balfaqih, 2022).

Another key technology revolutionizing automated supply chains is blockchain, which offers enhanced traceability, security, and trust in supply chain transactions. The decentralized nature of blockchain ensures that supply chain records are immutable and transparent, reducing fraud, errors, and inefficiencies. Additionally, smart contracts—self-executing agreements coded on blockchain platforms—will automate supplier transactions, procurement processes, and financial settlements, eliminate intermediaries and increasing operational efficiency. These advancements will not only streamline logistics operations but also provide greater accountability and compliance with regulatory requirements (Horenberg, 2017).

Warehousing will also undergo a significant transformation with the rise of hyperautomation, a concept that integrates AI, robotics, IoT, and digital twins to create fully autonomous warehouse operations. Digital twins—virtual models that replicate physical supply chain processes—will enable predictive maintenance, inventory optimization, and workflow automation. Additionally, edge computing will play a crucial role in warehouse automation by processing data locally at the warehouse level rather than relying on cloud servers (Balfaqih et al., 2021). This will allow for real-time decision-making, reducing latency and enhancing efficiency in warehouse operations. The convergence of these technologies will lead to the emergence of self-managed warehouses that operate with minimal human intervention.

Sustainability will be a core focus in the evolution of automated supply chains, with green and sustainable logistics automation playing a crucial role in reducing the environmental impact of logistics operations. AI and automation will optimize route planning to minimize fuel consumption and carbon emissions, contributing to more sustainable transportation networks. Additionally, logistics hubs will increasingly integrate renewable energy solutions, such as solar-powered warehouses and electric automated guided vehicles (AGVs), to enhance energy efficiency and reduce dependency on fossil fuels. As regulatory pressures and consumer demand for eco-friendly practices continue to grow, businesses will need to prioritize sustainability-driven automation to maintain competitiveness in the global supply chain landscape.

As these cutting-edge technologies converge, automated supply chains will become more intelligent, efficient, and sustainable. Businesses that invest in AI-driven optimization, blockchain-enabled transparency, hyperautomated warehousing, and environmentally conscious logistics will gain a significant competitive advantage in the Industry 4.0 era (Lamba & Singh, 2017).

Automation is redefining the future of logistics and supply chain management by reducing costs, improving efficiency, and enabling data-driven decision-making. In the era of Industry 4.0, supply chains are evolving into self-organizing, AI-powered ecosystems that are more resilient, agile, and adaptive to changing conditions. Yet, widespread adoption faces critical challenges, including the high costs of implementation, the need for workforce adaptation, and the growing risks of cybersecurity threats. To ensure successful integration, future research should prioritize AI-driven automation capable of predictive logistics, blockchain-based solutions that enhance supply chain transparency and security, collaborative models that strengthen the relationship between human operators and automated systems, and sustainable approaches that advance green logistics. By addressing these areas, businesses can fully leverage automation as a driver of competitive advantage, operational excellence, and long-term supply chain resilience.

CONCLUSION

The digital transformation of logistics and SCM has fundamentally reshaped the way businesses operate, driving efficiency, accuracy, and adaptability through the integration of Industry 4.0 technologies such as artificial intelligence, the Internet of Things, big data analytics, and automation. These innovations enable organizations to optimize delivery networks, anticipate disruptions, and implement contingency plans, while real-time analytics and digital twin technologies provide the ability to simulate scenarios, monitor performance, and support predictive decision-making in dynamic environments. Although the benefits of automation and digitalization are clear—ranging from cost reduction and faster operations to improved supply chain visibility—several obstacles remain, including high implementation costs, data security risks, and interoperability challenges between legacy and modern systems. The adoption of Logistics 4.0 also requires specialized expertise for system deployment, integration, and maintenance, making it particularly challenging for small and medium-sized enterprises despite the promise of long-term savings and efficiency gains. Nevertheless, the trajectory of supply chain management is moving toward increasingly self-optimizing systems, where machine learning, blockchain-enabled transparency, and sustainable automation solutions work together to create agile, service-oriented, and resilient supply networks. As these developments continue to evolve, future research should further explore practical applications of emerging technologies, best practices for integrating digital tools, and strategies for overcoming barriers to adoption, ensuring that businesses remain competitive in the fast-changing global economy.

REFERENCES

Alshibli, F., Alqarni, K., & Balfaqih, H. (2024). Analyzing the causes and impact of essential medicines and supplies shortages in the supply chain of the Ministry of health in Saudi Arabia: A quantitative survey study. *Informatics in Medicine Unlocked*, *45*, 101457.

Balfaqih, H. (2022, March). Artificial intelligence in logistics and supply chain management: A perspective on research trends and challenges. In *International Conference on Business and Technology* (pp. 1241-1247). Cham: Springer International Publishing.

Balfaqih, M., & Alharbi, S. A. (2022). Associated information and communication technologies challenges of smart city development. *Sustainability*, *14*(23), 16240.

Balfaqih, M., Balfagih, Z., Lytras, M. D., Alfawaz, K. M., Alshdadi, A. A., & Alsolami, E. (2023). A blockchain-enabled IoT logistics system for efficient tracking and management of high-price shipments: A resilient, scalable and sustainable approach to smart cities. *Sustainability*, *15*(18), 13971.

Balfaqih, M., Balfaqih, Z., Shepelev, V., Alharbi, S. A., & Jabbar, W. A. (2022). An analytical framework for distributed and centralized mobility management protocols. *Journal of Ambient Intelligence and Humanized Computing*, *13*(7), 3393–3405.

Balfaqih, M., Jabbar, W., Khayyat, M., & Hassan, R. (2021). Design and development of smart parking system based on fog computing and internet of things. *Electronics (Basel)*, *10*(24), 3184.

Barata, J., Rupino Da Cunha, P., & Stal, J. (2018). Mobile supply chain management in the Industry 4.0 era: An annotated bibliography and guide for future research. *Journal of Enterprise Information Management*, *31*(1), 173–192.

Bécue, A., Praça, I., & Gama, J. (2021). Artificial intelligence, cyber-threats and Industry 4.0: Challenges and opportunities. *Artificial Intelligence Review*, *54*(5), 3849–3886.

Belhadi, A., Kamble, S., Jabbour, C. J. C., Gunasekaran, A., Ndubisi, N. O., & Venkatesh, M. (2021). Manufacturing and service supply chain resilience to the COVID-19 outbreak: Lessons learned from the automobile and airline industries. *Technological Forecasting and Social Change*, *163*, 120447. PMID: 33518818

Choi, T. M., Kumar, S., Yue, X., & Chan, H. L. (2022). Disruptive technologies and operations management in the Industry 4.0 era and beyond. *Production and Operations Management*, *31*(1), 9–31.

Dallasega, P., Rauch, E., & Linder, C. (2018). Industry 4.0 as an enabler of proximity for construction supply chains: A systematic literature review. *Computers in Industry, 99*, 205–225.

Dorduncu, H., & Oruç, Z. S. (2021). Industry 4.0, digitalization, and big data: perspective of logistics and supply chain management. *Çukurova Üniversitesi Sosyal Bilimler Enstitüsü Dergisi, 30*(3), 170-180.

Douaioui, K., Fri, M., Mabroukki, C., & Semma, E. A. (2018, April). The interaction between industry 4.0 and smart logistics: concepts and perspectives. In *2018 international colloquium on logistics and supply chain management (LOGISTIQUA)* (pp. 128-132). IEEE.

Evtodieva, T. E. (2019). *Sustainable Growth and Development of Economic Systems. Contributions to Economics*. Springer., DOI: 10.1007/978-3-030-11754-2_16

Görçün. (2016). *Ömer Faruk. "Dördüncü endüstri devrimi endüstri 4.0."*. Beta Yayıncılık.

Govindan, K., Mina, H., & Alavi, B. (2020). A decision support system for demand management in healthcare supply chains considering the epidemic outbreaks: A case study of coronavirus disease 2019 (COVID-19). *Transportation Research Part E, Logistics and Transportation Review, 138*, 101967. PMID: 32382249

Gritsenko, S., Karpun, O., & Kolisnichenko, A. (2022). Digitalization of logistics company business process in conditions of industry 4.0. *Electronic scientific and practical publication in economic sciences.*

Grosse, E. H., Glock, C. H., & Neumann, W. P. (2017). Human factors in order picking: A content analysis of the literature. *International Journal of Production Research, 55*(5), 1260–1276.

Hofmann, E., & Rüsch, M. (2017). Industry 4.0 and the current status as well as future prospects on logistics. *Computers in Industry, 89*, 23–34.

Horenberg, D. (2017). *Applications within Logistics 4.0: A research conducted on the visions of 3PL service providers* (Bachelor's thesis, University of Twente).

Ivanov, D., Das, A., & Choi, T. M. (2018). New flexibility drivers for manufacturing, supply chain and service operations. *International Journal of Production Research, 56*(10), 3359–3368.

Junge, A. L., Verhoeven, P., Reipert, J., & Mansfeld, M. (2019). *Pathway of digital transformation in logistics: best practice concepts and future developments* (Vol. 8). Universitätsverlag der TU Berlin.

Kache, F., & Seuring, S. (2017). Challenges and opportunities of digital information at the intersection of Big Data Analytics and supply chain management. *International Journal of Operations & Production Management*, *37*(1), 10–36.

Kamble, S. S., Gunasekaran, A., & Gawankar, S. A. (2018). Sustainable Industry 4.0 framework: A systematic literature review identifying the current trends and future perspectives. *Process Safety and Environmental Protection*, *117*, 408–425.

Kayikci, Y. (2018). Sustainability impact of digitization in logistics. *Procedia Manufacturing*, *21*, 782–789.

Khayyat, M., Balfaqih, M., Balfaqih, H., & Ismail, M. (2024). Challenges and factors influencing the implementation of green logistics: A case study of Saudi Arabia. *Sustainability*, *16*(13), 5617.

Khayyat, M., Jambi, J., Kutbi, N., Garout, R., Al-Noumani, L., Asiri, R., & Balfaqih, M. (2025). Logistic caravan: A smart transportation system for green logistics and sustainable practices. *International Journal of Advanced and Applied Sciences*, *12*(6), 154–168.

Klumpp, M. (2018). Automation and artificial intelligence in business logistics systems: Human reactions and collaboration requirements. *International Journal of Logistics Research and Applications*, *21*(3), 224–242.

Lager, H., Virgillito, A., & Buchberger, T. P. (2021). Digitalization of logistics work: Ergonomic improvements versus work intensification. In *Digital supply chains and the human factor* (pp. 33–53). Springer International Publishing.

Lamba, K., & Singh, S. P. (2017). Big data in operations and supply chain management: Current trends and future perspectives. *Production Planning and Control*, *28*(11-12), 877–890.

Liao, Y., Deschamps, F., Loures, E. D. F. R., & Ramos, L. F. P. (2017). Past, present and future of Industry 4.0-a systematic literature review and research agenda proposal. *International Journal of Production Research*, *55*(12), 3609–3629.

Lopes de Sousa Jabbour, A. B., Jabbour, C. J. C., Godinho Filho, M., & Roubaud, D. (2018). Industry 4.0 and the circular economy: A proposed research agenda and original roadmap for sustainable operations. *Annals of Operations Research*, *270*(1), 273–286.

Nitsche, B., Straube, F., & Wirth, M. (2021, July). Application areas and antecedents of automation in logistics and supply chain management: A conceptual framework. In *Supply Chain Forum* [). Taylor & Francis.]. *International Journal (Toronto, Ont.)*, *22*(3), 223–239.

Öztuna, B. (2021). Logistics 4.0 and technologic applications. In *Logistics 4.0 and Future of Supply Chains* (pp. 9–27). Springer Nature Singapore.

Paprocki, W. (2016, May). How transport and logistics operators can implement the solutions of "Industry 4.0". In *TranSopot Conference* (pp. 185–196). Springer International Publishing.

Peters, M. A., Jandrić, P., & Hayes, S. (2022). Biodigital philosophy, technological convergence, and postdigital knowledge ecologies. In *Bioinformational philosophy and postdigital knowledge ecologies* (pp. 3–22). Springer International Publishing.

Rajković, T., Vasiljević, D., & Lečić-Cvetković, D. (2022, June). Logistics 4.0–Smart Transformation of Logistics and Supply Chain Management. In *International Symposium SymOrg* (pp. 386-402). Cham: Springer International Publishing.

Rennung, F., Luminosu, C. T., & Draghici, A. (2016). Service provision in the framework of Industry 4.0. *Procedia: Social and Behavioral Sciences, 221*, 372–377.

Sarker, I. H. (2021). Data science and analytics: An overview from data-driven smart computing, decision-making and applications perspective. *SN Computer Science, 2*(5), 377. PMID: 34278328

Sokolov, B., & Ivanov, D. (2015). Integrated scheduling of material flows and information services in industry 4.0 supply networks. *IFAC-PapersOnLine, 48*(3), 1533–1538.

Sundaramurthy, S. K., Ravichandran, N., Inaganti, A. C., & Muppalaneni, R. (2022). The future of enterprise automation: Integrating AI in cybersecurity, cloud operations, and workforce analytics. *Artificial Intelligence and Machine Learning Review, 3*(2), 1–15.

Tiwari, S., Wee, H. M., & Daryanto, Y. (2018). Big data analytics in supply chain management between 2010 and 2016: Insights to industries. *Computers & Industrial Engineering, 115*, 319–330.

Udeh, E. O., Amajuoyi, P., Adeusi, K. B., & Scott, A. O. (2024). The role of IoT in boosting supply chain transparency and efficiency. *Magna Scientia Advanced Research and Reviews, 11*(01), 178–197.

Vermesan, O., & Friess, P. (Eds.). (2022). *Digitising the industry internet of things connecting the physical, digital and VirtualWorlds*. CRC Press.

Wang, G., Gunasekaran, A., Ngai, E. W., & Papadopoulos, T. (2016). Big data analytics in logistics and supply chain management: Certain investigations for research and applications. *International Journal of Production Economics, 176*, 98–110.

Wang, Y., Feng, L., Chang, H., & Wu, M. (2017). Research on the impact of big data on logistics. In *MATEC Web of Conferences* (Vol. 100, p. 02015). EDP Sciences.

Winkle, H., & Zinsmeister, L. (2019). Trends in digitalization of intralogistics and the critical success factors of its implementation. *Brazilian Journal of Operations & Production Management*, *16*(3), 537–549.

Witkowski, K. (2017). Internet of things, big data, industry 4.0–innovative solutions in logistics and supply chains management. *Procedia Engineering*, *182*, 763–769.

Zhong, R. Y., Xu, X., Klotz, E., & Newman, S. T. (2017). Intelligent manufacturing in the context of industry 4.0: A review. *Engineering*, *3*(5), 616–630.

Chapter 6
Design and Analysis of a Solar-Assisted Combined Cooling, Heating, and Power System for Smart Cities:
Case Study From Doha

Abdulrazzak Akroot
Karabük University, Turkey

Akram A. Almohammedi
https://orcid.org/0000-0001-8476-0813
Abdullah Gül University, Turkey

Wadah Talal
https://orcid.org/0000-0002-1268-8944
Ministry of Industry and Minerals, Iraq

ABSTRACT

The rising demand for sustainable and energy-efficient solutions in urban areas has driven interest in renewable systems for smart cities. This chapter presents a solar-assisted combined cooling, heating, and power (SA-CCHP) system designed for Doha, Qatar, where high solar radiation and cooling needs prevail. Powered solely by a parabolic trough collector (PTC) field, the system delivers net power from 120 kW in winter to 195 kW in summer, with cooling loads of ~210–340 kW and heating loads of ~900–1450 kW. Increasing the superheating degree at the ORC turbine inlet enhances power and heating but reduces cooling, while raising

DOI: 10.4018/979-8-3373-0154-9.ch006

the pressure ratio (α) from 0.5 to 0.8 boosts net output and efficiency, cutting CO_2 emissions from 0.22 to 0.13 kg/kWh. Overall energy efficiency rises from 85% to 90% and exergy efficiency from 76% to 78.5%, while costs decline from $40/hr to $36/hr, confirming both environmental and economic viability. The study demonstrates the feasibility of solar-powered CCHP systems as scalable models for achieving clean energy goals in smart cities.

INTRODUCTION

The rapid growth of urban populations, particularly in regions characterized by extreme climatic conditions such as the Middle East, has intensified the demand for sustainable, efficient, and resilient energy systems (Chu et al., 2019; Salimi et al., 2022). Combined Cooling, Heating, and Power (CCHP) systems integrated with renewable energy sources have gained significant attention among the innovative solutions emerging to meet these needs (Eisavi et al., 2018; Wang et al., 2019a). Solar-assisted CCHP systems offer a promising approach to simultaneously address electricity, cooling, and heating demands while minimizing environmental impacts (Li et al., 2022; Wang et al., 2020; L. Zhang et al., 2019a).

Unlike conventional energy systems that dissipate a large proportion of input energy as unusable waste heat (Alsunousi & Kayabasi, 2024; Besevli et al., 2024; Mithat Delibaş et al., 2021), combined cooling, heating, and power (CHP) systems maximize energy utilization by capturing and repurposing this byproduct for practical applications such as space heating, hot water production, district heating networks, and even industrial thermal processes (K. Yang et al., 2019; Zhu & Chow, 2019). This integrated approach not only improves overall energy efficiency but also reduces dependence on separate, stand-alone systems for electricity and heating, thereby optimizing resource allocation. By harnessing the otherwise wasted thermal energy, CHP systems can achieve energy efficiencies as high as 80–90%, far surpassing the efficiency levels of traditional centralized power plants.

From an environmental perspective, CHP systems significantly advance sustainability goals by lowering greenhouse gas (GHG) emissions through efficient fuel utilization and the simultaneous production of electricity and useful heat (Fani & Sadreddin, 2017; Khan & Singh, 2024). This dual-output process reduces the need for additional fossil fuel combustion in conventional boilers, directly cutting CO_2 emissions and air pollutants such as NO_X and SO_2. These benefits align CHP deployment with global decarbonization strategies, climate action targets, and the growing demand for cleaner urban energy systems. Furthermore, integrating CHP with renewable sources such as biomass, solar thermal, or geothermal energy can enhance its environmental credentials even further, creating hybrid energy systems that are both efficient and low-carbon.

Economically, CHP adoption also presents clear advantages. By consolidating power and thermal energy production into a single process, organizations can lower operational costs, stabilize energy bills, and mitigate risks associated with fluctuating energy prices. This makes CHP particularly attractive for energy-intensive sectors such as manufacturing, healthcare, and large-scale commercial facilities, where uninterrupted energy supply is critical (Oliveira et al., 2002; Talal & Akroot, 2023; Almohammedi et al., 2023). In addition, institutions and municipalities benefit from the resilience of CHP systems, which can operate in island mode during grid outages, ensuring continuity of critical services—a feature especially valuable in the context of urban resilience and smart city development.

With advancements in smart grids, IoT-enabled monitoring, and AI-driven optimization, modern CHP systems are becoming more intelligent and adaptive (Mauladdawilah et al., 2025a). These innovations allow for dynamic load management, predictive maintenance, and real-time efficiency optimization, which further extend the economic and environmental benefits. As a result, CHP technology is increasingly recognized not only as a transitional solution bridging conventional and renewable systems but also as a cornerstone of future integrated energy infrastructures that prioritize efficiency, resilience, and sustainability (Abbas et al., 2023).

As global emphasis increasingly shifts toward cleaner and more efficient energy solutions, combined cooling, heating, and power (CCHP) systems have gained recognition as a versatile and impactful strategy to address the rising complexity of modern energy demands (Elsafi et al., 2025). These systems, by simultaneously delivering electricity, heating, and cooling from a single energy source, maximize resource utilization and minimize waste, making them particularly attractive for applications in smart cities and densely populated urban environments (Mauladdawilah et al., 2025b).

Wang et al. (2020) reviewed the integration of solar technologies into CCHP configurations, underscoring their capacity to deliver significant energy, economic, and environmental benefits. The authors emphasized that coupling solar energy with CCHP not only reduces fossil fuel dependence but also lowers greenhouse gas emissions, aligning with global decarbonization goals. However, they stressed that achieving large-scale commercial viability requires ongoing innovation in system design, improved operational optimization, and tailored applications for diverse climatic and urban contexts. In a similar vein, Hou et al. (2021) evaluated a solar-assisted CCHP system optimized using particle swarm optimization and demonstrated its practical advantages in a residential case study in Guangzhou. Their results revealed annual energy savings of 15,141.63 kWh and additional heat output of 170,183.57 MJ, with the system achieving a return on investment

(ROI) of 2.55 and a payback period of just over 11 years. Such findings highlight the promising economic feasibility of solar-assisted CCHP systems when properly optimized, reinforcing their potential role in delivering sustainable, cost-effective energy solutions for future urban developments.

Other researchers have also proposed innovative configurations to boost efficiency and reduce emissions. N. Zhang et al. (2018) presented a solar-assisted CCHP system with a chemically recuperated gas turbine and absorption chiller, achieving a solar-to-electricity efficiency of 26–29%, fossil fuel savings of 30.4%, and CO_2 emission reductions of 33% compared to conventional CCHP systems. L. Zhang et al. (2019b) developed a biogas and PV-driven RES-CCHP system optimized for rural applications, delivering 20.94% primary energy savings, 11.73% cost savings, and a 40.79% reduction in carbon emissions, validated in Jinan, China. Gao et al. (2023) investigated a solar-assisted CCHP system based on a reversible solid oxide fuel cell, reporting SOEC mode efficiencies of 19.0% (energy) and 20.5% (exergy), and SOFC mode efficiencies of 51.4% (electrical), 71.3% (energy), and 45.2% (exergy).

Experimental and modeling studies further demonstrate the versatility of solar-assisted CCHP designs. Cavalcanti et al. (2021) analyzed a solar-assisted trigeneration system with an absorption chiller, achieving a maximum cooling output of 264.15 kW and a COP of 1.2, noting that exergy efficiency decreased with rising dead state temperature (79.75% at 10°C to 17.57% at 30°C). Su et al. (2023) proposed a solar-assisted combined cooling and power system with energy storage and desulfurization for coal-fired plants, achieving power improvements of 138,976 kW, a 5.8% rise in efficiency, a refrigerating output of 91,698 kW, desulfurization capacity of 10,550 kg/h, and a payback period of 5.7 years. Similarly, Chen et al. (2023) presented a solar-assisted absorption–compression system for year-round heating and cooling, demonstrating a CO_2 increment of 45.8%, a power-saving ratio (PSR) of 31.4%, and an energy-saving ratio (ESR) of 18.9% in heating mode.

Hybrid and alternative configurations have also been explored. Li et al. (2022) optimized a hybrid CCHP system using solar thermal, PV, batteries, and microturbines, achieving 46.56% primary energy savings, 54.64% CO_2 reduction, and 78.51% energy efficiency. Akroot and Refaei (2025) compared solar-assisted CCHP systems using R245fa and butane. R245fa provided higher energy (86.89%) and exergy (16.26%) efficiencies and lower carbon emissions (0.195 kg CO_2/kWh), while butane offered a lower cost rate of $63.06/h, despite reduced performance. Ünal et al. (2018) experimentally studied a solar-assisted vertical ground source heat pump system in Mardin, Turkey, reporting 67.36% energy efficiency, 27.40% exergy efficiency, and an exergoeconomic factor of 60.51%. Finally, Wang et al. (2019b) proposed a solar thermal biomass gasification-based CCHP system, achieving average energy and exergy efficiencies of 56% and 28%, an energy ratio of solar to biomass of 0.19, and biomass savings of 9.22% (cooling mode) and 2.02% (heating mode).

This chapter contributes to the advancement of solar-based energy solutions by designing and analysing a purely solar-driven Combined Cooling, Heating, and Power (SA-CCHP) system tailored to the energy demands of smart cities, with Doha serving as a strategic case study due to its high solar irradiance and intense cooling needs. The originality of the work lies in three main areas: (1) the proposal of a solar-powered CCHP configuration using a parabolic trough collector (PTC) field; (2) the development of a comprehensive thermodynamic and exergy model to evaluate the system's seasonal performance, highlighting its ability to deliver stable cooling, heating, and power outputs under varying solar inputs; and (3) a parametric investigation into the effects of superheating temperature and pressure ratio on energy efficiency, exergy efficiency, cost rate, and CO_2 emissions, providing valuable insights for optimizing solar CCHP systems in arid urban environments.

Although previous studies have demonstrated the promising performance of solar-assisted CCHP systems, several limitations remain. Many research efforts focused primarily on specific case conditions without conducting comprehensive seasonal performance evaluations, limiting their applicability to regions with variable solar resources. Moreover, some models optimized individual components (such as the turbine or chiller) rather than the entire system holistically, resulting in narrow optimization scopes. Additionally, several studies emphasized thermodynamic analysis but offered limited economic assessments or cost evaluations, which are essential for real-world deployment. These gaps underline the necessity for integrated approaches that combine energy, exergy, economic, and environmental analyses, as pursued in the present work.

MODEL DESCRIPTION

To provide a clearer understanding of the operational framework and energy pathways, the proposed system's overall structure is first presented schematically. Figure 1 illustrates the main components and flow configuration of the proposed Solar-Assisted CCHP system. The system integrates renewable solar energy through a parabolic trough collector (PTC) field to drive a hybrid energy cycle capable of producing electricity, heating, and cooling. This system consists of a solar PTC Field, which concentrates solar energy to heat a thermal transfer fluid, which flows into a Thermal Storage Tank. This tank buffers energy, allowing continuous operation even under variable solar conditions. Heat recovery vapor generator (HRVG) receives high-temperature heat from the thermal storage tank and generates high-pressure vapor. The Organic Rankine cycle turbine (ORT) expands the vapor to produce electrical power via a generator (G). The Condenser 1 ($Cond_1$) recovers the remaining heat from the turbine exhaust, supplying heating output for domestic or district use. A fraction of the heat is directed to a Generator within an absorption chiller loop. The generator heats the working fluid, which flows through a Solution Heat Exchanger (SHEX) and Pressure Reducing Valve (PRV) to the Absorber. The working fluid is mixed and absorbed in the absorber, facilitating the cooling cycle. The refrigerant vapor is condensed in Condenser 2 ($Cond_2$), expanded through an Expansion Valve (EV), and evaporated in an Evaporator, providing cooling services.

Figure 1. Schematic diagram of the solar-assisted Combined Cooling, Heating, and Power (CCHP) system for smart cities

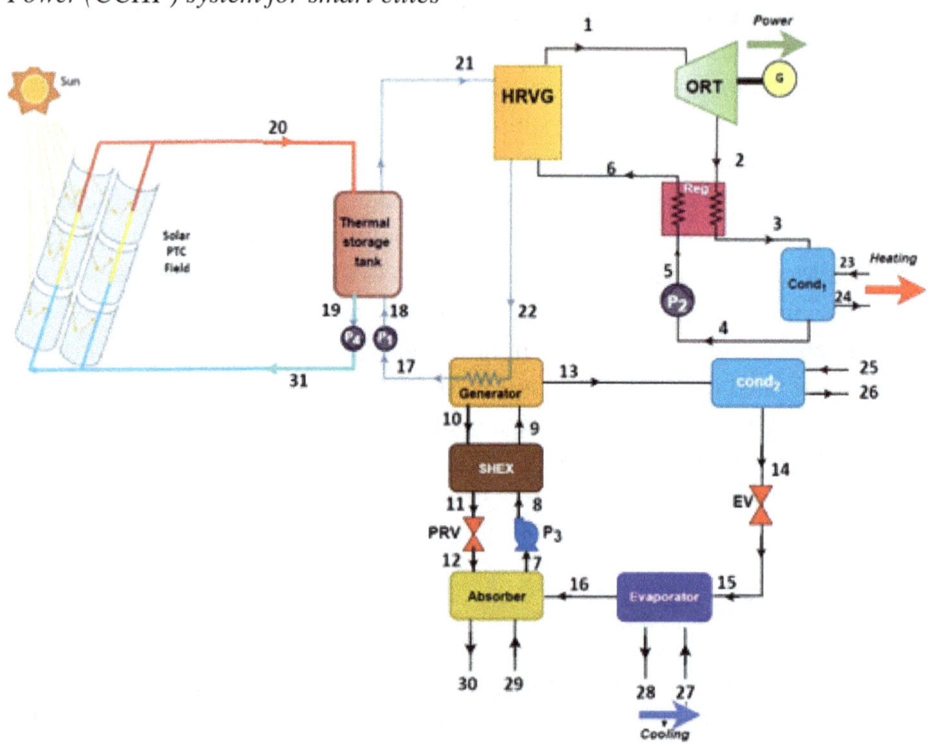

In the proposed model, butane (n-butane) was chosen as the working fluid for the organic Rankine cycle (ORC) system due to its favourable thermophysical properties. Butane has a low boiling point and low environmental impact, making it ideally suited for solar-powered applications. Its low critical temperature and pressure allow it to effectively utilize the thermal energy collected from parabolic trough collectors and transfer it via a heat exchanger. Furthermore, butane's relatively low cost and widespread availability support its selection. Pumps (P_1, P_2, P_3) facilitate fluid movement throughout the system, ensuring proper circulation between components. The integrated setup allows the efficient utilization of solar energy for simultaneous electricity production, heating, and cooling, making it highly suitable for smart city applications such as in Doha, where cooling demand is particularly high.

Table 1 presents the input parameters for modelling the SA-CCHP system. The following assumptions are considered in the SA-CCHP system modelling:

- All system components operate under steady-state conditions.
- Therminol-66 oil is employed as the heat transfer fluid in the solar energy subsystem to efficiently convey thermal energy from the solar collectors to the thermal storage tank, absorption refrigeration system (ARS) generator, and the HRVG.
- The expansion process across the throttling valve is assumed to be isenthalpic.
- The refrigerant vapor at the evaporator outlet and the liquid refrigerant at the condenser outlet are both considered saturated.
- The solution heat exchanger (SHEX) is modelled with an effectiveness of 0.72.
- The LiBr-H_2O solution exiting the ARS generator is assumed to be in a saturated state.
- Pressure losses in pipes and heat exchangers are neglected.

Table 1. Input data for modeling the suggested SA-CCHP system

Parameter	Value	Parameter	Value
Solar area	10000 m^2	Sun temperature	5770 K
Solar reservoir outlet temperature	245°C	Solar reservoir inlet temperature	128°C
Location	Doha, Qatar	Longitude (E)	51.531°E
Latitude (N)	25.2854°N	DNI	6.43 kWh/ m^2.day
DNI	6.43 kWh/ m^2.day	T_{amb}	30°C
Generator temperature (T_{15})	88°C	Absorber temperature (T_{12})	37°C
Turbine efficiency	90%	Pump efficiency	80%
LiBr solution strength	54%		

The fundamental equations governing the control volume's mass, energy, and exergy balances, excluding potential and kinetic energy variations, are presented below. These equations serve as the foundation for modelling the system's operation (Anvari et al., 2020; Javadi et al., 2021).

$$\sum \dot{m}_{in} - \sum \dot{m}_{out} = 0 \tag{1}$$

$$\dot{Q} - \dot{W} + \sum \dot{E}_{in} - \sum \dot{E}_{out} = 0 \quad (2)$$

$$\dot{E}x_{in} + \dot{E}x_{Q} - \dot{E}x_{out} - \dot{E}x_{W} - \dot{E}x_{dis} = 0 \quad (3)$$

Where *in* represents the control volume inlet, out the control volume exit, $\dot{m}$ the mass flow rate, $\dot{Q}$ the rate of heat transfer, $\dot{W}$ the network, h the enthalpy, $\dot{E}x_Q$ the exergy associated with heat transfer, $\dot{E}x_w$ the exergy associated with power, and $\dot{E}x_{dest}$ the exergy destruction within each component.

The exergy of a thermodynamic system exchanging heat with a reference environment is defined as follows (Alfaris et al., 2025; Mauladdawilah et al., 2025c):

$$\dot{E}x_{Q} = \dot{Q}_i\left(1 - \frac{T_0}{T_i}\right) \quad (4)$$

The exergy rate associated with work is equal to the energy transfer rate due to work (Khoshgoftar Manesh et al., 2022):

$$\dot{E}x_{W} = \dot{W} \quad (5)$$

Finally, the following performance indicators are introduced to evaluate the effectiveness of the SA-CCHP system. Overall output work can be represented as:

$$\dot{W}_{net} = \dot{W}_T - \dot{W}_{P1} - \dot{W}_{P2} - \dot{W}_{P3} - \dot{W}_{P4} \quad (6)$$

The heating and cooling loads obtained from the SA-CCHP plant are given as follows:

$$\dot{Q}_{\text{heating}} = \dot{Q}_{\text{Heater}} = \dot{m}_{23}\left(h_{24} - h_{23}\right) \quad (7)$$

$$\dot{Q}_{\text{cooling}} = \dot{Q}_{\text{Evaporator}} = \dot{m}_{27}\left(h_{27} - h_{28}\right) \quad (8)$$

While the overall efficiency can be calculated as:

$$\eta_{\text{AS-CCHP}} = \frac{\dot{W}_{net} + \dot{Q}_{heating} + \dot{Q}_{cooling}}{A_{Coll}.\eta_{coll}.DNI.N_{Coll}} \quad (9)$$

$$\psi_{\text{AS-CCHP}} = \frac{\dot{W}_{net} + \dot{EX}_{P,heating} + \dot{EX}_{P,cooling}}{\dot{Q}_{solar}\left(1 - \frac{T_0}{T_{sun}}\right)} \tag{10}$$

Exergoeconomic analysis assesses the cost flows, total cost rates, and specific costs linked to the system's exergy streams, including the costs associated with exergy destruction and exergoeconomic factors. In this context, the cost balance for the k-th component is formulated as shown in Equation (9) (Alfaris et al., 2024; K. Yang et al., 2024; X. Yang et al., 2024).

$$\sum\left(c_{in}\dot{\text{E}}\text{x}_{in}\right)_k + c_{q,k}\dot{\text{E}}\text{x}_{q,k} + \dot{Z}_k = \sum\left(c_{out}\dot{\text{E}}\text{x}_{out}\right)_k + c_{w,k}\dot{W}_k \tag{9}$$

Where c_{in} represents the unit cost of exergy for incoming streams, while c_q denotes the unit cost of exergy for outgoing streams, both expressed in \$/kJ or \$/kWh. The term $\dot{Z}_k$ corresponds to the cost rate associated with component *k*, which includes capital, maintenance, and operational expenses, typically measured in \$/s. For each component, $c_{q,k}$ defines the unit cost of the heat exergy entering the system, whereas $c_{w,k}$ refers to the unit cost of the work output. The actual work output of a component is denoted by $\dot{W}_k$, expressed in kW or kJ/s, and the exergy linked to the heat input is represented as $\dot{\text{E}}\text{x}_{q,k}$.

Integrating the AS-CCHP reduces reliance on fossil fuels and consequently lowers pollutant emissions. Since the proposed system does not involve direct fossil fuel combustion, only the CO_2 emissions associated with electricity consumption, primarily for operating equipment such as pumps, are considered. These emissions are calculated as follows (Akroot & Al Shammre, 2024):

$$\dot{m}_{\text{CO2}} = \gamma\left(\dot{W}_{P1} + \dot{W}_{P2} + \dot{W}_{P3} + \dot{W}_{P4}\right) \tag{10}$$

The emission conversion factor, denoted as γ, has a value of 0.968 kg/kWh.

RESULTS AND DISCUSSION

This section presents the performance evaluation of the proposed SA-CCHP system under varying operating conditions, highlighting how thermodynamic, economic, and environmental parameters respond to changes in superheating degree and pressure ratio. Seasonal and monthly variations in power, heating, and cooling outputs are analyzed in relation to Doha's solar irradiance profile, emphasizing the system's adaptability to local climate conditions. The discussion also considers cost-effectiveness, energy and exergy efficiencies, and CO_2 reduction potential, offering insights into how the system supports both technical and sustainability objectives. Finally, the proposed system is benchmarked against an existing solar-biomass CCHP configuration to validate its competitiveness and suitability for deployment in solar-rich urban environments.

Performance Analysis of the Proposed SA-CCHP System

To evaluate the thermodynamic behaviour of the proposed SA-CCHP system, the influence of superheating degree (ΔT) at the Organic Rankine Turbine (ORT) inlet was first analysed, as shown in Figure 2.

Figure 2. Effect of superheating degree at the ORT inlet on net power output, cooling load, and heating load in the SA-CCHP system

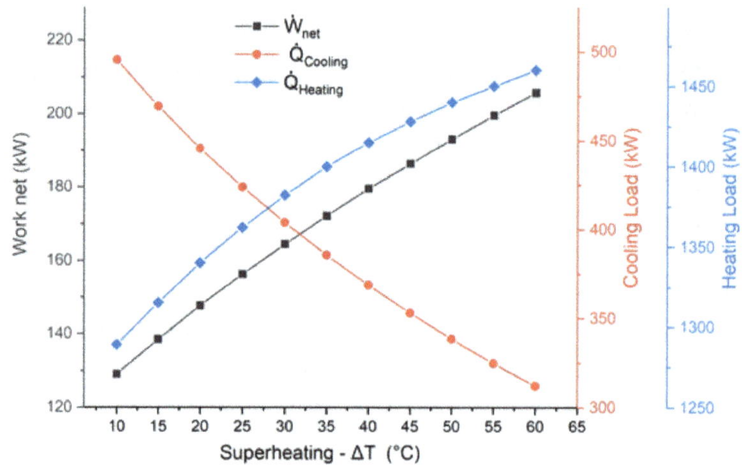

The findings present that increasing the superheating degree improves both electrical and heating outputs, but at the expense of cooling capacity. This trade-off suggests that operating at higher superheating temperatures is advantageous for systems prioritizing electricity and heating, such as those targeting district heating and smart grid integration in cities like Doha. However, for applications with a higher cooling demand (e.g., in extremely hot climates), a moderate superheating degree might be more appropriate for balancing power, cooling, and heating needs. As the superheating temperature increases from 10°C to 60°C, the net power output ($_{net}$) steadily increases, rising from about 130 kW to around 210 kW. This behaviour is expected, as higher superheating improves the expansion work potential in the turbine, thereby boosting power generation. On the other hand, the cooling load ($Q_{Cooling}$) shows a declining trend, decreasing significantly from around 510 kW at

10°C to about 310 kW at 60°C. This is likely because more available thermal energy is converted into mechanical work for power generation rather than being directed into the absorption cooling cycle. In contrast, the heating load ($Q_{Heating}$) increases with higher superheating degrees, rising from about 1260 kW to approximately 1470 kW. This indicates that higher inlet temperatures also enhance heat recovery for heating purposes, benefiting the thermal heating demands in the system.

Figure 3. Effect of superheating degree on economic, environmental, and thermodynamic performance indicators of the SA-CCHP system

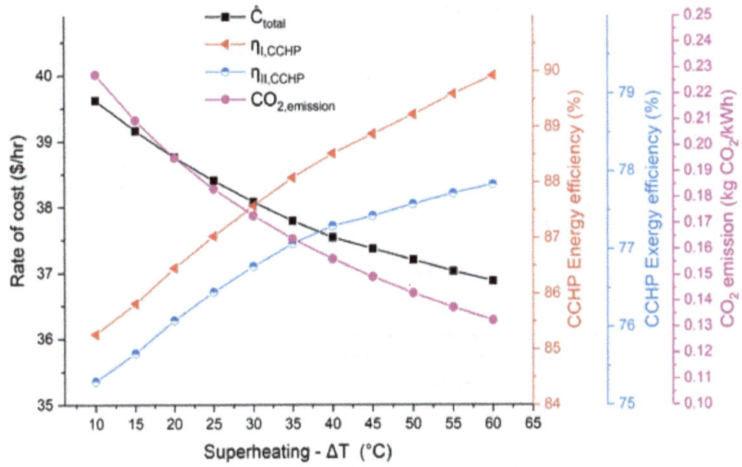

Figure 3 presents the influence of increasing superheating temperature (ΔT) at the inlet of the Organic Rankine Turbine (ORT) on four critical performance indicators of the SA-CCHP system: Total Cost Rate ($/hr), CCHP Energy Efficiency ($\eta_{I,CCHP}$), CCHP Exergy Efficiency ($\eta_{II, CCHP}$), and CO_2 Emissions (kg CO_2/kWh). The results reveal that increasing the superheating degree leads to simultaneous energy and exergy efficiency improvements, decreased system cost, and significantly reduced environmental impact. The cost rate decreases consistently as the superheating temperature increases from 10°C to 60°C, dropping from ~$40/hr to ~$36/hr. The energy efficiency increases steadily from approximately 85% to over 90% as ΔT increases. This enhancement is due to the increased thermodynamic potential of the working fluid, allowing more of the input thermal energy to be converted into useful outputs (power, heating, and cooling). It demonstrates a clear thermodynamic benefit of operating the ORT at higher superheating conditions. The exergy efficiency, which accounts for the quality of energy conversion, also improves with increasing ΔT, rising from about 75% to ~78.5%. This reflects more effective utilization of available energy to perform work and useful heating/cooling, reducing irreversibility within the cycle. The system's CO_2 emissions show a notable decline, from approximately 0.24 kg CO_2/kWh at 10°C to 0.11 kg CO_2/kWh at 60°C. This trend indicates that increased thermal efficiency at higher superheating results in less auxiliary system use, aligning the SA-CCHP system with sustainability and decarbonization goals.

Figure 4. Effect of pressure ratio (α) on net power output, cooling load, and heating load of the SA-CCHP system

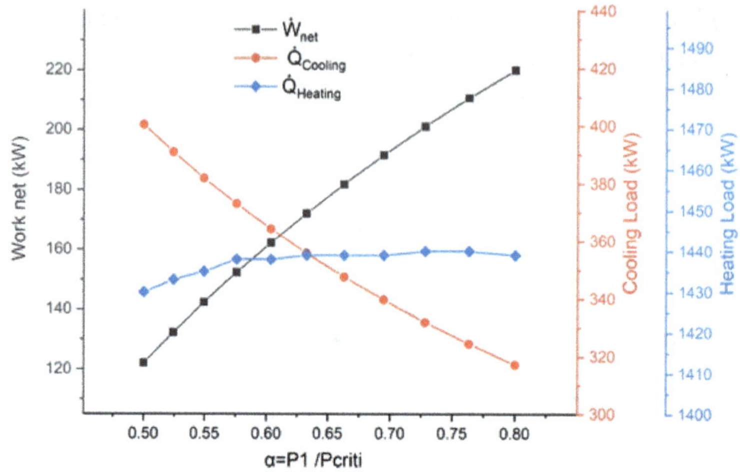

Figure 4 examines the influence of the pressure ratio (α)—defined as the ratio between the turbine inlet pressure (P_1) and the working fluid's critical pressure ($P_{critical}$)—on the key performance outputs of the SA-CCHP system. The figure highlights the importance of tuning the turbine intake pressure relative to the working fluid's critical pressure for achieving the desired performance outcome based on local energy needs and climate conditions. Higher pressure ratios improve electrical power generation at the expense of cooling capacity, while heating capacity remains relatively stable. In regions like Doha, where cooling demand dominates, a moderate pressure ratio (around $\alpha = 0.6$) may be preferable to balance sufficient cooling and acceptable power output. However, higher pressure ratios ($\alpha > 0.7$) are more advantageous when the system aims to maximize electricity production. The net power output increases significantly with an increase in α, from approximately 125 kW at $\alpha = 0.50$ to around 220 kW at $\alpha = 0.80$. This increase results from higher inlet pressures, enabling the Organic Rankine Cycle (ORC) turbine to perform greater expansion work and enhance power generation efficiency. The cooling load shows a declining trend as α increases. Initially, the cooling capacity is about 415 kW at $\alpha = 0.50$, but it falls to nearly 315 kW at $\alpha = 0.80$. This behavior suggests that as more energy is extracted as useful work (electricity), less thermal energy remains available for driving the absorption cooling cycle. Unlike the other two outputs, the heating load demonstrates a mild increase followed by stabilization. Heating output rises slightly from about 1410 kW to approximately 1450 kW as α increases, after which it plateaus.

Figure 5. Effect of pressure ratio (α) on economic, environmental, and thermodynamic performance indicators of the SA-CCHP system

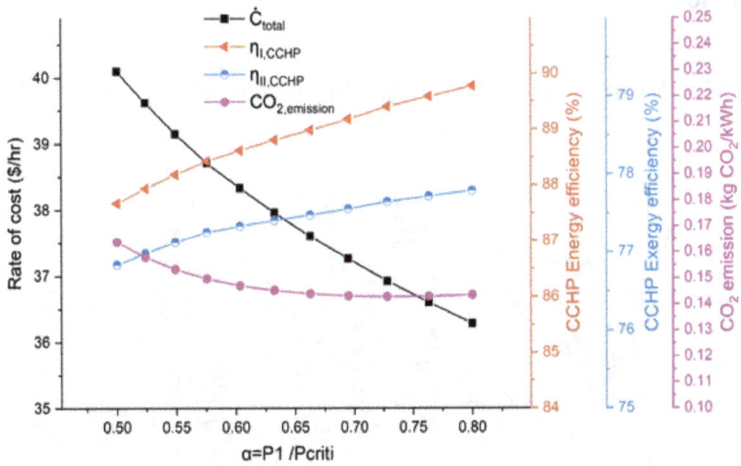

Figure 5 presents the variation of four critical performance indicators for the SA-CCHP system as a function of the pressure ratio (α). The findings demonstrate that optimizing the system around an α value near 0.75–0.80 offers economic savings, thermodynamic performance, and reduced environmental impact. However, cooling requirements should also be considered, as seen in the earlier thermal load figures. Higher pressure ratios reduce operational costs by increasing system output relative to input resources. Both energy and exergy efficiencies are enhanced with increased α, though the gain in energy efficiency is more pronounced. Lower CO_2 emissions at higher α levels highlight the SA-CCHP system's potential to meet green city and smart energy goals. The cost rate decreases sharply as the pressure ratio increases from 0.5 to 0.8. Specifically, the system cost drops from about $40/hr at α = 0.5 to around $36/hr at α = 0.8. This reduction is driven by enhanced thermodynamic performance at higher pressures, which increases net energy output and improves system cost-effectiveness.

First-law energy efficiency shows a steady improvement with increasing α. It rises from approximately 85% at α = 0.5 to about 90% at α = 0.8. Higher inlet pressures boost turbine performance and reduce energy losses, raising the overall thermal-to-useful energy conversion ratio. Second-law exergy efficiency also increases with α, though more moderately than energy efficiency. It grows from about 76% to approximately 78.5%. This indicates that system irreversibility decreases slightly as pressure rises, leading to a more efficient use of the input exergy (quality of energy). CO_2 emissions show a clear downward trend with increasing α. Emissions drop from around 0.22 kg CO_2/kWh at $\alpha = 0.5$ to approximately 0.13 kg CO_2/kWh at $\alpha = 0.8$. This outcome reflects the reduced auxiliary energy needs, positioning the system as a more environmentally sustainable solution when operated at higher pressure ratios.

Figure 6. Monthly average Direct Normal Irradiance (DNI) profile for Doha

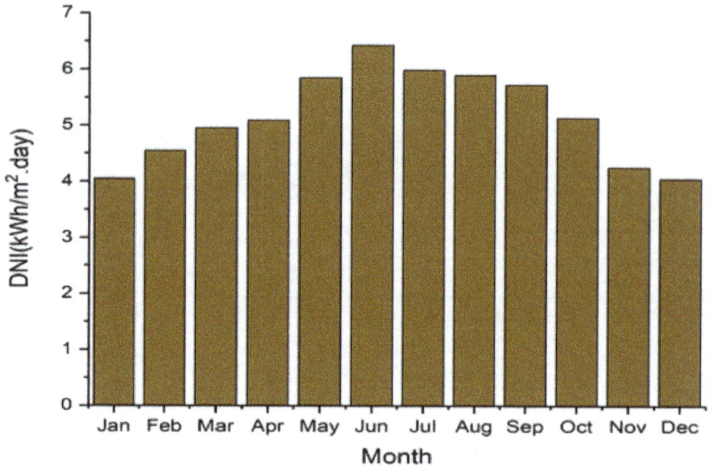

Figure 6 shows the monthly variation of the Direct Normal Irradiance (DNI), measured in kWh/m²·day, over a year in Doha, based on NASA meteorological data. DNI represents the amount of solar radiation received per unit area by a surface that is always held perpendicular (normal) to the rays of the sun, and it is a critical parameter for the performance of solar concentrating systems, such as the Solar PTC field used in the SA-CCHP system. The data presents that Doha enjoys excellent year-round DNI levels, especially from March to October, making it an ideal environment for solar-assisted systems like the SA-CCHP configuration. The thermal storage tank becomes important during lower DNI months (winter) to buffer energy and maintain system performance. Oversizing the solar field slightly could be a useful strategy to compensate for seasonal DNI variations without excessive cost. The SA-CCHP system can contribute significantly to sustainable energy targets and smart city energy management in Doha, given the strong solar resource.

The DNI values are the lowest during winter (December–February), ranging from about 4.0 to 4.5 kWh/m²·day. This is due to the lower solar altitude and shorter daylight hours. DNI levels rise significantly during spring (March–May), reaching around 5.0 to 5.8 kWh/m^2·day, indicating increasing solar potential as the sun's path rises in the sky. The maximum DNI occurs during the summer peak (June–August), with June showing the highest value of approximately 6.5–6.7 kWh/m^2·day. Although atmospheric dust (from desert conditions) can sometimes affect solar radiation during summer, the generally clear skies and long daylight hours dominate, leading to high solar resource availability. During the autumn (September–November), the DNI gradually decreases but remains relatively high (~5.5–4.5 kWh/m^2·day), supporting extended solar system operation throughout the year.

Figure 7. Monthly variation of net power output from the SA-CCHP system in Doha

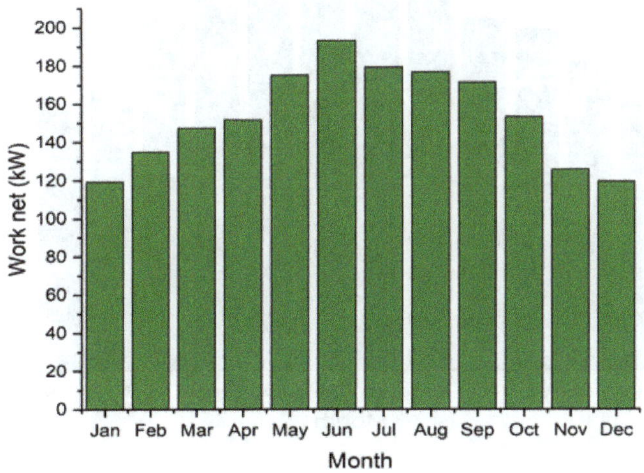

Figure 7 presents the monthly average net power output (in kW) generated by the SA-CCHP system throughout the year, based on Doha's local solar radiation conditions. The SA-CCHP system's power generation strongly correlates with solar resource availability, highlighting the importance of seasonal energy management strategies. During summer, the system can support peak electricity demand efficiently, which is especially critical in Doha's hot climate, where cooling needs are extremely high. In winter, supplementary energy strategies such as thermal energy storage or hybrid operation with other energy sources may be needed to maintain continuous service. This trend confirms the design suitability of the SA-CCHP system for regions with high solar potential like Doha, enabling renewable-based smart energy solutions for urban sustainability. The net power output is at its lowest in winter (December–February), ranging from 120 kW to 135 kW. This reduced output is consistent with the lower DNI observed during winter, as the system is primarily solar-driven. A progressive increase in net power generation is seen during spring months (March–May), reaching about 170–175 kW by May. This reflects the increasing solar intensity and longer daylight hours. The maximum power output occurs in June during Summer Peak (June–August), reaching approximately 195 kW. This corresponds to the highest solar irradiance conditions of the year. Although July and August show a slight decrease compared to June, the system maintains a very high-power output, benefiting from abundant solar energy. There is a gradual decline in the power output from September onwards. By November, net power drops to about 125 kW, following the seasonal decrease in solar energy availability.

Figure 8. monthly variation of total cost rate ($/hr) for the SA-CCHP system in Doha

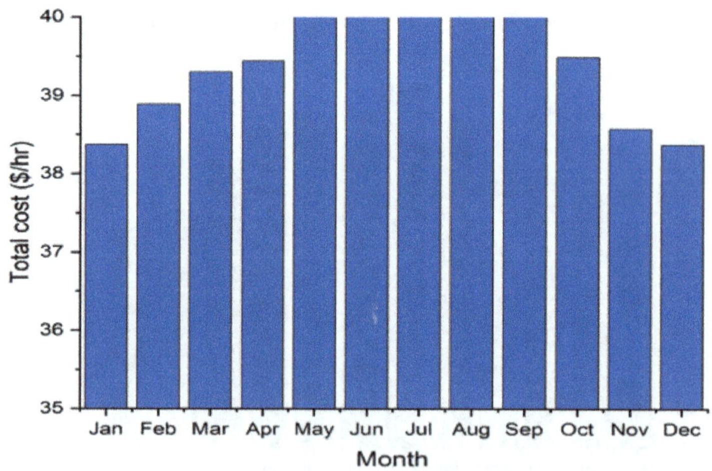

Figure 8 depicts the monthly changes in the total operating cost rate (in $/hr) of the SSA-CCHP system over a typical year in Doha. Although slight fluctuations occur, the operating cost remains relatively stable over the year (mostly between 38 and 38–40 $/hr), confirming the economic reliability of the SA-CCHP system for long-term urban deployment. The summer period shows how external factors like climate-driven cooling demand directly influence operational costs, despite favorable solar energy conditions. Integrating thermal storage or load management strategies could help reduce peak summer operational costs and enhance system cost-efficiency. During the winter, the total cost rate is at its lowest, around $38–38.5/hr. This is due to reduced energy demands (especially for cooling) and lower solar resource availability, but with efficient system management. A steady rise in cost is observed from March through May, reaching close to 39.5–40 $/hr. The increasing need for cooling and slightly improved energy outputs drives higher operational costs. The cost rate stabilizes at its maximum throughout the summer, around $40/hr. Despite high solar irradiance boosting energy outputs, very high cooling demands during

Doha's extremely hot summer significantly increase the system's energy use and auxiliary loads (such as pumps and cooling systems). After September, the total cost rate gradually declines, reflecting the reduction in cooling demand and overall load on the system.

Figure 9. Monthly variation of heating load supplied by the SA-CCHP system in Doha

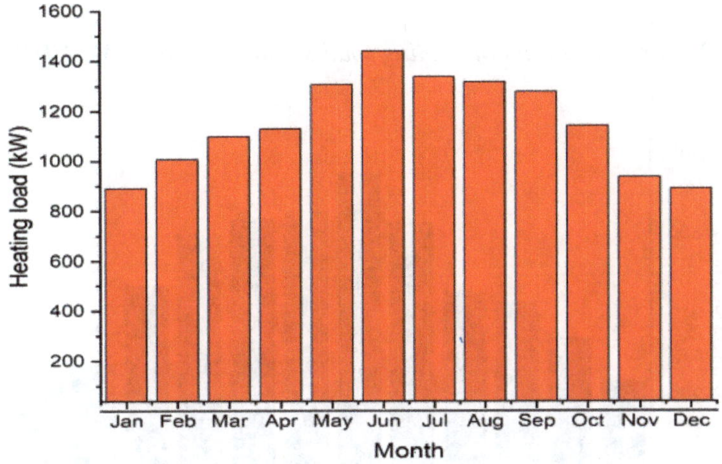

Figure 9 illustrates the monthly changes in the heating load (in kW) provided by the SA-CCHP system throughout the year under the climate conditions of Doha. This system's heating load reflects recovered thermal energy available for various uses, including hot water production, industrial processes, or district heating, not just space heating. During high solar input seasons (especially summer), the SA-CCHP system maximizes thermal energy recovery, resulting in higher heating load availability. In lower solar periods (winter), thermal recovery decreases, leading to a lower available heating load. The strong heating performance year-round highlights the effective thermal integration of the solar PTC field with the SA-CCHP system and its ability to operate flexibly under different seasonal conditions. While direct heating demand in Doha may be low in summer, the excess thermal energy can be diverted to desalination, industrial applications, or thermal storage, enhancing system value.

The heating load is relatively lower during winter, ranging from about 900 kW in January and December to about 1000 kW in February. Although this might seem counterintuitive compared to colder regions, heating demands in Doha's mild winter are modest compared to overall energy needs. A gradual rise in heating load is observed from March onwards, reaching approximately 1300 kW in May. This is likely associated with system heat recovery performance rather than external heating demands, reflecting the internal use of waste heat. During the peak summer months, the highest heating load values are observed, with a maximum around 1450 kW in June. End-user heating needs do not necessarily drive this high heating capacity, but rather the system's thermal energy management—recovering and utilizing high solar inputs. The heating load gradually declines, returning to around 950–1000 kW by November.

Figure 10. Monthly variation of cooling load supplied by the SA-CCHP system in Doha

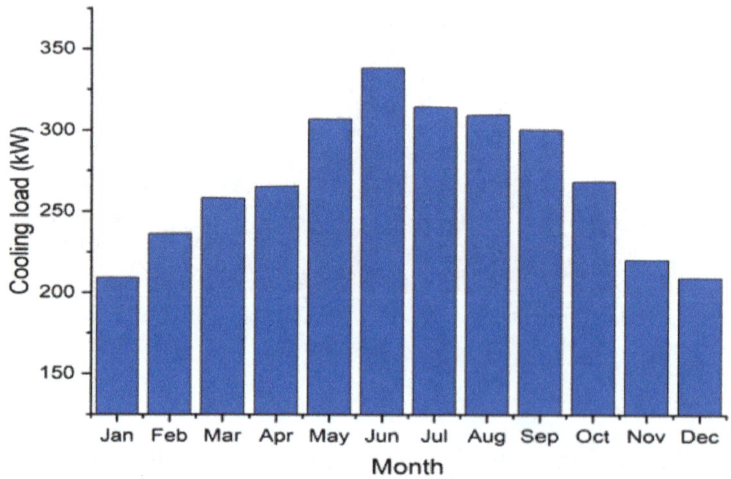

Figure 10 shows the monthly distribution of the cooling load (in kW) delivered by the SA-CCHP system over one year in Doha. The SA-CCHP system must adapt dynamically to large seasonal fluctuations in cooling load, with more than 60% variation from winter to summer. During summer, the system must prioritize cooling capacity, possibly with enhanced solar input and absorption chiller operation. As cooling demand drops, energy flows can be redirected during winter to maximize power and heating outputs. The cooling load is at its lowest during winter, ranging from ~210 kW in January and December to around 230 kW in February. This reflects the minimal cooling requirements during Doha's cooler season. The cooling load steadily increases during spring (March–May), reaching approximately 310 kW in May. Rising temperatures and extended daylight hours drive greater demand for indoor cooling. The maximum cooling load occurs in June, peaking at around 340 kW, followed closely by July and August. These months represent the hottest period in Doha, where cooling becomes a critical energy service for comfort and public health. A gradual decline in cooling load is observed as temperatures drop during the autumn (September–November), decreasing to around 270 kW in October and reaching ~210 kW again by December.

Comparative Evaluation with Existing Systems

A comparative evaluation was carried out to benchmark the performance of the proposed solar-assisted combined cooling, heating, and power (SA-CCHP) system against the solar-biomass CCHP configuration presented by Wang et al. (2019a). As shown in Table 2, the SA-CCHP system demonstrates superior energy efficiency at 58.74% compared to 56.0% in the reference system, while exergy efficiency is nearly equivalent, with values of 28.33% and 28.0%, respectively. These findings indicate that the proposed solar-assisted configuration is not only thermodynamically competitive but also capable of achieving incremental efficiency gains, which are particularly valuable in regions where maximizing renewable resource utilization is critical.

Table 2. Comparative performance evaluation between the proposed SA-CCHP system and solar-biomass CCHP system

Metric	Proposed SA-CCHP System	Solar-Biomass CCHP System (Wang et al., 2019a)
Energy Efficiency (%)	58.74	56.0
Exergy Efficiency (%)	28.33	28.0
CO_2 Emission Reduction (kg CO_2/kWh)	23.7	Not reported
Cost of Energy (USD/kWh)	0.087	Not reported

Beyond these metrics, the proposed system offers additional advantages that were not fully reported in the solar-biomass study. For example, the SA-CCHP system achieves a notable CO_2 emission reduction of 23.7 kg CO_2/kWh, aligning with global carbon mitigation goals and reinforcing its suitability for sustainable urban applications. Similarly, the system exhibits a favorable cost of energy at 0.087 USD/kWh, underscoring its economic viability in solar-rich regions such as Qatar where renewable energy is abundant and cooling demand is high. Unfortunately, a direct comparison with the solar-biomass system is limited by the absence of reported CO_2 and cost data in Wang et al. (2019a), making it difficult to establish a comprehensive cost-environmental performance baseline between the two systems.

Nevertheless, comparative evidence supports the feasibility and practicality of the proposed SA-CCHP system for deployment in arid and high-irradiance environments. While biomass integration can enhance system flexibility and provide a diversified renewable mix, the exclusive reliance on solar thermal energy in the SA-CCHP system offers a simpler, cleaner, and potentially more scalable solution for regions where biomass resources are scarce or logistically challenging to harness. Moreover, by minimizing reliance on biomass, the SA-CCHP system avoids potential drawbacks such as feedstock supply chain issues, land-use conflicts, and additional CO_2 emissions associated with biomass combustion. Overall, the analysis suggests that solar-assisted CCHP systems not only match but, in some aspects, outperform hybrid solar-biomass systems, particularly when designed for regions like Qatar with abundant solar energy and pressing sustainability targets.

CONCLUSION

This chapter presents a detailed thermodynamic, exergetic, and exergoeconomic evaluation of a SA-CCHP system designed specifically for smart cities such as Doha, which face high energy demands and possess abundant solar resources. Unlike conventional CCHP systems that rely on fossil fuels or hybrid configurations, the proposed system operates exclusively on solar thermal energy, utilizing parabolic trough collectors (PTCs). The model incorporates key components such as a thermal storage tank, heat recovery vapor generator (HRVG), ORC turbine, and absorption chiller, forming an integrated system capable of delivering electricity, cooling, and heating simultaneously. The results of seasonal simulations demonstrate that the proposed system can achieve a net power output ranging from 120 to 195 kW, depending on solar availability. The cooling capacity reaches a maximum of approximately 340 kW, while the heating load exceeds 1450 kW during peak months. System performance improves with higher superheating degrees and pressure ratios, where exergy efficiency increases from 76% to 78.5% and total cost rate drops from \$40/hr to \$36/hr. Additionally, CO_2 emissions, attributed only to auxiliary electricity consumption, decrease significantly with improved operating conditions, supporting the system's role in climate change mitigation. The parametric analyses reveal that increasing the superheating temperature increases net power and heating output but slightly lowers cooling capacity. Similarly, enhancing the pressure ratio improves overall system efficiencies and reduces environmental impact. These findings emphasize the system's adaptability to varying solar and demand conditions and its capability to provide year-round service without relying on fossil fuel inputs. Finally, the solar-powered SA-CCHP system proposed in this chapter offers a viable and sustainable solution for meeting the multi-functional energy demands of smart cities in hot, solar-rich environments. It effectively eliminates fossil fuel dependency, reduces CO_2 emissions, and supports the transition toward clean.

REFERENCES

Abbas, A. B., Almohammedi, A. A., Balfaqih, M., & Darshi, S. (2023, September). Conceptual Design of Wireless Smart Grid for the Optimization of Electric Transmission in Iraq. In *2023 3rd International Conference on Computing and Information Technology (ICCIT)* (pp. 577-580). IEEE.

Akroot, A., & Al Shammre, A. S. (2024). Techno-Economic and Environmental Impact Analysis of a 50 MW Solar-Powered Rankine Cycle System. *Processes (Basel, Switzerland), 12*(6), 1059. DOI: 10.3390/pr12061059

Akroot, A., & Refaei, M. (2025). Thermodynamic and Exergoeconomic Assessment of a Solar-Assisted Combined Cooling, Heating, and Power System in Antalya, Turkey. *Gazi Üniversitesi Fen Bilimleri Dergisi Part C: Tasarım ve Teknoloji*, 1.

Alfaris, A., Akroot, A., Alqaed, S., & Almehmadi, F. A. (2025). Performance analysis of integrated solar and natural gas combined cycle power plants in high solar potential regions. *Scientific Reports, 15*(1), 9181. PMID: 40097525

Alfaris, A., Akroot, A., & Deniz, E. (2024). The Exergo-Economic and Environmental Evaluation of a Hybrid Solar–Natural Gas Power System in Kirkuk. *Applied Sciences (Basel, Switzerland), 14*(22), 10113. DOI: 10.3390/app142210113

Almohammedi, A. A., & Yusof, M. H. M. bin Tuan Muda, S. R., Shepelev, V., Balfaqih, M., & Shamsuddin, S. (2023, October). Resistors Over Digital Analog Converter on PRBS Threat Generator with PLL Synchronizer: Comparison Between R2R & Weighted. In *2023 3rd International Conference on Emerging Smart Technologies and Applications (eSmarTA)* (pp. 1-6). IEEE.

Alsunousi, M., & Kayabasi, E. (2024). Techno-economic assessment of a floating photovoltaic power plant assisted methanol production by hydrogenation of CO2 captured from Zawiya oil refinery. *International Journal of Hydrogen Energy, 57*, 589–600. DOI: 10.1016/j.ijhydene.2024.01.055

Anvari, S., Mahian, O., Taghavifar, H., Wongwises, S., & Desideri, U. (2020). 4E analysis of a modified multigeneration system designed for power, heating/cooling, and water desalination. *Applied Energy, 270*(February), 115107. DOI: 10.1016/j.apenergy.2020.115107

Besevli, B., Kayabasi, E., Akroot, A., Talal, W., Alfaris, A., Assaf, Y. H., Nawaf, M. Y., Bdaiwi, M., & Khudhur, J. (2024). Technoeconomic Analysis of Oxygen-Supported Combined Systems for Recovering Waste Heat in an Iron-Steel Facility. *Applied Sciences (Basel, Switzerland), 14*(6), 2563. DOI: 10.3390/app14062563

Cavalcanti, E. J. C., Ferreira, J. V. M., & Carvalho, M. (2021). On the consideration of different dead states in the exergy assessment of a solar-assisted combined cooling, heat and power system. *Sustainable Energy Technologies and Assessments*, *47*, 101361. DOI: 10.1016/J.SETA.2021.101361

Chen, E., Zhao, Y., Wang, M., Bian, M., Cai, W., Li, B., & Dai, Y. (2023). Experimental investigation of a solar-assisted absorption-compression system for heating and cooling. *Solar Energy*, *257*, 18–33. DOI: 10.1016/J.SOLENER.2023.04.025

Chu, X., Yang, D., & Li, J. (2019). Sustainability assessment of combined cooling, heating, and power systems under carbon emission regulations. *Sustainability*, *11*(21), 5917.

Eisavi, B., Khalilarya, S., Chitsaz, A., & Rosen, M. A. (2018). Thermodynamic analysis of a novel combined cooling, heating and power system driven by solar energy. *Applied Thermal Engineering*, *129*, 1219–1229.

Elsafi, A., Almohammedi, A. A., Balfaqih, M., Balfagih, Z., & Sabri, S. (2025). Comparative analysis of maximum power point tracking methods for power optimization in grid tied photovoltaic solar systems. *Discover Applied Sciences*, *7*(9), 976.

Fani, M., & Sadreddin, A. (2017). Solar assisted CCHP system, energetic, economic and environmental analysis, case study: Educational office buildings. *Energy and Building*, *136*, 100–109. DOI: 10.1016/j.enbuild.2016.11.052

Gao, Y., Yao, W., Wang, J., & Cui, Z. (2023). Thermodynamic analysis of solid oxide fuel cell based combined cooling, heating, and power system integrated with solar-assisted electrolytic cell. *Journal of Thermal Science*, *32*(1), 93–108.

Hou, H., Wu, J., Ding, Z., Yang, B., & Hu, E. (2021). Performance analysis of a solar-assisted combined cooling, heating and power system with an improved operation strategy. *Energy*, *227*, 120516. DOI: 10.1016/J.ENERGY.2021.120516

Javadi, M. A., Jafari Najafi, N., Khalili Abhari, M., Jabery, R., & Pourtaba, H. (2021). 4E analysis of three different configurations of a combined cycle power plant integrated with a solar power tower system. *Sustainable Energy Technologies and Assessments*, *48*. Advance online publication. DOI: 10.1016/j.seta.2021.101599

Khan, Y., & Singh, P. K. (2024). Thermo-Economic and Environmental Assessment of a Novel SOFC-Based Hybrid Energy Generation System for Combined Cooling Heating and Power Generation. *Journal of Energy Resources Technology, Part A: Sustainable and Renewable Energy*, *1*(1). Advance online publication. DOI: 10.1115/1.4066110

Khoshgoftar Manesh, M. H., Mousavi Rabeti, S. A., Nourpour, M., & Said, Z. (2022). Energy, exergy, exergoeconomic, and exergoenvironmental analysis of an innovative solar-geothermal-gas driven polygeneration system for combined power, hydrogen, hot water, and freshwater production. *Sustainable Energy Technologies and Assessments, 51*(April 2021). https://doi.org/DOI: 10.1016/j.seta.2021.101861

Li, L. L., Ren, X. Y., Tseng, M. L., Wu, D. S., & Lim, M. K. (2022). Performance evaluation of solar hybrid combined cooling, heating and power systems: A multi-objective arithmetic optimization algorithm. *Energy Conversion and Management, 258*, 115541. DOI: 10.1016/J.ENCONMAN.2022.115541

Mauladdawilah, H., Balfaqih, M., Balfagih, Z., Gago, E. J., & Pegalajar, M. C. (2025b). Optimization of Photovoltaic Power Forecasting: A Comparative Study of Deep Learning Architectures, Optimization Techniques, and Evaluation Metrics. In *2025 22nd International Learning and Technology Conference (L&T)* (Vol. 22, pp. 109-114). IEEE.

Mauladdawilah, H., Balfaqih, M., Balfagih, Z., Pegalajar, M. D. C., & Gago, E. J. (2025a). Deep Feature Selection of Meteorological Variables for LSTM-Based PV Power Forecasting in High-Dimensional Time-Series Data. *Algorithms, 18*(8), 496.

Mauladdawilah, H., Gago, E. J., Pegalajar, M. C., & Balfaqih, M. (2025c). An Evaluation of Meteorological Variables Impact on Photovoltaic Power Generation Estimation Based on Deep Learning Model. In *2025 4th International Conference on Computing and Information Technology (ICCIT)* (pp. 638-642). IEEE.

Mithat Delibaş, H., Kayabaşı, E., Mithat Delibas, H., & Kayabasi, E. (2021). Environment and Economy Assessment of Waste Heat Recovery Technologies in Marine Industry. *The International Journal of Materials and Engineering Technology, 004*(2), 133–146. http://dergipark.gov.tr/tijmet

Oliveira, A. C., Afonso, C., Matos, J., Riffat, S., Nguyen, M., & Doherty, P. (2002). A combined heat and power system for buildings driven by solar energy and gas. *Applied Thermal Engineering, 22*(6), 587–593. DOI: 10.1016/S1359-4311(01)00110-7

Salimi, M., Hosseinpour, M., Mansouri, S., & Borhani, N., T. (2022). Environmental aspects of the combined cooling, heating, and power (CCHP) systems: A review. *Processes (Basel, Switzerland), 10*(4), 711.

Su, Z., Yang, L., Wang, H., Liu, S., Song, J., & Jin, X. (2023). Solar-assisted combined cooling and power system integrating energy storage and desulfurization for coal-fired power plants. *Thermal Science and Engineering Progress, 45*, 102110. DOI: 10.1016/J.TSEP.2023.102110

Talal, W., & Akroot, A. (2023). Exergoeconomic Analysis of an Integrated Solar Combined Cycle in the Al-Qayara Power Plant in Iraq. *Processes (Basel, Switzerland), 11*(3). Advance online publication. DOI: 10.3390/pr11030656

Ünal, F., Temir, G., & Köten, H. (2018). Energy, exergy and exergoeconomic analysis of solar-assisted vertical ground source heat pump system for heating season. *Journal of Mechanical Science and Technology*, *32*, 3929–3942.

Wang, J., Han, Z., & Guan, Z. (2020). Hybrid solar-assisted combined cooling, heating, and power systems: A review. *Renewable & Sustainable Energy Reviews*, *133*, 110256.

Wang, J., Ma, C., & Wu, J. (2019a). Thermodynamic analysis of a combined cooling, heating and power system based on solar thermal biomass gasification☆. *Applied Energy*, *247*, 102–115. DOI: 10.1016/J.APENERGY.2019.04.039

Yang, K., He, Y., Du, N., Yan, P., Zhu, N., Chen, Y., Wang, J., & Lund, P. D. (2024). Exergy, exergoeconomic, and exergoenvironmental analyses of novel solar- and biomass-driven trigeneration system integrated with organic Rankine cycle. *Energy*, *301*. Advance online publication. DOI: 10.1016/j.energy.2024.131605

Yang, K., Zhu, N., Ding, Y., Chang, C., Wang, D., & Yuan, T. (2019). Exergy and exergoeconomic analyses of a combined cooling, heating, and power (CCHP) system based on dual-fuel of biomass and natural gas. *Journal of Cleaner Production*, *206*, 893–906. DOI: 10.1016/j.jclepro.2018.09.251

Yang, X., Yan, Z., Bu, S., Li, W., Yu, N., Su, C., Lu, S., & Wang, X. (2024). Energy, exergy, exergoeconomic, economic, and environmental analyses and multi-objective optimization of a novel combined cooling and power system with dual-pressure Kalina cycle-absorption refrigeration. *Energy Conversion and Management*, *315*. Advance online publication. DOI: 10.1016/j.enconman.2024.118781

Zhang, L., Li, F., Sun, B., & Zhang, C. (2019a). Integrated optimization design of combined cooling, heating, and power system coupled with solar and biomass energy. *Energies*, *12*(4). Advance online publication. DOI: 10.3390/en12040687

Zhang, N., Wang, Z., Lior, N., & Han, W. (2018). Advancement of distributed energy methods by a novel high efficiency solar-assisted combined cooling, heating and power system. *Applied Energy*, *219*, 179–186. DOI: 10.1016/J.APENERGY.2018.03.050

Zhu, G., & Chow, T. T. (2019). Design optimization and two-stage control strategy on combined cooling, heating and power system. *Energy Conversion and Management*, *199*, 111869. DOI: 10.1016/J.ENCONMAN.2019.111869

Chapter 7
Utilizing IoT and AI to Enhance Urban Planning and Development

T. Venkat Narayana Rao
https://orcid.org/0000-0002-1996-1819
Sreenidhi Institute of Science and Technology, India

Dattaram Tellapati
https://orcid.org/0009-0005-4206-0874
Sreenidhi Institute of Science and Technology, India

Sangers Bhavana
https://orcid.org/0009-0005-6134-1081
Sreenidhi Institute of Science and Technology, India

Chenchu Swetha Swetha
https://orcid.org/0009-0005-1835-0679
Sreenidhi Institute of Science and Technology, India

ABSTRACT

A smart city is an urban area that uses advanced technology, data analytics, and connectivity to enhance the quality of life for its residents and reduce resource consumption. The IoT is a system of devices which can be accessed through internet and helps to collect and share data across number devices over internet. These innovative systems are fed lots of data and learn from patterns, make predictions, improve systems over time. Integrating IoT with machine learning combines data collection from IoT devices with advanced analytics. IoT sensors gather real-time data, which is then processed and cleaned for ML models. These models are trained to identify patterns, make predictions, and detect anomalies. This integration allows

DOI: 10.4018/979-8-3373-0154-9.ch007

predictive maintenance, and generate dynamic based on new data. The chapter focus on exploring smart transportation systems with energy management, environmental and sustainability with public safety and building management Citizen engagement. This would enable future trends in smart urban planning and Integrating IOT with Machine learning.

INTRODUCTION

The technological revolution of the late 20th and early 21st centuries has been pivotal in shaping the concept of smart urban planning. Smart cities are broadly defined as urban ecosystems that integrate data-driven decision-making, Internet of Things (IoT) technologies, and Artificial Intelligence (AI) to enhance operational efficiency, environmental sustainability, and social well-being. Early milestones include Singapore's deployment of sensor-based urban services and Barcelona's pioneering smart traffic management systems. Today, Geographic Information Systems (GIS) have become indispensable, enabling detailed spatial mapping and real-time data analysis to support effective planning (Balfaqih & Alharbi, 2022).

Modern urban development is guided by three core objectives: sustainability, resilience, and inclusivity. Concepts such as the "15-minute city" emphasize local self-sufficiency to reduce commuting distances and environmental emissions. Similarly, park-like urban models combine eco-friendly technologies with emergency preparedness systems to strengthen responses to environmental threats. Digital platforms further advance participatory planning by involving citizens in decision-making processes, ensuring urban policies align with diverse community needs.

Historically, urban planning has consistently adapted to technological innovation, social change, and economic transformation. From ancient city grids to today's AI-driven planning frameworks, cities have evolved toward greater efficiency and sustainability. The next stage in this evolution lies in climate-resilient digital cities, which integrate renewable resources and adaptive infrastructures to ensure equitable access to essential services and address future challenges (Hasan et al., 2023).

Smart transportation exemplifies this shift, as IoT devices combined with machine learning (ML) algorithms enable optimized traffic flow, improved public transit, and reduced congestion. Likewise, smart energy systems leverage real-time analytics to maximize efficiency while integrating renewable energy sources, advancing environmental goals. Environmental monitoring initiatives, using IoT-based sensors, track water and air quality, supporting proactive measures to protect ecosystems and safeguard public health (Balfaqih & Alharbi, 2022).

Public safety and emergency management are also enhanced through real-time analytics, smart surveillance, and responsive infrastructure systems, which optimize resource use and occupant well-being. At the same time, citizen engagement in governance fosters transparency and ensures inclusive participation in urban development decisions (Balfaqih et al., 2021).

Despite these advancements, challenges persist in applying IoT and ML technologies to urban planning. Key concerns include data privacy, algorithmic bias, and digital inequality, all of which must be addressed to build trust and guarantee equitable access. Intelligent governance and robust policy frameworks are therefore critical to realizing the full potential of smart city initiatives (Balfaqih & Alharbi, 2022).

This chapter maps the smart-city stack from the street to the control room and up to governance. It begins with smart transportation— Intelligent Transportation Systems (ITS) foundations, Singapore's congestion pricing and public-transit integration, and the security hardening required for connected mobility—then shows how IoT and ML reshape operations through V2X, real-time analytics, and the Amsterdam case, supported by rigorous data-collection pipelines. We then turn to smart energy management, detailing IoT-enabled monitoring and the cloud/edge ML workflows that drive forecasting, optimization, and demand response, before broadening to environmental monitoring and public safety, where sensor networks translate raw signals into decision support. The latter half interrogates implementation: data privacy and security, integration and data quality, algorithmic fairness, workforce capacity, and ethics in smart governance. Finally, we look ahead to digital-divide mitigation, planning innovations, cybersecurity and compliance, financing models, and the enabling roles of 5G, edge computing, and blockchain. Together, these sections provide a coherent roadmap for building cities that are intelligent, resilient, and trustworthy.

SMART TRANSPORTATION SYSTEMS

Smart mobility—often framed as ITS—applies IoT sensors, advanced communications, automation, high-speed networks, and big-data analytics to orchestrate the movement of people and goods. Beyond core tools such as in-vehicle navigation, traffic-signal control, container tracking, automated license-plate recognition, and speed enforcement, modern ITS fuse real-time, multi-source data to optimize flows and inform travelers. In doing so, they enhance network efficiency and safety, reducing human-error–related incidents and laying the groundwork for increasingly autonomous operations, (Khayyat et al., 2025).

Singapore illustrates this integrated approach. Constrained land and high demand led the city-state to pair supply-side efficiency with demand management. Since 1998, Electronic Road Pricing (ERP) has dynamically priced road use via roadside sensors and satellite support, shifting travel from peak periods and easing congestion on key corridors. A high-capacity Mass Rapid Transit (MRT) backbone, coordinated bus services with dedicated lanes, and sustained investment in walking and cycling infrastructure underpin a "car-lite" strategy. On the demand side, the Vehicle Quota System (VQS) and Certificate of Entitlement (COE) bidding limit private car growth and internalize the full cost of ownership, encouraging shared and public mobility (Calder, 2016).

Complementing these policies, Singapore deploys AI-enabled traffic management to retime signals, detect incidents, and deliver real-time updates to travelers. While short-lived congestion can still occur at construction zones or peak periods, the city continues to invest in electric mobility and autonomous-vehicle pilots to further decarbonize and de-risk the network. The result is a globally recognized model of coherent policy, technology integration, and continuous improvement in urban transport (Calder, 2016).

Expanded Security Strategies for the Connected Systems

Data integrity serves as the base requirement to achieve safety together with resilience in interconnected systems. The protection of data integrity uses hash values as mechanisms to verify unchanged data transfers through networks. The integration enables receivers to verify data because the verification process confirms the data remains complete without modification. Vehicle locations along with driver details require strict confidentiality measures as the primary precaution for data protection. Database security relies on encryption and VPNs as well as TLS (Transport Layer Security) to stop unauthorized parties from accessing sensitive information (ISO/IEC 17067:2013).

Strong authentication and authorization protocols play a key role within information security systems through the implementation of role-based access control (RBAC). The security mechanisms establish that permitted devices and approved users alone will receive permission to view and access particular functions and data points while blocking unauthorized persons from gaining access. The implementation of Message Authentication Codes (MACs) enables strong non-repudiation capability because parties become unable to decline their involvement in any communication exchange.

Service availability becomes critical since attacks and failures could occur. Redundancy mechanisms together with failover systems need implementation to maintain service availability despite adverse conditions. A real-time system for anomaly detection needs immediate implementation in the network infrastructure. This system operates at all times by utilizing machine learning or heuristic techniques to detect potential breaches while monitoring for irregularities in the system.

A critical component for system protection against attacked or compromised security involves continuous equipment updates accompanied by needed software patches. The maintenance of updated operating systems and firmware with additional software creates vulnerability solutions which makes it difficult for attackers to exploit the system. Replacement of physical security measures becomes vital for sections that remain exposed such as vehicles together with roadside infrastructure. Physical access prevention needs to be established together with equipment tampering protection strategies (Balfaqih et al., 2023).

The fundamental requirement for successful operations depends on system interoperability between different domains. Systems implementing defensive measures must adhere to pre-established standards including ISO or SAE guidelines in order to achieve smooth operation between different systems. Prioritize the protection of privacy in every operation. Organizations need to follow GDPR requirements and deploy data anonymization methods to handle personal data since this practice decreases the chances of leaking sensitive information (ISO/SAE 21434:2021).

Executing a complete incident response plan remains vital to handle possible security violations. A regular testing process should evaluate this plan to maintain its efficiency along with up-to-date content which enables coordinated responses for security failures. The creation of a security-conscious organization depends heavily on training its users. These educational services teach people to detect security risks while showing them the proper reaction steps which makes organizational systems more secure. These security measures unite into a comprehensive solution which protects networked systems by maintaining data integrity and confidentiality and providing resistance against cyber-attacks (Hartzog, 2018).

IoT in Transportation Systems

IoT technologies are transforming transportation by enabling seamless communication between vehicles, infrastructure, and users. Sensors and onboard communication devices allow real-time interaction with traffic systems, enhancing safety and operational efficiency. Data collected from cameras, radar, and inductive loops supports traffic optimization, while IoT-enabled smart signals adjust dynamically to conditions, reducing delays and improving flow. Mobile applications further enhance the user experience by providing real-time updates, optimized routes, and parking availability.

The evolution of Vehicle-to-Everything (V2X) communication expands this scope, linking vehicles with other vehicles (V2V), infrastructure (V2I), and pedestrians (V2P). These systems improve road safety, efficiency, and user convenience. Smart parking exemplifies this, reducing congestion and wait times through real-time space detection. Similarly, IoT-driven fleet management enables route optimization, fuel savings, and on-time deliveries, marking a fundamental shift toward safer, cleaner, and more efficient mobility systems (Balfaqih et al., 2017).

Integrating IoT with ML takes transportation innovation further. ML leverages historical and real-time data to predict traffic patterns, detect anomalies, and support proactive management. Computer vision and video analytics enhance recognition of traffic signs, vehicle movement, and incidents, ensuring faster responses. Algorithms optimize traffic flows and journey times, while NLP-powered assistants provide passengers with personalized, real-time information (Morozov et al., 2024). IoT-ML synergy supports demand forecasting, enabling smarter resource allocation in ride-sharing and public transit, ultimately contributing to sustainable mobility solutions (Balfaqih et al., 2021).

Amsterdam provides a global benchmark in smart transportation by combining sustainability with advanced digital infrastructure. Its strategy integrates electric mobility, autonomous vehicle development, and multimodal connectivity, supported by extensive cycling infrastructure and smart bike-sharing systems. Public transport is enhanced through smart trams, metro services, and buses linked via a Mobility-as-a-Service (MaaS) platform, offering seamless trip planning and ticketing. The city further reduces emissions by deploying electric and hydrogen-powered buses (Qonita et al., 2025).

At the core of Amsterdam's success is its open data platform, which fosters innovation and collaboration among researchers, businesses, and policymakers. Artificial intelligence is used for predictive traffic modeling, while blockchain and 5G enhance security, ticketing, and communication between vehicles, pedestrians, and infrastructure. Through its integrated, sustainable, and collaborative approach, Amsterdam has established itself as a global leader in smart urban mobility (Qonita et al., 2025).

Data Collection

The data collection phase within IoT-based Intelligent Transportation Systems (IoT-ITS) involves several critical management operations, including data cleaning, validation, anomaly detection, aggregation, transformation, and segmentation. These processes ensure that the diverse and large-scale datasets generated across the IoT-ITS framework are accurate, reliable, and ready for analysis. Because ITS applications rely heavily on continuous data streams, it is essential to define data requirements clearly to guarantee consistency and usability (Al-khatib et al., 2020).

Different categories of data are typically handled, such as background data, configuration data, generated data, daily operational inputs, real-time data from ITS devices, real-time outputs from ITS applications, and downstream data prepared for further processing. Data cleansing focuses on detecting and removing corrupt or inaccurate records, while eliminating redundant or irrelevant entries from communication channels. Validation follows to ensure accuracy, completeness, and usability of the datasets. Outliers or abnormal values are identified and removed to maintain integrity and support robust decision-making (Kolajo et al., 2019).

Data aggregation condenses and summarizes essential information, while transformation processes are applied when integration of heterogeneous data sources is required. For enhanced organization and accessibility, data partitioning techniques are used to segment datasets into manageable groups. Together, these operations form the foundation of effective data governance, enabling ITS to transform raw inputs into actionable insights for traffic optimization, safety management, and sustainable urban mobility (Kolajo et al., 2019).

SMART ENERGY MANAGEMENT SYSTEMS

Modern energy management solutions require immediate development because society demands more energy and faces environmental concerns about resources. SEMS powered by IoT and ML technology serves as leaders for this transformational energy process through new monitoring approaches optimized for energy management in various industries.

Role of IoT in Enhancing Smart Energy Management Systems (SEMS)

Smart Energy Management Systems (SEMS) derive much of their efficiency and sustainability from the integration of Internet of Things (IoT) technologies. IoT enables continuous monitoring of energy consumption through smart meters, allowing users to track usage patterns, detect inefficiencies, and identify opportunities for savings. A wide range of sensors monitor environmental conditions and operational parameters that directly influence energy demand, providing accurate data to support informed decision-making. With device interconnectivity, IoT enhances monitoring and control capabilities, enabling users to manage household appliances or industrial machines from a centralized platform. Real-time data streams further allow for automated efficiency adjustments, such as lighting control in unoccupied spaces or adaptive thermal regulation based on occupancy levels.

Beyond monitoring, IoT plays a vital role in predictive energy management. By analyzing historical consumption data, IoT devices can identify usage patterns and forecast future demand, improving distribution efficiency and facilitating peak load management. IoT also tracks renewable energy generation in real time, allowing seamless coordination between renewable sources and conventional systems. This integration enhances storage efficiency by optimizing the use of stored power during demand surges (Liu et al., 2020).

IoT technologies also strengthen demand response management by supporting dynamic pricing strategies that incentivize consumers to shift consumption during peak periods, thereby lowering costs and reducing grid stress (Balfaqih et al., 2021). Real-time data delivered through mobile apps and dashboards empowers customers to make informed decisions, while utilities benefit from improved distribution efficiency, fewer outages, and optimized supply-demand balancing. In IoT-enabled smart grids, bidirectional communication between consumers and providers enhances grid reliability, stability, and overall performance (Hasan et al., 2023).

Integration of IoT and Machine Learning in Energy Management Systems

The integration of the IoT and ML is transforming energy management systems by enabling advanced data processing, predictive analytics, and real-time optimization. This integration begins with cloud infrastructure, where vast amounts of data generated by IoT devices—including sensors, meters, and monitoring systems—are stored and processed. Continuous real-time data streaming, supported by communication protocols such as MQTT (Message Queuing Telemetry Transport) and CoAP (Constrained Application Protocol), allows organizations to analyze energy usage dynamically and respond rapidly, making management strategies more flexible and adaptive (Mazhar et al., 2023).

Before being used in ML model development, the collected data undergoes preprocessing and cleaning to ensure reliability. Normalization standardizes variables, while outliers and inconsistencies are removed to improve analytical accuracy. Feature engineering further enhances ML performance by extracting meaningful indicators such as peak demand periods, usage trends, and temperature fluctuations, which are critical for building predictive capabilities (Mazhar et al., 2023).

ML models developed from this processed data empower systems to perform predictive and optimization functions. Predictive analytics generate accurate forecasts of energy demand, consumption patterns, and peak load timings, enabling providers to allocate resources efficiently and correct operational deficiencies. Anomaly detection algorithms further enhance system reliability by identifying irregular consumption behaviors, equipment faults, or security risks before they escalate into major issues.

Energy distribution optimization is a key application of ML in IoT-enabled systems. Optimization algorithms analyze real-time consumption data to balance grid loads, reduce energy losses, and stabilize supply during peak demand. Through demand-side management strategies, supported by ML models, customers are incentivized to shift usage patterns, contributing to a more resilient and sustainable energy network (Nasir et al., 2021).

The integration of IoT and ML also supports renewable energy adoption. ML-based forecasting models, informed by historical data and weather conditions, predict the availability of renewable energy sources such as solar and wind, allowing operators to optimize energy allocation. In addition, ML algorithms applied to battery management systems enhance storage efficiency, minimize energy waste, and improve the reliability of renewable integration.

From the consumer perspective, IoT and ML enable the development of intelligent applications that provide users with easy access to real-time consumption data, personalized energy-saving recommendations, and remote device control via mobile platforms (Nasir et al., 2021). These applications improve user engagement and encourage sustainable energy behaviors.

A feedback mechanism between IoT and ML ensures continuous system improvement. As new data is collected, models are updated, enhancing prediction accuracy and optimization over time. This iterative learning process allows the system to become increasingly intelligent and efficient. However, security remains a critical requirement. Strong measures such as encryption, access control, and secure transmission protocols are essential to protect sensitive data and maintain user trust.

ENVIRONMENTAL MONITORING, AND PUBLIC SAFETY SYSTEMS

Environmental monitoring in smart cities relies heavily on IoT, which enables continuous data collection through a network of connected sensors. These sensors are deployed across diverse environments—such as irrigation systems, pipelines, water tanks, weather stations, marine applications, and industrial machinery—to measure critical parameters including temperature, humidity, water quality, leakage, and other physical conditions. Communication modules and edge computing ensure that this data is processed locally when needed and rapidly transferred to cloud or data centers for advanced analysis (Ullo & Sinha, 2020).

Automated monitoring systems extend these capabilities by detecting abnormal conditions and issuing real-time alerts via email or text messages. They can also generate service tickets or trigger shutdown procedures to prevent potential disasters. This combination of sensing, reporting, and preventive action makes IoT a cornerstone of proactive environmental management and sustainability strategies.

The deployment of IoT sensors worldwide has significantly advanced our ability to monitor and interpret climate change impacts. Networks of devices across urban and natural ecosystems provide high-resolution, real-time datasets that capture variations in temperature, pollution, and other environmental factors. These continuous data streams enable early identification of anomalies, allowing timely interventions to address ecological challenges. Documented applications include monitoring air and water quality, optimizing resource management, and protecting wildlife habitats, all of which demonstrate the versatility of IoT in environmental conservation.

A wide range of sensor types underpins IoT-based monitoring systems (Ullo & Sinha, 2020):

- Temperature sensors to track ambient conditions.
- Humidity sensors for air moisture levels.
- Proximity and motion sensors for detecting movement or occupancy.
- Light and pressure sensors for monitoring illumination and atmospheric or water pressure.
- Accelerometers to measure speed and direction.
- Gas sensors for detecting harmful emissions.

Urban ecosystems benefit particularly from pollution-monitoring sensors that measure CO_2, NO_2, PM2.5, and PM10 levels, providing city authorities with actionable insights to enforce emission controls, strengthen traffic regulations, and develop green infrastructure. Similarly, water quality sensors assess contaminants, pH, and temperature in lakes, rivers, and municipal supply systems, safeguarding communities from pollution and ensuring safe drinking water (Shepelev et al., 2023). Flood sensors in coastal cities provide critical alerts to reduce disaster risk, while soil moisture sensors support sustainable agriculture and urban landscaping by optimizing irrigation (Balfaqih, 2023).

Waste management has also been transformed by IoT-enabled systems. Smart bins equipped with fill-level sensors allow municipalities to optimize collection schedules, reduce fuel consumption, and improve recycling efficiency. Smart meters and energy sensors further enhance sustainability by monitoring real-time electricity consumption, supporting demand-side management, and integrating renewable sources such as solar and wind into the grid (Shepelev et al., 2023).

Beyond environmental sustainability, IoT sensors also strengthen public safety. AI-powered surveillance cameras with image recognition detect suspicious activities and assist in traffic monitoring and emergency coordination. Audio recognition systems can identify gunshots or other critical sounds, enabling authorities to respond quickly to threats (Morozov et al., 2024). Fire detection sensors monitor smoke and temperature changes, triggering early warnings to emergency services. In earthquake-prone areas, seismic sensors provide real-time alerts and can activate automatic shutdowns of pipelines, power systems, and transportation networks to reduce casualties (Almohammedi et al., 2023).

Additional innovations, such as motion-sensitive streetlights and panic button sensors in public areas, enhance safety by improving visibility, deterring crime, and providing rapid access to emergency assistance. Together, these technologies form a comprehensive ecosystem in which IoT-driven environmental monitoring, sustainability practices, and public safety measures converge to create smarter, more resilient cities.

Key Components of IoT-Based Environmental Monitoring for Decision Support

A typical IoT-based environmental monitoring system functions through four essential elements which help users make better decisions and extract vital information. Environmental monitoring serves as a fundamental system component through which sensors detect and send environmental conditions from agricultural fields as well as industrial locations and water management facilities. The data collection and transmission process is enabled through Digi XBee wireless modules together with connectivity gateways. The connected sensors operate in real time to gather information which becomes instantly ready for analysis and decisions.

The second element of this system concerns precise measurement of gathered data. The assessment of environmental impacts requires monitoring systems to measure particular indicators which detect water contamination and chemical spills together with potential equipment breakdowns. The collected data provides essential information which both industrial enterprises and local governments need for environmental footprint assessment and resource usage optimization and waste minimization. Data evaluation enables organizations to establish more sustainable practices while reaching better operational efficiencies and stopping environmental hazards (Shepelev et al., 2023).

Data cataloging functions as the vital solution to address the problems related to handling extensive environmental monitoring system data volumes. The Microsoft Planetary Computer operates as an environment-related data center that worldwide organizations use for storing and organizing environmental information. The analysis procedures require businesses and industries to maintain their gathered data in cloud storage structures together with data centers for convenient access. Organizations depend on structured data for extended environmental health pattern monitoring because it produces detailed environmental status reports.

The goal to extract actionable insights from data succeeds using advanced analysis procedures before integrating with cloud infrastructure. Microsoft Azure and Amazon Web Services together with IoT solutions analyze the data through powerful cloud platforms to generate valuable insights. The staff can use the system to receive early warning signals and track important monitoring points while taking steps to prevent potential threats. Organizations obtain practical information through insights which enables them to implement improved operational practices and effective environmental responses.

Linking of Wireless Devices Across an Industrial Area

The rising global demand for clean water underscores the critical role of technology in safeguarding resources, protecting human health, and ensuring environmental sustainability. Real-time operating system (RTOS)-based water quality monitoring systems equipped with IoT sensors provide continuous measurements of pollutants, pH levels, and dissolved oxygen. These systems are deployed across water treatment plants, irrigation networks, and industrial facilities, enabling early detection of hazardous substances and protecting both drinking water supplies and residential environments. Municipalities also utilize such systems to monitor groundwater, stormwater, and agricultural irrigation, thereby optimizing resource distribution, reducing pollution, and minimizing water waste—essential steps toward sustainable urban and rural water management (Shepelev et al., 2023).

Air quality monitoring is equally vital for protecting environmental and public health. Industrial operations and transportation systems release pollutants such as carbon monoxide, hydrocarbons, and methane, which degrade air quality and threaten human well-being. IoT-enabled air monitoring platforms provide real-time measurements of key pollutants—including CO, particulate matter (PM2.5 and PM10), and other harmful gases—allowing industries to mitigate their environmental footprint, municipalities to inform urban planning decisions, and automotive manufacturers to design vehicles with stricter emission controls. Smart traffic management further leverages IoT to reduce vehicular emissions, contributing to cleaner urban air. In agriculture, methane monitoring exemplifies how IoT supports environmental protection by enabling proactive emission management (Shepelev et al., 2023).

The integration of IoT with AI enhances environmental monitoring by transforming raw data into actionable insights. IoT sensor networks deliver continuous streams of real-time atmospheric data, while AI and machine learning analyze both historical and current datasets to identify long-term climate patterns and detect anomalies. These predictive capabilities allow policymakers and researchers to anticipate environmental risks and implement proactive measures against climate-related threats (Balfaqih & Alharbi, 2023). Advances in sensor accuracy, deployment strategies, and data analytics extend the reliability and lifespan of monitoring networks.

Together, IoT and AI provide powerful tools for climate resilience, enabling early detection of crises, supporting sustainable policy interventions, and strengthening society's ability to respond effectively to environmental challenges. As connected devices proliferate and technologies mature, their combined application will play an increasingly decisive role in shaping a sustainable and resilient future.

IOT AND MACHINE LEARNING IN SMART GOVERNANCE

As governments adopt IoT, AI, and other advanced technologies to improve efficiency and service delivery, critical challenges emerge around data privacy and security, data management and integration, and algorithmic bias and fairness. Protecting sensitive citizen information, ensuring data accuracy and interoperability, and addressing biases in automated decision-making are essential to maintaining public trust and achieving equitable governance outcomes. The following subsections examine these challenges in detail, outlining risks, regulatory responses, and the safeguards needed to create transparent, secure, and accountable smart governance systems.

Data Privacy and Security

Smart governance enhances efficiency and public service delivery by leveraging digital technologies such as IoT, AI, blockchain, and big data analytics. However, these advancements also introduce significant risks to personal data protection and system security. Safeguarding identity and maintaining robust cybersecurity measures are central to preserving public trust, particularly as governments handle vast volumes of sensitive information (Balfaqih & Alharbi, 2023).

Data privacy lies at the core of smart governance, ensuring the ethical and secure management of citizens' personal information across collection, storage, and processing. Digital platforms—including e-governance portals, surveillance systems, digital identity solutions, and AI-driven decision-making tools—require access to extensive datasets containing personally identifiable information (PII), biometric data, and behavioral records (Balfaqih & Alharbi, 2023). Without proper safeguards, mishandling this information can lead to identity theft, financial crimes, unauthorized surveillance, and violations of civil rights.

To address these challenges, regulatory frameworks such as the EU's General Data Protection Regulation (GDPR) and California's Consumer Privacy Act (CCPA) establish strict standards for responsible data handling. Governments must implement transparent protocols for data usage, secure user consent, and provide mechanisms for individual control (Balfaqih & Alharbi, 2023). Yet, integrating these requirements into legacy digital infrastructure remains complex and resource-intensive.

Smart governance systems are also exposed to escalating cyber threats, including hacking, ransomware, and disinformation campaigns. Sensitive data—from healthcare and financial records to law enforcement and national security information—makes government systems attractive targets for both state-sponsored and independent cybercriminals. Breaches not only compromise security but also erode public trust and result in significant financial losses.

IoT devices and cloud-based platforms pose particular vulnerabilities due to weak security configurations. Surveillance systems, traffic management platforms, and connected devices can be exploited if not adequately protected. Similarly, the integration of AI-driven governance tools introduces risks linked to algorithmic bias, opaque decision-making, and unethical data use, particularly in areas such as predictive policing and social welfare allocation (Balfaqih & Alharbi, 2023).

To mitigate these risks, governments must enforce comprehensive data protection policies aligned with GDPR, CCPA, and related standards. Regular audits, independent assessments, and strict guidelines for data collection, storage, and sharing are essential. Encryption of all communication channels, databases, and identity systems provides a first line of defense, while anonymization and pseudonymization offer additional safeguards by masking sensitive details during data analysis (Lartey & Law, 2025).

Investments in robust cybersecurity infrastructure are equally critical. Governments should deploy firewalls, intrusion detection systems, and threat intelligence solutions, while adopting zero-trust security models that require continuous identity verification. Routine updates, vulnerability assessments, and penetration testing further strengthen system resilience.

Citizen engagement is another key element of effective data governance. Public awareness campaigns, digital literacy initiatives, and transparent communication about data practices help individuals understand risks and adopt responsible sharing behaviors. Including citizens in policymaking fosters trust and ensures democratic oversight of digital governance systems.

Finally, AI applications in governance must embed transparency, accountability, and fairness. Explainable AI (XAI) approaches are necessary to make automated decision-making processes understandable to the public. Ethical frameworks should guide the design and deployment of AI to minimize bias, ensure accuracy, and protect citizens' rights, and reinforcing the legitimacy of smart governance systems (Lartey & Law, 2025).

Data Management and Integration

The rapid expansion of IoT devices generates vast volumes of digital assets, including highly sensitive information such as medical records, financial data, and real-time location details. Inadequate data management and weak storage protocols pose significant risks, as breaches can result in identity theft, financial fraud, and unauthorized surveillance. As personal data becomes increasingly accessible, the potential for misuse grows, underscoring the urgent need for robust security and privacy standards.

Cybersecurity risks are amplified as IoT devices extend their network connections. Many of these devices are deployed with insufficient security safeguards, leaving critical infrastructure, essential services, and vendor data vulnerable to cyberattacks. Exploiting such weaknesses not only disrupts operations but also erodes public trust, thereby discouraging the broader adoption of smart governance technologies (Balfaqih & Alharbi, 2023).

To address these risks, organizations and governments must comply with data protection regulations such as the EU's General Data Protection Regulation (GDPR), the California Consumer Privacy Act (CCPA), and similar frameworks worldwide. Compliance requires clear policies on data collection, storage, and sharing, but interpreting and implementing these frameworks remains complex and resource-intensive. Failure to comply can result in severe financial penalties, reputational damage, and a loss of public confidence. Strict adherence to these standards is essential to foster trust and establish an interoperable framework that allows disparate IoT devices and platforms to communicate effectively, improving the overall efficiency of smart governance initiatives (Lartey & Law, 2025).

Another key challenge lies in ensuring the accuracy and reliability of IoT-generated data. The performance of ML algorithms is highly dependent on the quality of training datasets. However, IoT devices—particularly sensors—are prone to malfunctions, calibration errors, and faulty readings. Poor-quality data can distort analyses, leading to flawed insights, inaccurate forecasts, and misguided policy decisions. As highlighted by Lartey & Law (2025), establishing common data frameworks and validation mechanisms is critical for producing effective and efficient governance outcomes.

Maintaining data quality requires continuous device maintenance, calibration, and the implementation of rigorous validation procedures. Data cleaning methods must be employed to eliminate outliers, noise, and inconsistencies that could compromise analytical accuracy. By ensuring high-quality inputs, governments and organizations can develop reliable ML models that support sound governance, equitable resource allocation, and evidence-based decision-making.

Algorithm Bias and Fairness

A major challenge of applying ML in public governance lies in addressing algorithmic bias. The accuracy of ML models depends entirely on the quality of their training data; if biases exist within these datasets, they can become amplified in the system's predictions. This risk is particularly concerning when algorithms influence critical governance functions such as law enforcement, social welfare, or resource distribution. For instance, biased algorithms in policing may result in disproportionate surveillance of specific communities, reinforcing racial, socio-economic, and gender inequalities. Similarly, in welfare programs, bias can cause discriminatory resource allocation, reducing accessibility for vulnerable groups. Mitigating these risks requires training datasets that are representative across all demographics, along with independent audits to evaluate and ensure algorithmic fairness. Transparency in algorithm design and deployment is essential, as it enables the early detection and correction of unintended biases before they cause harm (Zheng et al., 2024).

Another critical concern is the opacity of many ML models. Deep learning and other advanced algorithms often function as "black boxes," where even developers lack a full understanding of how decisions are made. This lack of interpretability poses serious challenges in governance, where decisions can directly affect citizens' rights, access to services, and personal security. Without clear explanations, accountability and public trust are undermined. Citizens may doubt the fairness and legitimacy of automated decisions that influence their lives. To address this, governments must prioritize the development and adoption of explainable AI (XAI), ensuring that decision-making processes are transparent, interpretable, and aligned with democratic values (Balfaqih & Alharbi, 2023).

Accountability within algorithmic decision-making further complicates governance. In traditional systems, responsibility can be clearly attributed to government officials or policymakers. However, in ML-powered systems, accountability becomes diffuse, raising the question of whether developers, data scientists, policymakers, or system operators should bear responsibility for adverse outcomes such as wrongful detentions, biased service delivery, or flawed policy recommendations. Without well-defined accountability frameworks, resolving issues becomes difficult, leading to confusion among stakeholders and erosion of public trust. To ensure fairness and ethical operation, governments must establish clear accountability mechanisms that assign responsibility, enforce oversight, and maintain transparency across the lifecycle of algorithmic governance systems (Zheng et al., 2024).

FUTURE TRENDS IN SMART URBAN PLANNING

Smart urban planning faces a range of interconnected challenges that extend beyond technology deployment to encompass social equity, financial feasibility, cybersecurity, workforce preparedness, and ethical responsibility. Addressing the digital divide is essential to ensure inclusivity and equitable access to smart city benefits, while emerging IoT and ML-driven trends reshape planning processes and urban services. However, the reliance on digital technologies amplifies cybersecurity and data privacy risks and raises concerns about the financial viability and sustainability of large-scale projects. At the same time, future technologies such as 5G, edge computing, and blockchain are redefining urban infrastructure, yet their effective integration depends on overcoming skill gaps and workforce development barriers. Finally, strong ethical frameworks are required to guide the use of data, AI, and machine learning in governance. Together, these dimensions highlight the complex landscape of smart city development, underscoring the need for holistic strategies that balance innovation with inclusivity, resilience, and trust.

Digital Divide and Inclusivity

One of the primary barriers to the successful adoption of smart city initiatives is the digital divide, which highlights the gap between technology users and non-users. As smart city planning becomes increasingly reliant on digital infrastructure, there is a risk that marginalized communities may be excluded from its benefits (Hidayat et al., 2024). Ensuring equitable digital access must therefore be a central priority for governments and urban planners. Policies should guarantee affordable internet connectivity for all citizens while offering specialized support for vulnerable groups to enable their effective participation in governance and community life (Lartey & Law, 2025). Complementary digital literacy programs are equally important to empower residents to fully engage with and benefit from smart city systems.

Community participation plays a critical role in addressing these challenges. When local residents are actively involved in smart city planning, they help identify barriers to technology use and ensure that solutions are aligned with their real needs. This inclusive, participatory approach fosters public trust, which is essential for the long-term success of smart city initiatives.

At the same time, challenges such as engagement fatigue must be carefully managed. Citizens can become overwhelmed if repeatedly asked to provide feedback across multiple digital platforms, which can reduce their willingness to engage with governance processes. To overcome this, urban planners must adopt communication strategies that are efficient, targeted, and respectful of citizens' time, ensuring opportunities for meaningful participation without overburdening communities. In this way, smart city projects can secure genuine public involvement, safeguard investments, and create fairer outcomes for all stakeholders (Hidayat et al., 2024).

Future Trends in Smart Urban Planning

The integration of IoT and ML is reshaping urban planning by enabling greater citizen participation and more responsive governance. Smart platforms that collect real-time feedback from residents, combined with advanced data analysis, allow cities to design policies and services that are more inclusive and aligned with community needs. These digital channels bring citizens closer to the decision-making process, fostering open governance and stronger community engagement (Kalfas et al., 2025).

Crowdsourcing platforms represent a key development in this context, empowering residents to propose solutions for local challenges and contribute actively to urban governance. By harnessing collective intelligence, such systems generate innovative ideas that directly address population needs. Alongside this, predictive analytics powered by ML plays a pivotal role in enhancing urban service delivery. Real-time data analysis supports traffic optimization, reducing congestion, improving public transport efficiency, and enabling proactive route and signal management.

In the energy sector, predictive modeling enhances smart building performance by analyzing usage patterns to optimize consumption, delivering both economic savings and environmental benefits. The integration of IoT with ML is particularly impactful in smart grid development, where embedded sensors provide accurate demand forecasts that improve distribution efficiency and minimize energy waste (Balfaqih & Alharbi, 2023).

Looking ahead, the convergence of IoT and ML will also strengthen urban resilience against climate change-related challenges. By leveraging data-driven insights, city planners can make more informed decisions on infrastructure design, disaster preparedness, and resource allocation. This ensures the development of urban systems that are not only more adaptive and sustainable but also capable of withstanding future environmental uncertainties.

Cybersecurity Risks and Data Privacy

Smart urban planning relies extensively on digital technologies, data analytics, and connected networks to enhance efficiency, sustainability, and quality of life. However, this reliance also introduces significant cybersecurity and privacy risks. Modern smart cities collect vast amounts of personal information through surveillance cameras, IoT sensors, traffic management systems, and mobile applications. While these technologies support safety, mobility, and energy efficiency, they also expose citizens to potential privacy breaches, unauthorized surveillance, and cyberattacks. To address these risks, regulatory frameworks such as the General Data Protection Regulation (GDPR) in the European Union and the California Consumer Privacy Act (CCPA) in the United States establish essential safeguards to balance innovation with citizens' privacy rights (Sindhuja et al., 2025).

Unauthorized data collection and constant surveillance remain among the most pressing concerns. Permanent monitoring through IoT sensors, biometric systems, and AI-driven analytics often provides tangible benefits for governance but risks infringing upon civil liberties. Weak or inconsistent data governance may lead to misuse of personal data by governments or private entities, enabling mass surveillance, discriminatory profiling, or commercial exploitation. GDPR requires explicit user consent and grants individuals the right to control their personal data, while the CCPA provides mechanisms for residents to opt out of data sharing and request data deletion—thereby enhancing transparency and trust (Sindhuja et al., 2025).

At the same time, cybersecurity vulnerabilities present serious challenges for smart cities. Critical infrastructure—such as smart grids, public transport systems, and emergency response networks—becomes an attractive target for malicious actors. Cyberattacks, including ransomware and data breaches, can disrupt essential services and expose sensitive information such as addresses, biometric records, healthcare files, and financial data (Sindhuja et al., 2025). To counter these threats, GDPR obliges organizations to adopt robust security practices, conduct periodic risk assessments, and provide prompt breach notifications. Similarly, the CCPA requires businesses to safeguard consumer data and grants individuals the right to pursue legal action in cases of negligent data handling.

Data interoperability and sharing among governments, private companies, and third-party providers further amplifies risks. Without strict protocols, personal data may be leaked, sold to advertisers, or exploited for political profiling. GDPR mandates secure data exchange practices and informed user consent, while CCPA emphasizes transparency in corporate data usage and gives individuals the right to restrict or prevent data sharing.

Another challenge arises from the use of AI in governance, particularly in predictive policing, traffic control, and resource allocation. If algorithms are trained on biased datasets, they can perpetuate racial, gender, or socio-economic discrimination. Facial recognition technologies, for example, have been criticized for inaccurate and biased results, sometimes leading to wrongful profiling or detentions. To mitigate these risks, organizations must ensure transparency, fairness, and accountability in AI systems. Under GDPR, individuals are entitled to contest automated decisions that significantly impact them, while CCPA provides mechanisms to challenge profiling and targeted advertising practices (Balfaqih & Alharbi, 2023).

Ensuring cybersecurity and privacy by design is critical for smart urban development. Security frameworks must be embedded from the earliest stages of system design, supported by encryption, anonymization, and continuous identity verification. Regular audits, staff training, vulnerability assessments, and collaborative partnerships with cybersecurity experts are essential to strengthening resilience against evolving threats. Ultimately, the success of smart cities depends on prioritizing the protection of citizen data. By aligning with GDPR, CCPA, and similar international standards, urban planners and governments can foster innovation while safeguarding privacy, thereby creating intelligent and secure cities that maintain public trust.

Financial Feasibility, Funding Models and Cost-Benefit Analyses

For moderately developed cities, integrating digital technologies, data analytics, and eco-friendly infrastructure is central to improving efficiency, sustainability, and quality of life. However, deploying smart city programs requires substantial investment, making financial feasibility a critical factor in urban development. Local governments, private financiers, and stakeholders must carefully evaluate funding strategies and conduct thorough cost-benefit analyses (CBA) to ensure the sustainability of such initiatives (Turečková & Nevima, 2020).

A project's financial viability depends on balancing costs, projected revenues, and long-term operational savings. This requires a comprehensive review of capital expenditures—including infrastructure, platforms, and technological systems—alongside recurring operational and maintenance expenses. Municipal authorities must also account for long-term budgetary sustainability to avoid financial strain. Effective feasibility studies help governments identify risks, predict demand, and determine the most suitable financing models (Turečková & Nevima, 2020).

Funding for smart urban projects comes from diverse sources. Public financing remains foundational, derived from municipal budgets, national appropriations, tax revenues, and government grants. Tools such as smart city bonds are increasingly used to raise capital for large-scale infrastructure. Yet public funding alone is often insufficient due to fiscal constraints and competing priorities (Balfaqih & Alharbi, 2023).

As a result, public-private partnerships (PPPs) have become a common model, enabling governments to leverage private sector expertise and financing while reducing public financial burdens. PPPs have successfully supported smart transportation systems, smart grids, and digital infrastructure projects. However, designing agreements that distribute risks and benefits fairly between stakeholders remains a key challenge. In parallel, private investments—particularly from technology firms and venture capital—play an essential role in piloting IoT-based city management, AI-driven traffic control, and other smart infrastructure systems. Regulatory frameworks are needed to ensure these contributions serve public interests while fostering innovation and efficiency (Turečková & Nevima, 2020).

International institutions such as the World Bank, Asian Development Bank (ADB), and European Investment Bank (EIB) also provide vital financial support for smart urban development in emerging economies. Their grants and loans prioritize sustainability-focused projects, including renewable energy networks, transportation infrastructure, and water resource management. To secure such funding, cities must present robust development plans, financial sustainability assessments, and measurable socio-environmental outcomes (Lark, 2020).

A cost-benefit analysis (CBA) is indispensable in evaluating the financial and economic viability of smart urban initiatives. CBAs weigh capital, operational, and maintenance costs against anticipated benefits such as improved efficiency, environmental sustainability, and social well-being. Costs typically include infrastructure deployment, IoT integration, smart grids, software platforms, and security measures, while benefits may extend to reduced emissions, energy savings, increased property values, and job creation in technology sectors. Projects with clear returns on investment (ROI), shorter payback periods, and strong long-term benefits are more attractive to investors and policymakers (Turečková & Nevima, 2020).

Smart city development also entails broader social and environmental considerations. Poorly planned projects may cause displacement, rising living costs, or privacy risks linked to data collection. Conversely, well-designed initiatives foster inclusivity, environmental sustainability, and economic growth. For instance, smart traffic management systems can reduce vehicle emissions, fuel consumption, and congestion, producing both financial savings and environmental gains. Similarly, integrating renewable energy with efficient building practices and waste management automation helps cities align with global climate goals while enhancing community health. Despite its promise, financial planning for smart cities must account for uncertainties such as rapid technological obsolescence, cybersecurity risks, and regulatory shifts. Proactive risk assessments and contingency planning are therefore essential to minimize potential losses. Sustainable financing requires balancing economic returns with principles of equity, resilience, and environmental responsibility.

Future Technologies

The future of smart urban planning is increasingly shaped by emerging technologies such as 5G networks, edge computing, and blockchain, which together provide the foundation for faster connectivity, enhanced efficiency, and secure urban ecosystems.

5G networks play a central role in enabling ultra-fast, low-latency communication, surpassing previous generations by supporting real-time interaction between billions of IoT devices, autonomous systems, and urban infrastructure. These capabilities are critical for applications such as intelligent transportation, public safety, healthcare, and energy-efficient buildings. In transportation, 5G enables autonomous vehicles to interact seamlessly with traffic signals, pedestrian zones, and other vehicles, while traffic management centers leverage predictive analytics for congestion reduction. Similarly, smart grids benefit from 5G-enabled real-time electricity monitoring, which enhances demand-response coordination and overall energy efficiency. Public safety and healthcare also advance through rapid data transmission: emergency services receive instant video streams from drones and surveillance systems, while telemedicine platforms deliver remote diagnostics, patient monitoring, and even robotic-assisted surgeries at minimal latency (Balfaqih & Alharbi, 2023).

Edge computing complements 5G by processing data closer to its source, minimizing delays associated with cloud-based systems. Localized processing at IoT nodes, gateways, and micro-data centers enhances responsiveness and reduces bandwidth consumption. This capability is particularly effective in real-time traffic management, where connected vehicles and smart traffic lights require immediate decision-making to optimize flows and support emergency vehicle routing. In environmental monitoring, edge computing accelerates analysis of air quality, noise levels, and water conditions, enabling faster responses to toxic leaks or flooding events. It also improves urban services such as smart waste management, where sensor-enabled bins optimize collection routes and reduce operational costs, and smart buildings, where on-site analysis of HVAC data enhances energy efficiency (Puliafito et al., 2021).

Blockchain technology introduces security, transparency, and trust into smart urban ecosystems through its decentralized, tamper-resistant ledger. It protects sensitive urban data—including personal identity records, financial transactions, and surveillance data—by ensuring secure storage and preventing unauthorized alterations. In governance, blockchain supports tamper-proof voting systems, transparent land registries, secure property transactions, and identity verification for safe access to public services. In the energy sector, blockchain facilitates decentralized power markets where prosumers trade surplus renewable energy directly through automated smart contracts, promoting sustainability and reducing reliance on centralized utilities (Balfaqih & Alharbi, 2023).

The integration of 5G, edge computing, and blockchain creates a robust technological foundation for smart cities. Together, they enable applications such as autonomous transportation, AI-driven urban planning, decentralized governance, and resilient infrastructure. For instance, in mobility, 5G provides real-time connectivity, edge computing ensures immediate processing of traffic data, and blockchain secures transactions and data exchange between authorities and mobility providers. In healthcare, the same triad supports remote patient monitoring, real-time medical data processing, and secure electronic health records.

Skill Gaps and Workforce Development

The deployment and long-term sustainability of IoT and ML systems in government face significant challenges due to a shortage of specialized technical expertise. Public institutions often struggle to recruit personnel with the necessary skills to manage, integrate, and maintain these advanced technologies. This skills deficit threatens not only the implementation of smart governance initiatives but also their continuous evolution and adaptation. To address this issue, public agencies must prioritize investments in workforce development, ensuring employees acquire the technical competencies required to support IoT and ML-based projects effectively.

Bridging this skills gap requires comprehensive training and capacity-building programs tailored for government employees. These programs should combine technical proficiency in IoT devices and ML algorithms with broader digital literacy and data management skills. As Balfaqih and Alharbi (2023) emphasize, strengthening digital competencies within the public sector is fundamental for ensuring the successful adoption and long-term sustainability of smart governance systems.

The integration of IoT and ML also demands organizational change, requiring adjustments in business processes, workflows, and institutional mindsets. Resistance from public-sector employees—often stemming from fears of job displacement, difficulty in adapting to new tools, and uncertainty about evolving responsibilities—can hinder adoption. Overcoming these challenges calls for structured change management strategies that combine clear communication, participatory decision-making, and continuous training. By addressing employee concerns and equipping staff with the right tools and knowledge, governments can facilitate smoother transitions and promote more effective integration of smart governance technologies (Balfaqih & Alharbi, 2023).

Ethical Considerations

The integration of IoT and ML into governance introduces critical ethical challenges, particularly regarding the extensive collection and use of citizen data. Protecting privacy must remain a core priority, requiring robust frameworks to ensure sensitive information is handled responsibly and not exploited in ways that undermine individual rights. Embedding ethical principles into the design and deployment of smart governance systems ensures that technology serves the public good while safeguarding fundamental liberties.

Public trust depends on transparency and accountability. Governments must adopt clear policies that disclose how data is collected, processed, and used, as well as the logic and potential consequences of algorithmic decisions. Mandatory transparency in algorithmic operations and data governance enhances citizen confidence and reinforces the legitimacy of smart governance initiatives (Balfaqih & Alharbi, 2023).

Developing ethical AI requires structured guidelines that promote fairness, accountability, and inclusivity. Governments should collaborate with developers, ethicists, and civil society to design frameworks that minimize algorithmic bias and prevent discriminatory outcomes. As Rawindaran (2023) highlight, regular performance audits and impact assessments are essential to detect and correct issues before they cause harm. Ensuring voluntary and informed consent for data collection is equally critical. Citizens must be fully aware of what data is gathered, how it will be used, and the associated risks, empowering them to make informed decisions about participation in smart city initiatives.

Data security further reinforces ethical governance. Many IoT devices remain vulnerable to cyberattacks, exposing personal data to unauthorized access. To mitigate this risk, governments must prioritize investments in advanced cybersecurity infrastructure, including strong encryption, frequent audits, and continuous system updates. Such measures not only protect citizen data but also ensure the integrity and resilience of IoT-enabled governance systems (Rawindaran, 2023).

CONCLUSION

The chapter analyzes IoT and Machine Learning as agents of change for smart urban system development through six key areas such as smart transportation and energy management and environmental monitoring alongside public safety and infrastructure management and citizen engagement. Through combined implementation of these technologies residents can expect better quality of life and enhanced efficiency as well as improved sustainability. Besides the numerous opportunities that exist there are major challenges together with ethical issues which demand proper attention. Extensive obstacles emerge from privacy and security concerns about data alongside biases in algorithms together with public trust issues and the digital divide which require proper management. The implementation of smart city initiatives depends on transparent government actions and sincere community engagement to achieve widespread trust in addition to equal access. The integration of IoT with Machine Learning into smart urban systems will develop cities that powerfully respond to community needs and citizen dreams while making cities smarter thus facilitating urban sustainability and equity.

REFERENCES

Al-khatib, A. A., Mohammed, B., & Abdelmajid, K. (2020). *A survey on outlier detection in Internet of Things big data. Big Data-Enabled Internet of Things*. IET.

Almohammedi, A. A., Balfaqih, M., Nahas, S., Bokhari, A., & Alqudsi, A. (2023). Design and Implementation of IoT-Enabled Intelligent Fire Detection System Using Neural Networks. In *International Conference on AI and Mobile Services* (pp. 63-70). Cham: Springer Nature Switzerland.

Balfaqih, M. (2023). Design and Implementation of IoT-enabled Intelligent Irrigation System Using Machine Learning. In *2023 28th Asia Pacific Conference on Communications (APCC)* (pp. 436-441). IEEE.

Balfaqih, M., & Alharbi, S. A. (2023). Associated information and communication technologies challenges of smart city development. *Sustainability*, *14*(23), 16240.

Balfaqih, M., Alharbi, S. A., Alzain, M., Alqurashi, F., & Almilad, S. (2021). An accident detection and classification system using internet of things and machine learning towards smart city. *Sustainability*, *14*(1), 210.

Balfaqih, M., Balfagih, Z., Almohammedi, A. A., & Alfawaz, K. M. (2023). A Smart and Privacy-Preserving Logistics System Based on IoT and Blockchain Technologies. In *2023 1st International Conference on Advanced Innovations in Smart Cities (ICAISC)* (pp. 1-5). IEEE.

Balfaqih, M., Ismail, M., Nordin, R., & Balfaqih, Z. A. (2017). 802.21-assisted distributed mobility management solution in vehicular networks. *IEEE Access : Practical Innovations, Open Solutions*, *5*, 9518–9532.

Calder, K. E. (2016). *Singapore: Smart city, smart state*. Brookings Institution Press.

Hartzog, W. (2018). *PrivacyOs Blueprint: The Battle to Control the Design of New Technologies*. Harvard University Press.

Hasan, M. K., Habib, A. A., Islam, S., Balfaqih, M., Alfawaz, K. M., & Singh, D. (2023). Smart grid communication networks for electric vehicles empowering distributed energy generation: Constraints, challenges, and recommendations. *Energies*, *16*(3), 1140.

Hidayat, M. N. F., Baharun, H., Aisyah, E. N., Zaini, A. W., Sanjani, M. A. F., & Hasanah, R. (2024, October). Bridging the Digital Divide: The Role of Public Relations in Enhancing Digital Inclusivity. In *2024 10th International Conference on Education and Technology (ICET)* (pp. 59-66). IEEE.

ISO/IEC 17067:2013, Conformity assessment — Fundamentals of product certification and guidelines for product certification schemes

ISO/SAE 21434:2021 for Road Vehicles—Cybersecurity Engineering

Kalfas, D., Kalogiannidis, S., Spinthiropoulos, K., Chatzitheodoridis, F., & Ziouziou, E. (2025). Enhancing Predictive Urban Planning in European Smart Cities Through AI-Driven Digital Twin Technology: A Case Study of Greece. *Urban Science (Basel, Switzerland)*, *9*(7), 267.

Khayyat, M. M., Jambi, J., Kutbi, N., Garout, R., Al-Noumani, L., Asiri, R., & Balfaqih, M. (2025). Logistic caravan: A smart transportation system for green logistics and sustainable practices.

Kolajo, T., Daramola, O., & Adebiyi, A. (2019). Big data stream analysis: A systematic literature review. *Journal of Big Data*, *6*(1), 1–30.

Lartey, D., & Law, K. M. (2025). Artificial intelligence adoption in urban planning governance: A systematic review of advancements in decision-making, and policy making. Landscape and Urban Planning, 258, 105337.

Lark, J. (2020). Uneven innovation: The work of smart cities. Columbia University Press.

Liu, C.. (2020). Smart Energy Systems: A Comprehensive Review of IoT-Enabled Solutions for Sustainable and Optimized Energy Usage. *Energy Reports*, *6*, 1342–1355.

Morozov, V., Petrov, A. I., Shepelev, V., & Balfaqih, M. (2024). Ideology of urban road transport chaos and accident risk management for sustainable transport systems. *Sustainability*, *16*(6), 2596.

Nasir, T., Bukhari, S. S. H., Raza, S., Munir, H. M., Abrar, M., Muqeet, H. A. U., & Masroor, R. (2021). Recent Challenges and Methodologies in Smart Grid Demand Side Management: State-of-the-Art Literature Review. *Mathematical Problems in Engineering*, *2021*(1), 5821301.

Puliafito, A., Tricomi, G., Zafeiropoulos, A., & Papavassiliou, S. (2021). Smart cities of the future as cyber physical systems: Challenges and enabling technologies. *Sensors (Basel)*, *21*(10), 3349. PMID: 34066019

Qonita, M., Rachmawati, R., & Rijanta, R. (2025). Assessing Amsterdam's urban mobility through the lens of smart mobility goals. *Case Studies on Transport Policy*, *20*, 101443.

Rawindaran, N. (2023). Legal Considerations and Ethical Challenges of Artificial Intelligence on Internet of Things and Smart Cities. In *Data Protection in a Post-Pandemic Society: Laws, Regulations, Best Practices and Recent Solutions* (pp. 217–239). Springer International Publishing.

Mazhar, T., Irfan, H. M., Haq, I., Ullah, I., Ashraf, M., Shloul, T. A., Ghadi, Y. Y., Imran, & Elkamchouchi, D. H. 2023. Analysis of Challenges and Solutions of IoT in Smart Grids Using AI and Machine Learning Techniques:Review, A. . *Electronics (Basel)*, *12*(1), 242.

Shepelev, V., Glushkov, A., Slobodin, I., & Balfaqih, M. (2023). Studying the relationship between the traffic flow structure, the traffic capacity of intersections, and vehicle-related emissions. *Mathematics*, *11*(16), 3591.

Sindhuja, M., Alice, K., Nancy, A. M., & Vinoth, N. A. S. (2025). Cybersecurity Strategies for Smart Cities. In *Harnessing AI in Geospatial Technology for Environmental Monitoring and Management* (pp. 77–98). IGI Global Scientific Publishing.

Turečková, K., & Nevima, J. (2020). The cost benefit analysis for the concept of a smart city: How to measure the efficiency of smart solutions? *Sustainability*, *12*(7), 2663.

Ullo, S. L., & Sinha, G. R. (2020). Advances in smart environment monitoring systems using IoT and sensors. *Sensors (Basel)*, *20*(11), 3113. PMID: 32486411

Zheng, Y., Hao, Q., Wang, J., Gao, C., Chen, J., Jin, D., & Li, Y. (2024). A survey of machine learning for urban decision making: Applications in planning, transportation, and healthcare. *ACM Computing Surveys*, *57*(4), 1–41.

Chapter 8
Centering Ground Water Scarcity:
Ensuring Futuristic Sustainable Farming With Blockchain and IoT

Bhupinder Singh
https://orcid.org/0009-0006-4779-2553
Sharda University, India

Christian Kaunert
https://orcid.org/0000-0002-4493-2235
Dublin City University, Ireland & University of South Wales, UK

Arunima Shastri
https://orcid.org/0000-0002-6260-1253
Christ University, Bangalore, India

Saurabh Chandra
https://orcid.org/0000-0003-4172-9968
Bennett University, Greater Noida, India

ABSTRACT

Depletion of groundwater is a major threat to agricultural sustainability, especially in and around irrigated areas. This chapter discusses monitoring water usage, soil moisture and environmental using sensors powered by IoT and how such technologies offers real-time insights into water usage for efficient management of water. It also examines decentralized smart contracts that can promote fair water use rights and incentive farmers' conservation behaviours. Combining the adaptation of technological innovation and sustainable water management, the paper offers policy prescriptions to climate proof agriculture. With means of interdisciplinary

DOI: 10.4018/979-8-3373-0154-9.ch008

views, practical applications and policy suggestions, this book snapshots the agro-environmental management policy and the smart irrigation systems towards the protection of water resources, and future generations of smart farming applications worldwide.

INTRODUCTION

Escalating groundwater depletion has become a binding constraint on food security and ecosystem stability, particularly in regions where extraction outpaces recharge (Flores et al., 2021). Addressing this deficit requires a shift from schedule-based watering to demand-driven irrigation guided by data.

The Internet of Agricultural Things (IoAT) provides that data layer. Networks of in-field sensors measure soil moisture, temperature, and nutrient levels; weather stations supply local forecasts; and aerial drones equipped with infrared payloads detect crop stress before yield losses occur (Voulgaris et al., 2019). Fused in real time, these streams enable precision irrigation—delivering water to the root zone only when and where plants need it—reducing withdrawals, conserving aquifers, and raising productivity (Kaur et al., 2023). Decision engines can automatically adjust valve timings and flow rates, while farmer dashboards surface anomalies and recommendations.

Blockchain augments this stack with integrity, traceability, and automated control. Because device readings are immutably recorded on a decentralized ledger, stakeholders can verify water use histories, audit compliance, and curb data tampering across the supply chain (Gupta et al., 2020). Smart contracts operationalize policy: for example, an agreement can trigger or halt irrigation when on-chain soil-moisture thresholds are reached, translating agronomic rules into verifiable, low-friction actions that save additional water (Siwach et al., 2022). The same mechanisms support transparent allocation of water rights and targeted incentives for conservation.

Real-world deployment is advancing but not trivial. Up-front costs for sensors, connectivity, edge devices, and ledger integration remain high for smallholders; rural areas may lack reliable power and broadband; and governance frameworks for data ownership, privacy, and blockchain use are still maturing (Rudrakar & Rughani, 2022). Here, smart-city and regional infrastructure can play a catalytic role: smart grids and distributed renewables power field devices; reclaimed wastewater expands non-potable irrigation supplies; and intelligent logistics move perishables faster, cutting losses and emissions (Rakhee & Elayidom, 2023).

Despite these hurdles, the convergence of IoAT, drones, weather analytics, and blockchain is well placed to deliver water-smart, climate-resilient agriculture. With supportive policy and fit-for-purpose financing, these technologies can underpin equitable allocation, verifiable conservation, and sustained yields under increasing climatic stress (Kumari & Sharma, 2023; Kaur et al., 2023; Gupta et al., 2020).

SMART AGRICULTURE AND WATER MANAGEMENT FOR SUSTAINABLE FUTURES

Over the last couple of years, it has been seen that Blockchain Technology IoAT are coming together to bring a revolution in agricultural practices mainly focusing on resource optimization, transparency and sustainability. With the unprecedented effect of simple data storage, blockchain hardware technology provides an immutable decentralized ledger that records what data is held on it in a traceable and verifiable way to enable supply chain level transactions from grower to consumer. Conversely, in the field of changing how we do agriculture, IoAT can allow us to monitor our data out in the field with sensors and devices connected over a network covering everything from soil health to crop conditions as well local weather patterns and water needs (Singh, 2023). The complementary strengths of Blockchain and IoAT therefore make for a powerful combination, improving precision farming, minimizing wastage of resources and allowing data-driven decision-making that improves efficiency and sustainability. These two together provide great benefits, especially when it comes to the challenges of groundwater depletion and sustainable food security.

Global Groundwater Scarcity: Current and Future Statistics

The scarcity of ground water emerged as a major global problem affecting agriculture, ecosystem and human population. Groundwater is critical for human communities, serving as the daily water source for about 2.5 billion people and contributing to around 40% of global irrigation needs according to United Nations data. Yet, many areas have imported water bottles filled with groundwater by the tens of billions and poorly managed extractions are drawing down aquifers from regions hungry for liquidity (both agricultural and orgasmic), e.g. India, China and the United States. Studies on the other hand show groundwater tables in Indo-Gangetic plains and North China Plains are plummeting at rates as high as one meter per annum. According to the World Resources Institute, more than 33 countries including heavyweights in agriculture such as India and the U.S. will face extremely high levels of water stress by 2040, threatening food production and heightening the risk of droughts (Ahamed et al., 2023).

The outlook for groundwater availability is growing more bleak, especially in negative climate change and human (population) growth/consumption scenarios. Groundwater extraction is already too high, and it might be exhausted within just some decades in many of the arid and semi-arid areas where people have relied on this resource during droughts (Guven & Parlak, 2022). Since global food production relies heavily on access to stable water supplies, up to 50% of the world's food supply may be at risk from extreme water scarcity by 2050. Consequently, the depletion of groundwater needs to be tackled in order to promote sustainable agriculture and maintain water availability for future generations.

Technologies for Sustainable Agriculture

With the challenge of groundwater drought and other resource challenges, sustainable farming technologies have become a major focus area including smart irrigation, precision farming and IOT based solutions. Such technologies facilitate resource-efficient usage, generate low waste, and improve crop production while minimizing the ecological footprint. Smart irrigation involves sensor based systems to monitor soil moisture, local climate, and crop water usage requirements. For example, smart irrigation gives water only when and where it is needed, which can decrease water consumption by up to 50% compared to traditional methods (Food and Agriculture Organization FAO). By connecting these systems to IoAT networks, it is possible to monitor and operate them remotely and change irrigation agendas according to the changing weather. Sophisticated systems such as drip and sprinkler irrigations with live readings save a lot of water and provide optimum moisture in soil for the growth of plants (Ravindran, 2022).

Sustainable farming technologies that are revolutionizing agriculture is precision farming. Using satellites, drones, and IoT sensors, precision farming offers farmers rich field data on soil composition, as well as crop health and nutrient profiles. Precision agriculture allows farmers to apply fertilizers, pesticides, and water only in the necessary amounts in specific places of a certain field while minimizing waste and environmental effects (Strobel, 2020). This is where IoAT comes to the rescue by providing a foundational layer of data collection, storage and analysis which empowers farmers to make evergreen decisions on crop management, resource allocation and pest control.

IoAT-enabled solutions have moved beyond field management to cover wider agricultural sustainability issues. IoAT devices enable farmers to monitor the health of their livestock, along with feed levels and environmental conditions continuously so they can not only manage and maintain higher standards for animal welfare but also identify early signs of a disease. The purpose of cultivation with these types of references is to increase the quality and safety of agricultural products, while minimizing greenhouse gas emissions by improving resource use efficiency and reducing unnecessary input (Vangipuram et al., 2022).

Interface of Smart Cities and Agriculture

Realizing the global trend of growing urbanization, Smart Cities have emerged through mobile and wireless technologies such as IoT (Internet of Things), AI (Artificial Intelligence), Blockchain etc to economize Urban Infrastructure, improve quality of life for urban dwellers and contribute to Environmental Sustainability. Smart cities can serve as a paradigm shift for agriculture by enabling the integration of urban infrastructure and sustainable farming practices. The interplay between smart cities and agriculture occurs along soft lines, with urban farming skills, food distribution optimization, and water sustainability being just a few dimensions of this interaction (Chattopadhyay et al., 2022).

Urban agriculture is a critical component of smart cities, which help cities become food producers and less reliant on agricultural products from rural areas. Innovations such as vertical farming, hydroponics, and aeroponics enable urban centers to produce vegetables and fruits in controlled environments sometimes there are even no additional horticulture efforts from within the city itself. Using Blockchain and IoAT, urban agriculture can be a data driven process to manage resources, where this in turn leads the ability of bringing full transparency on the production process (Lamtzidis & Gialelis, 2018). IoAT sensors such as measuring light, temperature and humidity in a vertical farm to ensure crops grow properly, and blockchain to guarantee that the food they supply to consumers has been grown sustainably or sourced from its origin can all be examples.

The important place in which these two groups interact is the logistics and distribution of food within cities. IoAT systems can keep track of the inventory, how long something is going to take to expire, and what conditions something is being transported in to ensure the food is stored or transported properly. Tracking produce from the farm all the way to your table, blockchain can help reduce food fraud while empowering consumers to make informed decisions about what they eat (Bera et al., 2023). Municipalities and city planners can leverage this system to better plan logistics by making food transportation more efficient indirectly alleviating traffic congestion caused by transporting foods, as well as reduce carbon emissions from the transport.

Water management is yet another important sector where the overlaps between smart cities and agriculture can assist in a sustainable manner. With advanced IoAT, urban areas can utilize it along with the blockchain technology in the water infrastructures to control and lessen the usage of this essential resource as both help in reducing waste of it and encourage urban farming. For example, IoAT sensors could track how much water is used, where, and if there are any leaks as well as manage distribution as per demand. Data regarding water usage can be transparently recorded on a blockchain which is crucial for ensuring equitable distribution and efficient allocation of resources in times of scarcity (Dinc et al., 2019). It can also be used as reclaimed water from urban areas and supply a sustainable resource of irrigation water to urban farms and green zones in the context of smart cities. Smart cities and agriculture can also help promote a circular economy where resources are reused and recycled to reduce waste and minimize environmental impact. Linking urban farms with their organic waste from urban areas, and then composting such waste to be deployed as fertilizer in the farmer communities effectively closes the loop; thereby reducing dependency on chemical fertilizers. This process can be monitored and recorded with the help of IoAT and blockchain technology to ensure that composting and waste recycling is efficient, safe and transparent.

The integration of blockchain technology and IoAT systems with sustainable agricultural practices can thus become a remedy for groundwater scarcity band-aids, Intensive resource use barriers and food security challenges (Bodkhe et al., 2020). Adopting smart irrigation, precision farming as well as sustainable urban agriculture will facilitate in making our food systems more resilient and sustainable. The provided transparency and traceability are invaluable when talking about food security and accuracy-related issues at the consumer end, while IoAT will promote optimization in on-farm resource usage by linking local requirements with supply-side offerings, thus paving the path for smarter cities atop urban farms powered by Blockchain and IoAT. Collectively, they provide a vision for a sustainable tomorrow, where agriculture and urbanity can co-exist and evolve to sustainably meet the needs of present and future generations.

BLOCKCHAIN IN AGRICULTURAL SUPPLY CHAIN

Blockchain technology has emerged as a revolutionary force in agricultural supply chains, fundamentally changing the manner in which food products are tracked, managed and secured from farm to fork. In agriculture, product traceability, water resources management, and data privacy are all of high importance, making the decentralized ledger system that blockchain affords transformative in providing an immutable and transparent record of transactions. Use of the technology in agriculture solves age-old issues including food counterfeiting, wastage of resources and inefficiencies while encouraging transparency and accountability amongst all parties involved (Bayih et al., 2022). Whether it is regarding improving traceability and transparency, handling water resources or maintaining privacy in data, we see blockchain helping towards a healthier and fairer agriculture ecosystem. It has been explained how blockchain can help drastically improve agricultural supply chain with better transparency, manage water resources and facilitate secure data sharing.

Blockchain technology is changing this landscape by providing a highly reliable tracking mechanism from farm to market, which provides transparency and accountability in agriculture, helps rebuild consumer trust and create strong deterrents against food fraud. Which in turn blockchain for water management facilitating more efficient irrigation practices and equitable allocation of a critical resource (Verma, et al., 2022). But widespread adoption of blockchain in agriculture requires the consideration of privacy, security and ethical challenges to enable farmer-centric applications realizing economic benefits for all relevant stakeholders without the trade-off of individual data privacy ownership or environmental sustainability. The embedding blockchain with care and continuing to come up with new solutions, it can foster an agricultural ecosystem that is more resilient, transparent, and sustainable than ever before.

Traceable and Transparent Blockchain

Blockchain technology is heralded as a game-changing technology that forces traceability and transparency where it seems to be most important in the agricultural supply chain, which is essential to building trust. Blockchain creates an immutable digital ledger of every transaction or step along the journey from farm to market, allowing stakeholders to authenticate the origin, quality and movement of products at every stage (Ferrag & Shu, 2021). That is an especially useful capability for agriculture, where consumers, retailers and regulators are demanding more information about how food items originate and the practices behind their production.

For example, instead of relying on thousands of audits to verify the exact situation around each crop and grain shipment a farmer can directly input their data onto the blockchain with all relevant information such as planting dates, pesticides use, and harvesting conditions recorded. Subsequent participants in the supply chain processors, distributors, retailers and so on then update that ledger with information, giving you a digital record from end to end of where the product has gone. Such transparency reduces the risks of food fraud and counterfeit products as consumers can select what they buy based on verified information on product origins and quality. to connect the nodes, for instance with long-distance freight transportation and even if would all be online, this would not only simplify companies but also retailers through the features of traceability provided by blockchain, which allows them increase the assurance in the labour processes of their suppliers and proves compliance to safety standard key for avoiding unnecessary recalls (Phang et al., 2023).

Blockchain traceability along the agricultural supply chain enables adherence to sustainability practices as well. For example, if a farm is making sustainable or organic claims, those can be recorded and verified on the blockchain, providing consumers with proof of responsible environmental care. Companies who are looking to soon meet sustainability goals and align global standards of eco-friendly practices believe that this ability to audit these practices is imperative. Blockchain that encourages transparency and traceability is expected to eliminate malpractices by providing higher confidence in the agricultural value chain (Bala & Kaur, 2022).

Blockchain for Efficient Water Resource Management

Agriculture is a huge consumer of our precious water with water scarcity or droughts being one of the main challenges affecting agriculture. Blockchain technology has potential use cases in irrigation, water use and conservation management by tracking the allocation of resources efficiently. Together with Internet of Things (IoT) sensors and smart irrigation systems, blockchain can enable the creation of a decentralized and transparent ledger of water usage to monitor consumption, identify overconsumption, and even promote sustainable resource management (Dwivedi et al., 2021; Podder et al., 2021).

Sensors placed on farmlands for monitoring soil moisture levels and weather and crop water requirements can obtain such data in real time. That data can then be connected to a blockchain solution, which logs the amount of water used per farm. Farmers can utilize this data to irrigate according to the data– water only when and where it is needed. In addition, blockchain fosters traceability and validation of water used for agricultural production, showing responsible irrigation practices. This involves keeping a record, which is critical to comply with regulatory reporting requirements, obtaining conservation subsidy payments or certifying crops produced under sustainable practices.

Blockchain can also be used to enable water-sharing contracts between farmers or communities. Fair and efficient allocation of water resources in regions experiencing water scarcity is critical. Blockchain captures, shares and monitors distributions of water transparently so that under any circumstance key stakeholders will have access to a clear overview in real-time (Balfaqih et al., 2023a). This transparency decreases the chances of disputes regarding water rights, promotes trust between stakeholders and facilitates equitable sharing of resources (Chauhan & Dwivedi, 2022). Farmers who manage to bring their water usage below a defined ceiling could receive blockchain-powered credits or rewards in other forms, providing strong incentives that further promote sustainability. Blockchain can simplify the process of integrating data between different water management systems of municipal water utilities, agricultural cooperatives, environmental agencies. As it provides a single, secure ledger, blockchain facilitates collaboration across organizations towards water conservation projects and allows for better coordination of efforts and planning of shared resources. With blockchain being a decentralized technology, the risk of data tampering is also minimized, thus making sure that all records pertaining to water consumption are transparent and true, which can be crucial during droughts or shortages.

Data Privacy- Security and Ethical considerations

Although blockchain technology can give agriculture a lot of impetus, it also raises great concerns about data privacy, security and ethical issues. The idea behind blockchain is transparency that once we add any information to the ledger, everyone who is a participant of the network will see this information. Although this openness is necessary to ensure traceability and trust, it can risk disclosing sensitive information, such as farm-specific information, proprietary practices or personal information about farmers and workers. Maintaining the privacy and confidentiality of data on blockchain systems is vital for protecting the rights of all stakeholders involved in blockchain, notably small farmers at risk of being exploited through data.

These privacy worries call for blockchain-based agricultural frameworks to ensure selective access to data and give users control over information disclosure. One method is through the implementation of private or permissioned blockchains where open access to specific information is limited to those who have been authorized as participants. With this model, farmers can manage who sees each data entry (a specific yield or soil health metrics) while still maintaining the overall transparency of the supply chain. Furthermore, cryptographic methods like zero-knowledge proofs can offer a solution for verifying data without needing to disclose the underlying details, thereby ensuring a compromise between transparency and privacy.

Ethical challenge related to blockchain for agriculture is the risk of data monopolization and use by big companies. As agricultural data sources grow in value, smaller farmers (particularly from developing regions) may lose their ability to control that data to larger agribusinesses or technology corporations responsible for administering blockchain platforms. Such power asymmetry may lead to farmers having to cede data in return for little or no compensation, benefit, and thus reduces their agency from the agricultural bio-economy. A similar message can be seen in using blockchain platforms in agriculture where ethical practice of data governance needs to become a priority so that not only is ownership and accessibility equitably distributed, but also that small farmers are able to control their information as creators.

The other aspect is safety and this is a crucial one, blockchain does not allow data read or write access to be freely modified without proper permission for a region (part of the input); however, there exists a chance that this is compromised due to cyberattacks or hacking of agricultural data recorded on it at the level IoT devices. The consequences of a breach can be severe as it may impact entire supply chains, affecting not only businesses and consumers but the farmers that produce food as well. Implementing robust security protocols such as encryption and multi-factor authentication on the blockchain platforms and IoT devices helps in protecting agricultural data minimizing threats from cybersecurity actors. This approach would also require some data redundancy strategy and occasional security audits to ensure the integrity of blockchain systems in agriculture.

There are environmental and ethical implications arising from blockchain! It is well documented that blockchain systems, and in particular public clones based on proof-of-work protocols, are energy-intensive. That energy consumption could, ironically, actually harm the environment with agricultural operations that are increasingly integrating blockchain to drive sustainability. A solution for this is that agricultural blockchain networks can use energy-efficient consensus mechanisms for example proof-of-stake or proof-of-authority to minimize environmental damage.

INTERNET OF AGRICULTURAL THINGS AND SMART IRRIGATION

IoAT is a modern agricultural revolution where numerous connected devices and other emerging technologies to Components of IoT in Agriculture are embraced for optimization. One of its most effective applications is smart irrigation which utilizes IoAT to tackle water management, resource use and crop production issues. IoAT such as Sensor and drone systems with data management system allows farmers for specific irrigation which lead to water savings, yield increases, and sustainable agricultural practices. IoAT supports farmers in their decision-making process by enabling the collection and processing of vast amounts of data through which resources could be allocated efficiently resulting in optimal utilization whilst improving productivity.

Elements of IoAT in Agriculture

The driving force behind IoAT is the combination of IOT devices like sensors and drones along with sophisticated data management systems in farming. These devices serve as the foundation for precision farming by gathering real-time data from on-farm areas and sending it to centralized platforms through various communication technologies for analysis. It is this data about soil status, moisture content, temperature and humidity that is important for knowing how much water the crops need and sensors play a great role in collecting this data. These sensors are usually buried in the soil or hanging with plants to continuously monitor environmental and soil variables. They facilitate quick response to changing conditions and adjustment of irrigation schedules by providing real-time information.

High resolution imagery and crop health over wide areas are becoming available, enabling the use of drones. Drones are equipped with cameras and sensors that capture high-resolution visual data allowing farmers to identify crop stress signs, insects and moisture differences. Using aerial surveys, drones provide accurate maps of soil variability and crop health in order to carry out precise interventions such as targeted irrigation treatments. Processing the huge data derived from sensors and drones needs excellent data management systems. These platforms pull information from multiple data sources and deliver useful knowledge to farmers such as predictive analytics and automated irrigation control. These systems include cloud-based platforms to store data regularly in a secured manner, access it remotely and analyze with the help of machine learning algorithms. With this data management infrastructure, farmers can handle their irrigation systems and other agronomic operations for better work.

Smart Irrigation Methods Backed By IoAT Perspective

IoAT is transforming irrigation practices that have remained largely unchanged for centuries by introducing intelligent, data-driven systems that optimize water usage and improve crop yields. A machine learning (ML)-based irrigation framework, integrated with IoT technologies, has been developed to support precision agriculture and sustainable farming. These systems rely on IoT devices such as soil moisture sensors, temperature and humidity monitors, and weather forecasting modules to continuously capture real-time and historical data about environmental conditions and crop requirements. The collected data are transmitted to centralized cloud servers or edge-computing platforms, where ML algorithms—ranging from decision trees and support vector machines to neural networks—analyze patterns across variables such as soil type, crop species, climatic fluctuations, and growth stages (Balfaqih, 2023).

Once the optimal irrigation schedule is determined, the system actuates irrigation through smart valves or automated actuators, ensuring that water is delivered precisely when and where it is needed. This minimizes waste, reduces runtime, and prevents both over- and under-irrigation. Remote accessibility via mobile applications or web dashboards gives farmers greater flexibility to monitor and control their fields. Predictive maintenance alerts can further reduce downtime by detecting potential equipment failures in advance. As the ML models are continuously retrained with new data, the system evolves over time, making more accurate and adaptive irrigation decisions. In doing so, it not only mitigates water scarcity but also enhances crop productivity and strengthens resilience against climate variability. By integrating

AI, IoT, and automated control mechanisms, smart irrigation transforms traditional methods into highly efficient, climate-responsive practices suitable for both small-holder farms and large-scale enterprises, supporting sustainable rural development and equitable access to advanced technologies (Balfaqih, 2023).

Central to IoAT-enabled irrigation are three interrelated techniques: precision irrigation, soil moisture sensing, and climate-smart systems. Precision irrigation employs IoT sensors to measure the exact water needs of crops at specific points within a field, adjusting delivery in real time to prevent overuse or shortages. This targeted approach reduces waste while maintaining optimal soil humidity for healthy growth, and advanced systems can even align irrigation schedules with weather forecasts to avoid unnecessary watering. Soil moisture sensing adds another critical layer, where sensors monitor hydration thresholds and automatically trigger irrigation when soil becomes too dry. This not only conserves water but also prevents nutrient leaching that can harm soil health and water quality. Climate-smart systems go a step further by integrating weather forecasts into irrigation planning. Anticipating rainfall, temperature shifts, or drought conditions allows these systems to adapt watering schedules proactively, reducing runoff and ensuring more sustainable use of water resources.

Like other smart agricultural systems, IoAT irrigation frameworks rely heavily on data collection, processing, and analytics. Sensors continuously gather information about soil conditions, weather, and crop health, which is then processed using big data platforms and AI models. These analytical tools reveal patterns and long-term trends that may not be visible through human observation alone, such as shifts in rainfall patterns or correlations between crop type and irrigation efficiency. By learning from large and diverse datasets, ML algorithms refine irrigation strategies over time, customizing recommendations for specific crops, environments, and growth stages. This ability to adapt dynamically enables farmers to better align their practices with resource availability, environmental constraints, and productivity goals, ultimately leading to smarter water management, higher yields, and more resilient food systems.

Benefits of Smart Irrigation with IoAT

IoAT-based smart irrigation systems offer significant benefits, including higher crop yields, improved resource efficiency, and substantial water savings. This sub-section highlights these advantages as illustrated in Figure 1.

Higher Output

With optimized irrigation, IoAT allows crops to get the perfect amount of water at the right time resulting in healthier plants and higher growth rates. Uniform and exact irrigation leads to increased production since crops are neither stressed by irrigating less than needed nor injured due to excessive irrigation. Through improved water management, crops can grow in climate-adverse conditions that would repress their growth in the absence of such management, especially if they are to be cultivated in areas with water scarcity or drought.

Resource Efficiency

Systems based on IoAT considerably improve resource utilization. Farmers conserve water, energy and fertilizers through irrigation that lowers waste and delivers the right amount of water more accurately. It even applies to other resources like pesticides because precision irrigation will be better synchronized with pest management. Farmers can cut costs of resource consumption by optimizing the irrigation schedules as per real-time data, which helps in making farming an economically sustainable practice with reduced operational costs.

Water Preservation

The biggest benefit that smart irrigation powered by IoAT provides is water conservation. In the context of water scarcity that the world has been facing ever more seriously, these technologies can help save a great deal of water without sacrificing yield or even gain increases. Sensors and data analytics monitor irrigation schedules but allow farmers to use the water only at times and in areas that require it, leading Iran to make up about 30–50% of "water saving volume". This contributes by conserving important water resources, most especially where the climate has a character of arid and semi-arid.

IoAT and smart irrigation methods are changing the game for the agriculture domain by enhancing the efficiency of water consumption, saving resources, and increasing food production. Farmers can use these sensors, drones, and data management systems to make automated irrigation decisions from real-time data. With the escalating global issue of water scarcity, IoAT provides an approach to managing our water resources sustainably and in turn helps foster food security even under uncertain climate change. The applications of IoAT are disrupting traditional farming practices and laying the foundation for a more climate-smart, resilient way of agriculture.

Figure 1. Benefits of smart irrigation with IoAT

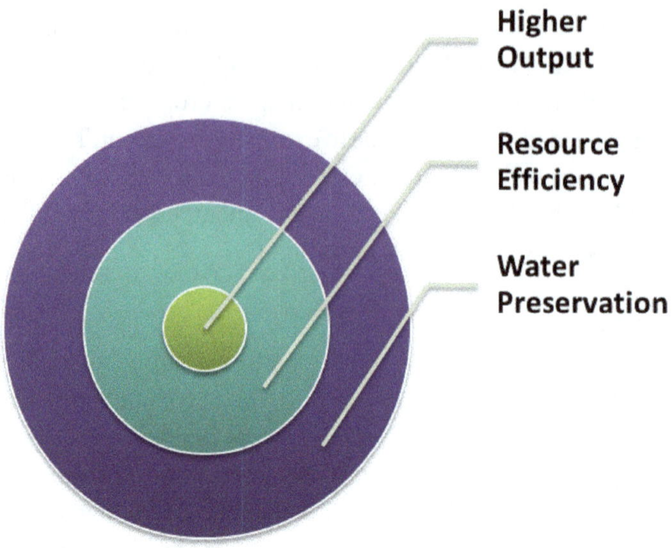

CHALLENGES IN ADOPTING BLOCKCHAIN AND IOAT IN AGRICULTURE

In agriculture, the innovation of advanced technologies such as Blockchain and IoAT hold an exceptional opportunity for optimizing agricultural activities, increasing productivity, and assuring sustainable use. But the power of these smart city systems has yet to be realized, and a few hurdles must be cleared before they will cement their place in urban infrastructure. From technology and infrastructure requirements to regulatory issues as well as cost challenges and the willingness of farmers to adopt new technologies, these barriers abound. Moreover, the environmental and social costs associated with widespread adoption of technology must also be managed so that the siphoning off of profit does not mean a degradation to bottom-line services. Figure 2 shows the challenges that will be discussed in this section.

Figure 2. Challenges in adopting blockchain and IoAT in agriculture

Technological Limitations and Infrastructure

This refers to the vast technological and infrastructure needs, especially in rural and remote locations counties of lab when adopting Blockchain, but also IoAT on agri enterprise level. Coherence: Consistent, Reliable Connectivity and Access to Power with IoAT comes the ability to turn a lot of interconnections in each manufacturing process into tangible promises, but fulfilling those promises does depend on continuous access to high-quality connectivity and reliable power sources. Issues with reliable internet connectivity and electricity infrastructure in many rural places, especially developing ones remain a major barrier for effective application of IoAT devices such as sensors, drones and automated irrigation systems.

Real-time analytics and decision-making require constant data streaming from IoAT devices to cloud platforms. Traditional devices perform their best only through well-established internet access, otherwise it would delay or distort information processing. In addition, the extensive deployment of Blockchain needs enough computing and storage capacity which may not appear in rural areas easily. Such infrastructural constraints raise the cost of technology adoption and may widen the gap between rural and urban farming practices, with people in rural areas often being at a disadvantage when it comes to using modern agricultural technologies.

Apart from connectivity and power issues, most farmers have no technical know-how to operate IoAT devices or Blockchain systems. The lack of knowledge is another barrier to adoption as growers require training in how to use the technology properly. Tackling these sustainability issues will need investments in the technology infrastructure, tech education, and support systems to make these technologies relevant and applicable for rural farmers.

Regulatory, Legal and Ethical Issues

Blockchain and IoAT uptake in agriculture pose a number of regulatory, legal and ethical challenges. A clear regulatory framework is essential for the effective integration of blockchain technology, which allows a decentralized and transparent system to record transactions time (Balfaqih et al., 2023a). The unique nature behind Blockchain and IoT technologies is beyond scope of existing laws and regulations which clearly create loopholes in governance along enforcement. In fact, Blockchain provides either immutable transactions or data on the ledger or decentralized storage of those records, a characteristic that may be at odds with current regulations needing centralized visibility & control (Balfaqih et al., 2023b) . When it comes to agricultural supply chain IoAT, there are concerns over the ownership, sharing and liability surrounding this data. This can cause uncertainty to farmers, producers and consumers and can prevent the adoption of these technologies due to a lack of clear guidelines/ legal standards.

Data privacy is one of the major issues facing on IoAT and Blockchain in agro industry. IoT devices and platforms could be at risk of breaches that expose vast, new volumes of personal and financial information. Agricultural data can be sensitive, such as details on farm operations and crop yields, or water usage history; farmers therefore need to feel confident that their data will be protected from unauthorized access and improper use. We need to talk about the ethical implications of data ownership and privacy, particularly in rural communities in which concerns regarding trust may be a barrier to the uptake of this technology.

Cost, Maintenance and Farmer Preparedness

The price of carrying out Blockchain and IoAT solutions in farming is still high, particularly for smallholder growers. The upfront cost of IoT devices, sensors, drones, and the associated infrastructure elements (internet connection and electricity to run) can be huge. Moreover, farmers have to keep in mind all the ongoing maintenance costs that come with it including software updates, hardware repairs, and troubleshooting. This isn't financially possible for all farmers, especially those whose margins are tight enough as is or who have no access to financing.

These technologies come with high initial and maintenance costs that can prevent small-scale and subsistence farmers from adopting them. Although smart farming solutions can yield benefits and save costs in the long run, it can take heavy toll financially on farmers up front. Farmers may also not have access to financing or government subsidies to help them cover the costs associated with adopting these technologies. The other dilemma is whether farmers will be ready and willing to utilize a new technology. Traditional agricultural communities are typically skeptics of Blockchain and IoAT, as many farmers. Transitioning from traditional farming techniques to cutting-edge mechanization necessitates a change in both mindset and practice – understandably quite challenging for farmers who have spent decades growing food the old way. Finally, limited awareness and knowledge on the potential advantages of these technologies may limit any speed in their adoption. Tackling such resistance and underuse is not just a matter of funding; both training programs and dissemination campaigns should be used in these nations to show what productivity, sustainability, or profitability can be gained from using these technologies.

Environmental and Social Impacts

However, the large, scaled usage of Blockchain and IoAT in agriculture may also have wider environmental and social consequences. Evolution of IoAT in Smart Farming On the environmental side, IoAT-assisted smart farming can help make farming more sustainable (through things like precision irrigation and resource management), however deploying billions of sensor devices can also have a negative ecological impact. However, the impact of e-waste may escalate to pollution based on how it manages in their production phase and waste disposal. The power usage of IoT devices and Blockchain networks could contribute to the carbon footprint of agriculture, having a larger impact in areas with high energy consumption or dependence on non-renewable electricity sources.

Socially, technological change in agriculture could create a divide between technology-rich large commercial farms and smallholder or subsistence farmers. It is important to note that while larger farms may have the capital to adopt IoAT and Blockchain technologies more easily, smaller farms will face challenges in keeping pace, creating widening disparity between access to 21st century farming implements. Such a split may further reinforce social inequalities in rural societies and discourage the building of y inclusive and equitable agricultural systems. For another, the emphasis on tech could have a negative effect by further removing labor from the equation and thus decreasing job opportunities in agricultural communities. The adoption of drones and IoT systems is likely to reduce manual labor in several instances, leading to job losses especially in rural regions where agriculture serves as a primary source of livelihood.

Blockchain and IoAT do indeed highlight enormous opportunities in transforming agriculture but their adoption challenges should not be underestimated. For broader adoption of these solutions in agri-practices technological limitations, regulatory issues, cost and maintenance barrier and on-farm readiness of farmers towards new technologies shall be addressed. In addition, carefully considering the environmental as well as social impacts of their large-scale use can avert unintended consequences. Addressing these barriers will necessitate a collaborative approach by governments, the private sector, and farming communities to foster an enabling environment for the sustainable and inclusive development of smart agriculture technologies.

SMART CITY INFRASTRUCTURE SUPPORT FOR SUSTAINABLE AGRICULTURE

Although the world is experiencing an unprecedented wave of urbanization, incorporating agriculture into smart city systems has become a key element for sustainable urban development. Smart city technologies such as IoT, Artificial Intelligence (AI) and Blockchain are critical tools for urban-agricultural symbiosis improvement. By utilizing these technologies, the connection between metropolitan areas and rural farming communities can be further bridged to make urban agriculture more efficient, sustainable and robust. Thus, consumers benefit from smart city infrastructure by being able to optimize sustainable agricultural production with access to necessary resources-energy, water and efficient logistics systems-necessary for a wider urban mobilization of agriculture initiatives.

Smart cities use modern advanced technologies to make life easier and a lot more convenient. The role of such technologies in urban agriculture includes IoT, AI and Blockchain. Sensors, drones, and automated irrigation systems are among the types of IoT devices that gather data in real time to help farmers use resources more efficiently while also scouting their crops for pest or disease outbreaks. By analyzing this data, AI predicts trends and optimizes the planting schedule as well as crop farming resources that need to be deployed for optimal productivity whilst minimizing wastage. It guarantees transparency, security and traceability of all agricultural transactions and data creating trust and accountability between different players across the supply chain. The synergy between smart cities and agriculture facilitates better resource utilization, which can lead to climate-smart urban areas that fulfill both their environmental sustainability objectives and their local food systems.

However, the main challenge in supporting IoAT in urban agriculture is a sustainable and reliable energy supply. Smart city infrastructure techniques such as smart grids play a fundamental role in providing energy supply to IoAT systems. They leverage AI and IoT in order to make optimal use of production plants, including renewable energy sources like solar and wind power. Smart grids can guarantee an uninterrupted supply of energy to the energy-dependent IoT devices which are important parts of urban agriculture such as sensors and automatic irrigation systems, a seamless smart grid would ensure that there is environmentally conscious energy supply for urban agriculture. It also helps farmers save costs as we are promoting the used of renewable energy to contribute in achieving even bigger sustainability goals by the city level. Moreover, smart energy management solutions can help integrate additional energy storage systems to provide backup power supply in a reliable way even during peak demand hours or when the renewable sources are not available. The combination of smart grid and IoAT not only facilitates the idea of smart city, but also promotes sustainability in agriculture as well as urban infrastructure, thereby utilizing energy efficiently.

Waste and water management are efficient elements of sustainable urban agriculture. Smart city system can therefore maximize the efficiencies of recycling waste, reusing wastewater and conserving water. This is a critical factor, particularly for urban areas where water scarcity is an issue as IoAT devices can monitor the usage of water and with this data in hand, irrigation systems can be optimized to curb over-irrigation and wastage. In addition, urban farmers can irrigate with treated water using wastewater treatment and reuse systems made possible by smart city infrastructure that helps to ensure freshwater resources are not over-utilized. Sensor technology can also be integrated to maintain the soil moisture by watering crops and saving water resources when needed.

The other area of agriculture influenced by smart city technologies is waste management. In smart cities, the collection and processing of agri-waste would be efficient and processed to obtain resources like compost or bioenergy. Such an ecosystem of IoT-driven waste management system can monitor the flow of waste, securing its processing in a greener way. This reduces the quantity of organic waste being dumped by landfills and provides inputs for urban farming in the form of compost or biogas resulting in enhanced soil quality, thus having a positive impact on agriculture as it mitigates the requirement of synthetic fertilizers.

Timely and cost-effective transport and logistics ensure urban markets have fresh local agricultural produce. Smart cities have intelligent logistics and transportation systems which are IoT, AI and blockchain enabled, that ensure the optimal movement of agricultural goods. IoT-based sensors monitor the temperature, humidity and state of a particular agricultural product during transport to keep it fresh and consumable. This is very vital for perishables like fruits, vegetables, and dairy with cold chain concerns. They can even lend support in the adoption of last-mile delivery solutions that can help reduce food wastage and make distribution more convenient in urban areas. Alternative delivery methods such as Electric vehicles (EV) and Autonomous delivery drones are being tested as sustainable solutions to the traditional delivery method, in order to help lower the carbon emission produced by transporting agricultural products. Furthermore, they carry out demand projection and route optimization with the help of AI algorithms so that transportation costs are kept low, and food wastage is reduced.

The latter ensures that the final retail without compromising on the visibility of each stage and security, which is getting an added boost by blockchain to make logistics and cold chain management at ease for agricultural products. Farmers, retailers, etc all stakeholders can have real-time access to information regarding the status (Goes out of stock or expiry) and location (Where the goods are kept) significantly improves coordination between farmers and consumers resulting in reduced inefficiencies. Smart city should help to support local food systems, by enabling agricultural products safe and timely delivery which is the key stone in the overall sustainability of urban agriculture.

Smart City infrastructure is the cornerstone of sustainable agriculture when used with IoT technologies, AI technology enables organizations to run high-performance and deep predictive analyses on large volumes of information in real time, Blockchain can provide data access over a secure ledger and smart grids will support the successful rollout of energy efficient initiatives directly into communities. These technologies support urban agriculture in improving efficiency, resource-efficiency, and resilience against climate change factors while also aiding agricultural incorporation into urban life. Smart cities, by tackling major issues such as energy management, waste and water management or logistics could help create a conducive environment for sustainable agriculture (Ordonez García & Martos Nunez, 2023). The interrelationship between smart cities and agriculture provides a compelling opportunity for us to develop more sustainable and livable cities that can promote food security, environmental sustainability, and economic diversity.

INFORMATION AND COMMUNICATION TECHNOLOGIES CHALLENGES OF SMART CITY DEVELOPMENT

Smart cities heavily depend on Information and Communication Technologies (ICT) to provide efficient infrastructures, on-line services, and citizens participation. Nevertheless, deployment and management of these technologies have several challenges. One of the big hurdles is to make systems and devices from different vendors talk to one another. Data privacy and cybersecurity are also big issues – because smart cities generate so much sensitive personal and operational data, they become prime targets for hackers. Moreover, it may be difficult and resource-demanding to scaleup and the integration of legacy systems into state-of-the-art ICT platforms can be challenging. Limited network connectivity and available bandwidth, in particular in outskirts, or in dense areas, also severely restrict the possibility of real-time communication and control. No less essential an issue is the access gap, which when it comes to equipping the citizens can result in social exclusion. The very high costs of implementation as well as the absence of a set of common policies may cause project delays and fragmented urban solutions. However, sustainability, technological adaptability, long term maintenance in the fast changing ICT landscape is a strategic challenge. There are challenges such as coordinated planning, capacity in terms of infrastructure, policy framework for inclusiveness, stakeholder engagement in order to take full advantage of the potential of ICT in the development of the smart city.

CONCLUSION

The combination of modern technologies like Blockchain, IoAT and smart city infrastructure has a game-changing ability to address sustainable agriculture. This helps to not only reduce resource consumption like water and energy but also increase the transparency and traceability of agricultural supply chains. Combined, these technologies facilitate better farming methods, thus allowing farmers to tackle major challenges such as groundwater depletion, food security and sustainable agriculture. There is still a long way to go in this area. IoT devices and AI algorithms are constantly improving, offering increasingly high-resolution, predictive capabilities to help manage agricultural resources. The continuous development of blockchain technology will continue to help ensure the integrity and trustworthiness of agricultural data and transactions along supply chains. However, successful implementation of these technologies demands cost-effective solutions to infrastructure hurdles and regulatory frameworks especially in rural or developing countries. The role of these technologies will continue to evolve in relationship to sustainable policy frameworks that protect our ecosystems, data rights and human right across smart cities. Once these solutions are perfected, they shall allow for more resilient, efficient and sustainable agricultural systems that will provide food with health urban populations while allowing them to continue chasing high quality of life in their natural environment.

REFERENCES

Ahamed, N. N., Karthikeyan, P., Velliangiri, S., & Gunjan, V. K. (2023). Blockchain Security for the Internet of Things. In *Modern Approaches in IoT and Machine Learning for Cyber Security: Latest Trends in AI* (pp. 75–92). Springer International Publishing.

Bala, K., & Kaur, P. D. (2022, January). Changing trends of blockchain in IoT: Benefits and challenges. In *2022 12th International Conference on Cloud Computing, Data Science & Engineering (Confluence)* (pp. 324-329). IEEE.

Balfaqih, M. (2023, November). Design and Implementation of IoT-enabled Intelligent Irrigation System Using Machine Learning. In *2023 28th Asia Pacific Conference on Communications (APCC)* (pp. 436-441). IEEE.

Balfaqih, M., Balfagih, Z., Almohammedi, A. A., & Alfawaz, K. M. (2023b). A Smart and Privacy-Preserving Logistics System Based on IoT and Blockchain Technologies. In *2023 1st International Conference on Advanced Innovations in Smart Cities (ICAISC)* (pp. 1-5). IEEE.

Balfaqih, M., Balfagih, Z., Lytras, M. D., Alfawaz, K. M., Alshdadi, A. A., & Alsolami, E. (2023a). A blockchain-enabled IoT logistics system for efficient tracking and management of high-price shipments: A resilient, scalable and sustainable approach to smart cities. *Sustainability*, *15*(18), 13971.

Bayih, A. Z., Morales, J., Assabie, Y., & de By, R. A. (2022). Utilization of internet of things and wireless sensor networks for sustainable smallholder agriculture. *Sensors (Basel)*, *22*(9), 3273. PMID: 35590963

Bera, S., Dey, T., Mukherjee, A., & Buyya, R. (2023). E-CropReco: A dew-edge-based multi-parametric crop recommendation framework for internet of agricultural things. *The Journal of Supercomputing*, 1–35.

Bodkhe, U., Tanwar, S., Parekh, K., Khanpara, P., Tyagi, S., Kumar, N., & Alazab, M. (2020). Blockchain for industry 4.0: A comprehensive review. *IEEE Access : Practical Innovations, Open Solutions*, *8*, 79764–79800.

Chattopadhyay, P., Patel, H. P., & Parmar, V. (2022, August). Internet of Things (IoT) in Smart Agriculture. In *2022 3rd International Conference on Electronics and Sustainable Communication Systems (ICESC)* (pp. 536-540). IEEE.

Chauhan, N., & Dwivedi, R. K. (2022, April). Designing a secure smart healthcare system with blockchain: a review. In *2022 1st International Conference on Informatics (ICI)* (pp. 214-219). IEEE.

Dinc, E., Kuscu, M., Bilgin, B. A., & Akan, O. B. (2019). Internet of everything: A unifying framework beyond internet of things. In *Harnessing the Internet of Everything (IoE) for Accelerated Innovation Opportunities* (pp. 1–30). IGI Global.

Dinc, E., Kuscu, M., Bilgin, B. A., & Akan, O. B. (2019). Internet of everything: A unifying framework beyond internet of things. In *Harnessing the Internet of Everything (IoE) for Accelerated Innovation Opportunities* (pp. 1–30). IGI Global.

Dwivedi, S. K., Roy, P., Karda, C., Agrawal, S., & Amin, R. (2021). Blockchain-based internet of things and industrial IoT: A comprehensive survey. *Security and Communication Networks*, *2021*, 1–21.

Ferrag, M. A., & Shu, L. (2021). The performance evaluation of blockchain-based security and privacy systems for the Internet of Things: A tutorial. *IEEE Internet of Things Journal*, *8*(24), 17236–17260.

Flores, A., Morales, A., Campos, G., & Gelso, J. (2021, December). Energy Efficiency Using IOTA Tangle for Greenhouse Agriculture. In *Annual International Conference on Information Management and Big Data* (pp. 122-138). Cham: Springer International Publishing.

Gupta, D., Bhatt, P., & Bhatt, S. (2020, December). A game theoretic analysis for cooperative smart farming. In *2020 IEEE International Conference on Big Data (Big Data)* (pp. 2350-2359). IEEE.

Guven, I., & Parlak, M. (2022, November). Blockchain, AI and IoT Empowered Swarm Drones for Precision Agriculture Applications. In *2022 IEEE 1st Global Emerging Technology Blockchain Forum: Blockchain & Beyond (iGETblockchain)* (pp. 1-6). IEEE.

Kaur, U., Kaur, M., & Khang, A. (2023). Futuristic Technologies in Agriculture Challenges and Future Prospects. *Handbook of Research on AI-Equipped IoT Applications in High-Tech Agriculture*, 347-364.

Kumari, A., & Sharma, I. (2023, October). Augmentation of Internet of Things Security and Privacy with IOTA Tangle Network. In *2023 4th IEEE Global Conference for Advancement in Technology (GCAT)* (pp. 1-5). IEEE.

Lamtzidis, O., & Gialelis, J. (2018, December). An IOTA based distributed sensor node system. In *2018 IEEE Globecom Workshops (GC Wkshps)* (pp. 1-6). IEEE.

Ordonez García, B. J., & Martos Nunez, M. V. (2023). Blockchain in Agriculture: A PESTELS Analysis JAVIER.

Phang, S. K., Chiang, T. H. A., Happonen, A., & Chang, M. M. L. (2023). From Satellite to UAV-Based Remote Sensing: A Review on Precision Agriculture. *IEEE Access : Practical Innovations, Open Solutions.*

Podder, A. K., Al Bukhari, A., Islam, S., Mia, S., Mohammed, M. A., Kumar, N. M., & Abdulkareem, K. H. (2021). IoT based smart agrotech system for verification of Urban farming parameters. *Microprocessors and Microsystems*, *82*, 104025.

Rakhee, M., & Elayidom, S. (2023, January). Blockchain Technology as a Solution for Vulnerabilities in Internet of Things-A Survey. In *2023 Advanced Computing and Communication Technologies for High Performance Applications (ACCTHPA)* (pp. 1-6). IEEE.

Ravindran, M. (2022). Energy Efficient Clustering for Precision Agriculture using LEACH Protocol. *NeuroQuantology : An Interdisciplinary Journal of Neuroscience and Quantum Physics*, *20*(7), 2702.

Rudrakar, S., & Rughani, P. (2022). *IoT based Agriculture (IoTA).* Architecture, Cyber Attack, Cyber Crime and Digital Forensics Challenges.

Singh, B. (2023). Federated Learning for Envision Future Trajectory Smart Transport System for Climate Preservation and Smart Green Planet: Insights into Global Governance and SDG-9 (Industry, Innovation and Infrastructure). *National Journal of Environmental Law*, *6*(2), 6–17.

Siwach, V., Sehrawat, H., & Dahiya, O. (2022, October). Hybrid Blockchain Security Mechanism for protection of data related to Precision Agriculture. In *2022 10th International Conference on Reliability, Infocom Technologies and Optimization (Trends and Future Directions)(ICRITO)* (pp. 1-7). IEEE.

Strobel, G. (2020). Agriculture in the Era of the Internet of Things: An Information System Architecture for Smart Agriculture.

Vangipuram, S. L., Mohanty, S. P., Kougianos, E., & Ray, C. (2022). G-DaM: A Distributed Data Storage with Blockchain Framework for Management of Groundwater Quality Data. *Sensors (Basel)*, *22*(22), 8725. PMID: 36433322

Verma, A., Bhattacharya, P., Madhani, N., Trivedi, C., Bhushan, B., Tanwar, S., & Sharma, R. (2022). Blockchain for industry 5.0: Vision, opportunities, key enablers, and future directions. *IEEE Access : Practical Innovations, Open Solutions*, *10*, 69160–69199.

Voulgaris, S., Fotiou, N., Siris, V. A., Polyzos, G. C., Jaatinen, M., & Oikonomidis, Y. (2019). Blockchain technology for intelligent environments. *Future Internet*, *11*(10), 213.

Chapter 9
Are You Ready for Your Future Home?
Modeling the Intention to Purchase Homes in Smart Cities

Aqilah Yaacob
https://orcid.org/0000-0002-4287-1977
Taylor's University, Malaysia

Jen Ling Gan
https://orcid.org/0000-0002-1684-1440
Taylor's University, Malaysia

Kim Leng Khoo
https://orcid.org/0000-0002-0329-1451
Taylor's University, Malaysia

ABSTRACT

Despite the growing interest in smart cities development and their potential benefits, the existing research lacks empirical studies into the nuanced factors shaping purchase intentions of homeowners in the 21st century. Adopting Value-based Adoption Model (VAM), hypotheses were developed and tested. This study addresses the gap and is among the first to empirically utilize perceived price value as a mediator; and provides insights into how perceptions of affordability and value-for-money influence consumers in developing countries. By utilizing PLS-SEM the findings indicate that the relationship among perceived environmental initiatives, convenience, IT resources and behavioral intention to purchase homes in smart cities are mediated by perceived price value, while perceived privacy and perceived security did not. These findings offer city officials an approach to gauging Malaysian' inten-

DOI: 10.4018/979-8-3373-0154-9.ch009

tions to purchase homes in smart cities as well as identify those factors critical to developing a successful smart-city strategy effectively cater to the evolving needs of future homeowners.

INTRODUCTION

In recent years, the idea of future homes in smart cities has gained significant interest worldwide as urban neighborhoods grapple with the concepts of sustainability, efficiency, and livability. With technological advancements and the progression towards UNSDGs on affordable and clean energy (SDG7) and sustainable cities and communities (SDG11), the integration of smart technologies and sustainability into urban development has become a key focus. As Malaysia Madani initiatives geared towards transforming cities with principles of environmental sustainability, social inclusivity, economic vitality, and good governance, the Malaysian government has been actively pursuing the Smart City Framework, national Internet of Things (IoT) Strategic Roadmap, and MyDigital Blueprint. Within this domain, the decision to purchase a home in a smart city represents a significant investment for individuals, influenced by a multitude of factors ranging from technological advancements to socio-economic considerations.

In order to understand the demand towards the seamless integration of technology into various aspects of urban life, including transportation, energy management, healthcare, and governance, it is crucial to examine the factors that influence the intention to purchase a home in smart cities, in which the current chapter is aimed at addressing. By understanding these determinants, policymakers, urban planners, and real estate developers can tailor strategies to enhance the attractiveness and accessibility of smart city living, thereby fostering sustainable urban development and improving quality of life.

Even though there are few studies pertaining to smart cities, most of them are related to hearing aids in smart cities, artificial intelligence (Herath and Mittal, 2022), smart home energy technology, technology sensors (Balfaqih et al., 2023), smart mobility (Leow and Krishnaswamy, 2021), and eco-friendly smart home services (Jabbar et al., 2019). Despite the growing emphasis on smart city development in Malaysia, limited empirical research has been conducted to understand the factors driving individuals' intentions to purchase homes within these urban ecosystems. This is the substantial gap in which the current study will address. The knowledge gap hampers efforts to design targeted interventions that cater to the diverse needs and preferences of potential homebuyers.

Additionally, utilizing price value as a mediator in this chapter is crucial as it provides insights into how perceptions of affordability and value-for-money influence consumers in developing countries. Using price value as a mediating variable, this research seeks to address this gap by identifying and analyzing the key factors within these key research questions: (1) to assess the impact of "Sacrifice" component in VAM, such as perceived privacy and security on the intention to purchase a home in smart cities; and (2) to examine the influence of "Benefit" component in VAM, such as environmental initiatives, convenience, and IT resources on homebuyers' inclination towards acquiring residential properties within smart urban environments.

The rest of the paper is organized as follows. Section 2 presents the hypothesis building, Section 3 describes the chapter methodology, and Section 4 reports the results of the empirical analysis. Section 5 concludes the discussion and suggests recommendations to decision-makers and construction industry regulators.

LITERATURE REVIEW

The Value-based Adoption Model (VAM) offers a valuable theoretical lens for understanding how individuals make decisions about adopting new technologies. Unlike models that focus solely on perceived usefulness or ease of use, VAM emphasizes a value-maximization perspective in which consumers weigh the expected benefits against the perceived sacrifices before committing to adoption (Kim, Chan, & Gupta, 2007). This framework is particularly relevant in the context of smart-city living, where advanced technologies—ranging from IoT-enabled services to digital governance platforms—promise significant lifestyle improvements but also introduce concerns around privacy, cost, and security. Within this framework, the consumer's ultimate intention to adopt is shaped not only by the benefits on offer but also by their perception of whether the sacrifices are acceptable. Crucially, perceived value functions as the mediating factor that translates these evaluations into behavioral intentions.

In the case of smart-city housing decisions, sacrifices are conceptualized as potential risks, particularly related to data privacy and digital security. Smart homes and urban systems rely heavily on the continuous collection of residents' personal and behavioral data, from energy usage to mobility patterns. While this data fuels greater efficiency and convenience, it simultaneously raises concerns about surveillance, misuse, and breaches of confidentiality (Vimalkumar et al., 2021). Prospective buyers therefore assess not just the functional advantages of smart-city technologies but also the level of trust they can place in the system's ability to safeguard their information. Studies such as Dogra and Adil (2024) highlight that privacy concerns are no longer secondary considerations; instead, they directly shape adoption decisions in technology-driven environments.

Conversely, benefits in the VAM framework extend across environmental, technological, and lifestyle dimensions. For instance, smart cities often integrate environmental initiatives such as renewable energy systems, green building practices, and efficient resource management, which align with growing consumer preferences for sustainability. Likewise, the convenience of seamless connectivity, automation of daily routines, and access to integrated IT resources enhances the attractiveness of smart-city living. These benefits collectively create a strong value proposition for potential homebuyers, particularly when they are positioned as contributing to both personal comfort and broader societal goals such as sustainability and efficiency.

The role of perceived price value as a mediator is especially important in emerging markets and developing countries. Affordability determines whether the balance of benefits and sacrifices translates into actual purchase intention. Wirth et al. (2020) argue that individuals are often willing to tolerate higher costs or potential risks if they believe the benefits—such as enhanced privacy protections or sustainable infrastructure—justify the expenditure. In this context, the present chapter applies VAM to hypothesize that privacy and security concerns, when adequately addressed, enhance the perceived value of smart-city living. Similarly, benefits such as environmental initiatives, convenience, and IT resources are expected to positively influence intention, provided that consumers perceive the financial investment as reasonable.

Taken together, this expanded application of VAM highlights the interplay between technology, trust, and affordability in shaping consumer behavior. By framing adoption as a value-driven decision process, VAM provides a comprehensive way to explain why some individuals readily embrace smart-city housing while others remain hesitant despite the availability of advanced features. The integration of sacrifices (privacy and security) and benefits (environmental initiatives, convenience, and IT resources) mediated by perceived price value thus forms the conceptual foundation of this chapter's hypotheses on home purchase intention in smart cities.

Hypothesis 1: Perceived Price Value mediates the relationship between Perceived Privacy and Behavioral Intention.

Other than the privacy of the data, potential residents are concerned about the security of the data as well. It is defined as how much the users believe that the system would transmit private data without any leakage (Merhi et al., 2019). With cyber security issues, residents might be hesitant to stay in smart cities as they do not want their data to be exposed. It is challenging for the residents to perceive the price value if they fail to perceive the sense of security living in a smart city which requires high expenses (Jiang, Jun, & Yang, 2016). When the users perceive that the benefits do not match the high cost that they are bearing, their intention to purchase a home in smart cities will decrease (Han & Han, 2023). This shows that price value might play an intervening role in affecting the residents' behavioral intentions in a smart city. Therefore, hypothesis 2 is proposed as follows:

Hypothesis 2: Perceived Price Value mediates the relationship between Perceived Security and Behavioral Intention.

Smart cities have always been linked to sustainability and environment initiatives (Shamsuzzoha et al., 2021). Residents who are concerned about the environment tend to pay higher attention to any initiatives that protect the environment (Ogiemwonyi et al., 2023). In other words, green consumers are willing to pay more for a green service. If they find the value in green products. Consequently, if a project of smart cities is able to save the environment, green consumers tend to have higher intention to purchase a home in smart cities as they perceive the price value in producing green outcomes. Therefore, this chapter proposed:

Hypothesis 3: Perceived Price Value mediates the relationship between Perceived Environmental Initiatives and Behavioral Intention.

Other than a well-structured mobility system, residents also aim for convenience in their residence area. With most things that can be assessed or completed using high technology in smart cities, convenience levels could be increased among the users. Other than the physical convenience experienced by the users, Hayat (2016) also highlighted 'convenience' as saving energy and cost consumption, which in turn improves their life quality in smart cities. The pleasant experience of convenience can indirectly affect the intention of purchasing among the users by recognizing the value of spending in smart cities (Lin et al., 2019). Hence, this chapter proposed:

Hypothesis 4: Perceived Price Value mediates the relationship between Convenience and Behavioral Intention.

Information technology (IT) plays a major role in implementing smart cities. Smart devices, smart transportation, or any smart facilities require IT resources to run them, which involves high costs for good maintenance (Balfaqih & Alharbi, 2022). It was highlighted by Ahad et al. (2020) that smart cities are not possible if the nation does not have a pool of experts and residents who are technology-savvy. In other words, if potential users of smart cities do not perceive support in terms of IT resources, the price value will decrease as they do not have the literacy in IT. Therefore, this chapter proposed:

Hypothesis 5: Perceived Price Value mediates the relationship between IT resources and Behavioral Intention.

Based on the hypothesis development above, the present chapter has developed a conceptual framework, indicated in Figure 1.

Figure 1. Conceptual framework

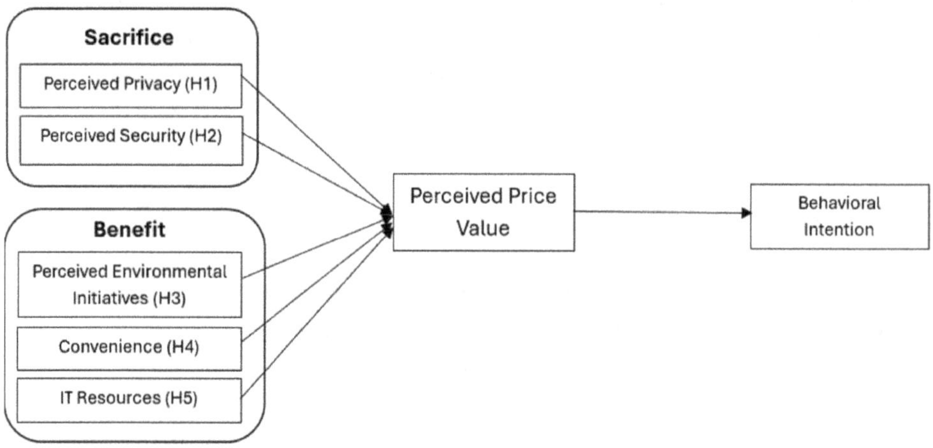

DATA COLLECTION AND MEASURES

To understand the relationships between the proposed variables, this chapter used quantitative methods through the survey questionnaire method. The scale items used in the chapter were derived from existing literature. The data was collected through a snowball sampling method, with 261 complete questionnaires received from diverse respondents. The sample comprised individuals of different ages and genders, with 48% male and 52% female respondents. Among the respondents, 72% belonged to the respondents aged 18-29 years old, 15% were between 30-39 years old, 8% were between 40-49 years old, 3% were between 50-60 years old, and 2% were 61 years old and older. Table 1 presents the demographic profile of respondents for this chapter.

Table 1. Demographic profile

Demographic	Details	Frequency (%)
Gender	Male	126 (48.3%)
	Female	135 (51.7%)
Age	18-29 years old and above	187 (71.6%)
	30-39 years old and above	39 (14.9%)
	40-49 years old and above	22 (8.4%)
	50-60 years old and above	7 (2.7%)
	61 years old and above	6 (2.3%)
Marital Status	Single	195 (76.2%)
	Married	62 (23.8%)
Education	High School or lower	31 (11.9%)
	O-level	4 (1.5%)
	A-level	7 (2.7%)
	Diploma	60 (23%)
	Bachelor's Degree	125 (47.9%)
	Master's Degree	24 (9.2%)
	Doctorate Degree	10 (3.8%)
Profession	Housewife/Stay at Home Husband	11 (4.2%)
	Private Sector Employee	102 (39%)
	Public Sector Employee	23 (8.8%)
	Retiree	7 (2.7%)
	Student	107 (41%)
	Unemployed	4 (1.5%)
	Others	7 (2.7%)
Monthly Income	Less than RM1500	100 (38.3%)
	RM1500-RM3000	49 (18.8%)
	RM3001-RM4500	46 (17.6%)
	RM4501-RM6000	20 (7.7%)
	RM6001-RM7500	13 (5.0%)
	RM7501-RM9000	13 (5.0%)
	RM9001 and above	9 (3.4%)
	Others	11 (4.2%)

Source(s): Authors' own illustration

In this chapter, measurement items of construction were adopted and adapted from the established literature. Respondents were instructed to answer the questions based on their intention to purchase homes in a smart city. We have adapted perceived privacy and perceived security from the work of Chatterjee and Kar (2018) with 4 and 6 items, respectively. Perceived environment initiatives were adapted from Giffinger and Gudrun (2010) and Kumar and Dahiya (2017) with 5 items. Next, price value was adapted from Baudier, Ammi, and Deboeuf-Rouchon (2020) with 3 items and IT resources using a scale of three items adapted from Pratama and Imawan (2019). convenience and behavioral intention were adapted from Venkatesh, Thong, and Xu (2012) with 4 and 3 items, respectively.

EMPIRICAL RESULTS

In this chapter, the primary analytical method used to test the proposed research model is partial least squares equation modeling (PLS-SEM). The preference for PLS-SEM was based on several reasons proposed by Hair et al. (2019). Firstly, this chapter did not attempt to confirm a theory but to analyze the proposed model based on the literature. Secondly, PLS-SEM is suited to handling non-normal data and datasets with a small sample size.

Common Method Bias

In survey-based research, one of the recurring concerns is the potential for common method variance (CMV), which arises when the measurement method rather than the constructs of interest accounts for observed correlations among variables. This issue can lead to inflated or misleading relationships, thereby compromising the validity of the findings. Since this study relied on self-reported data collected from a single instrument, it was essential to examine whether CMV posed a threat to the reliability of the results.

To address this concern, a collinearity diagnostic test was carried out, in line with methodological recommendations in the literature. Specifically, the study assessed variance inflation factors (VIFs) by introducing a randomly generated dependent variable and checking for multicollinearity among the predictors. According to Kock (2015), a VIF value greater than 3.3 suggests the potential presence of method bias, whereas values below this threshold indicate that CMV is unlikely to significantly distort the model.

The results of this diagnostic, summarized in Table 2, demonstrate that all VIF scores for the study's constructs—including behavioral intention, convenience, environmental initiatives, IT resources, perceived price value, perceived privacy, and perceived security—were comfortably below the recommended threshold. The highest observed value was 2.424 for convenience, which still lies well within acceptable limits. This finding confirms that collinearity among indicators was not problematic, and therefore, CMV does not present a substantial risk in this dataset.

By confirming the absence of significant common method bias, the study ensures greater confidence in the robustness of its structural model and the accuracy of its path relationships. This diagnostic step strengthens the internal validity of the analysis and provides reassurance that the reported associations among constructs reflect genuine relationships rather than artifacts of the data collection method.

Table 2. Collinearity test

	Variance inflation factors (VIF)
Behavioral intention	1.765
Convenience	2.424
Environment initiatives	1.729
IT resources	1.490
Perceived price value	1.745
Perceived privacy	1.910
Perceived security	2.134

Measurement Model

Before testing the hypothesized structural relationships, it was necessary to evaluate the measurement model to ensure that the constructs were both statistically reliable and theoretically valid. Following the guidelines of Hair et al. (2019), the assessment was conducted in a two-step process that focused on the reliability, convergent validity, and discriminant validity of the constructs. This procedure is critical because it provides assurance that the latent variables are being measured accurately and consistently, thereby lending credibility to subsequent structural model analysis.

The first step involved an evaluation of internal consistency reliability. Both Cronbach's Alpha and composite reliability (CR) were used as indicators, with the generally accepted cut-off being 0.70. As reported in Table 3, all constructs exceeded this threshold, confirming that the scale items consistently represented their underlying constructs. This suggests that the measurement instrument demonstrated strong reliability and can be considered dependable for further analysis.

Next, the study examined convergent validity, which evaluates the extent to which items that are theoretically related actually converge to represent the same construct. This was tested by analyzing the outer loadings of the items and the average variance extracted (AVE) for each construct. The findings revealed that the outer loadings were above 0.70, while all AVE values were greater than 0.50. These results indicate that the constructs were able to capture more than half of the variance in their indicators, thereby meeting the minimum requirements for convergent validity and confirming that the items adequately reflected their intended concepts.

Finally, discriminant validity was assessed to determine whether the constructs were truly distinct from one another. This was achieved using the heterotrait-monotrait ratio of correlations (HTMT), a robust criterion proposed by Henseler, Ringle, and Sarstedt (2015). The HTMT values for all construct pairs were well below the threshold of 0.85, demonstrating that the constructs were not excessively correlated and that each represented a unique dimension of the conceptual model.

The results provide strong evidence that the measurement model is both reliable and valid. The high levels of internal consistency, coupled with satisfactory convergent and discriminant validity, confirm that the survey instrument successfully captured the intended constructs. With the measurement model validated, the study was able to proceed confidently to the evaluation of the structural model, where the hypothesized relationships between constructs were tested.

Table 3. Measurement model

Construct	Item	Loading	Cronbach's alpha	Composite reliability	Average variance extracted (AVE)
Perceived Privacy	PP1: I believe that the information of the residents should not be disclosed to unwanted authorities or personals	0.799	0.862	0.904	0.702
	PP2: I believe that citizens' image, voice, etc. should not be distributed without the consent of that citizen to some other unwanted person or authority	0.834			
	PP3: I believe that confidentialities of information of the residents should be remain protected	0.844			
	PP4: I believe that information should not be linked to any department without the proper approval from the concerned resident	0.872			

continued on following page

Table 3. Continued

Construct	Item	Loading	Cronbach's alpha	Composite reliability	Average variance extracted (AVE)
Perceived Security	PS1: I believe that the security aspect of the IT-enabled system should not be compromised under any circumstances	0.671	0.853	0.890	0.578
	PS2: I believe that the perceived threat of data stolen from the IT system should be at minimal	0.641			
	PS3: I believe that the people working to secure the IT-enabled services of the smart city should be adequately trained	0.813			
	PS4: I believe that the residents should be adequately train and aware of how to use the IT services safely and securely	0.750			
	PS5: I believe that the residents should be well protected from any security threat	0.822			
	PS6: I believe that the IT-enabled services provided to the residents of smart city should have a high degree of security features which can keep the digital services fully secured	0.840			
Perceived Environment Initiatives	PES1: I believe that development of homes in smart city supports pollution reduction	0.843	0.899	0.925	0.712
	PES2: I believe that development of homes in smart city provides environmental protection	0.892			
	PES3: I believe that development of homes in smart city provides provision of sewerage and waste water treatment	0.796			
	PES4: I believe that development of homes in smart city adheres to the green practices and recycling of solid waste	0.841			
	PES5: I believe that development of homes in smart city promotes sustainable resource management	0.844			

continued on following page

Table 3. Continued

Construct	Item	Loading	Cronbach's alpha	Composite reliability	Average variance extracted (AVE)
Convenience	C1: I believe that homes in smart cities and its connected services would be a good assistant in my daily life.	0.838	0.906	0.934	0.781
	C2: I believe that homes in smart cities and its connected services would help me save useful time in my daily life.	0.934			
	C3: I believe that homes in smart cities and its connected services would make my everyday life easier.	0.869			
	C4: I believe that homes in smart city and its connected services would increase my efficiency in my daily life	0.891			
IT Resources	ITR1: I believe that there is sufficient IT infrastructure to support smart city implementation	0.890	0.886	0.929	0.815
	IT2: I believe that there is sufficient maintenance of IT infrastructure to support smart city implementation	0.921			
	ITR3: I believe that there is sufficient IT expertise to implement the smart city programme	0.896			
Perceived Price Value	PV1: I believe that homes in smart city are reasonably priced	0.808	0.824	0.895	0.740
	PV2: I believe that homes in smart city are a good value for the money	0.886			
	PV3: I believe that at the current price, homes in smart city provides a good value	0.884			
Behavioral Intentions	BI1: I intend to live in a home in a smart city in the future	0.908	0.853	0.911	0.773
	BI2: If I had access to a home in a smart city, I intend to live in one	0.848			
	BI3: The probability that I buy a home in a smart city in the future is high	0.882			

Source(s): Authors' own illustration

Table 4. Heterotrait-monotrait ratio (HTMT)

Construct	1	2	3	4	5	6	7
1. Behavioral Intentions							
2. Convenience	0.717						
3. Perceived Environment Initiatives	0.477	0.632					
4. IT Resources	0.408	0.438	0.488				
5. Price Value	0.551	0.544	0.517	0.607			
6. Perceived Privacy	0.279	0.506	0.389	0.178	0.146		
7. Perceived Security	0.383	0.607	0.522	0.234	0.145	0.812	

Source(s): Authors' own illustration

Table 5. Direct effects

	Path coefficient	Std error	*t*-value	*p*-value	2.5%	97.5%	f^2	Remark
Perceived Privacy → Perceived Price Value	0.31	0.069	0.440	0.660	-0.098	0.173	0.001	Not significant
Perceived Security → Perceived Price Value	-0.261	0.077	3.403	0001	-0.428	-0.125	0.050	Significant
Perceived Environmental Initiatives → Perceived Price Value	0.216	0.080	2.713	0.007	0.056	0.371	0.046	Significant
Convenience → Perceived Price Value	0.349	0.081	4.287	0.000	0.185	0.501	0.111	Significant
IT Resources → Perceived Price Value	0.337	0.058	5.783	0.000	0.224	0.451	0.149	Significant
Price Value → Behavioral Intentions	0.196	0.058	3.374	0.001	0.078	0.310	0.040	Significant
Perceived Privacy → Behavioral Intentions	-0.064	0.068	0.939	0.348	-0.195	0.072	0.004	Not significant
Perceived Security → Behavioral Intentions	0.196	0.073	0.916	0.360	-0.085	0.202	0.003	Not significant
Perceived Environmental Initiatives → Behavioral Intentions	0.014	0.065	0.217	0.828	-0.106	0.146	0.000	Not significant
Convenience → Behavioral Intentions	0.504	0.070	7.213	0.000	0.377	0.646	0.220	Significant
IT Resources → Behavioral Intentions	0.045	0.063	0.714	0.476	-0.074	0.176	0.002	Not significant

Source(s): Authors' own illustration

Structural Model

To rigorously examine the mediating role of perceived price value, this chapter employed a bootstrapping procedure involving 5,000 resamples, which is widely regarded as a robust non-parametric method for testing indirect effects. Bootstrapping is particularly suitable in structural equation modeling because it does not rely on assumptions of normality and generates confidence intervals directly from the empirical resampling process, thereby enhancing the accuracy of mediation testing.

The analysis revealed that the indirect paths associated with H3, H4, and H5 were statistically significant. Specifically, the mediation effects for environmental initiatives ($\beta = 0.042$; $p < 0.05$), convenience ($\beta = 0.068$; $p < 0.05$), and IT resources ($\beta = 0.066$; $p < 0.05$) were validated. Importantly, the confidence intervals for these paths excluded the value of zero, which strengthens the evidence that perceived price value acts as a genuine mediator in these relationships, rather than the results being attributed to random sampling variation (Nitzl, Roldan, & Cepeda, 2016).

The finding indicates that individuals' behavioral intentions to purchase homes in smart cities are not solely influenced by the direct benefits of environmental initiatives, convenience, or IT resources, but also by how these factors shape their perceptions of affordability and value-for-money. In methodological terms, this demonstrates the effectiveness of applying mediation analysis in complex adoption models such as the VAM, where psychological assessments of value function as key mechanisms linking perceived benefits to behavioral intentions.

Table 6. Mediation analysis

Relationship	Indirect effect	Std error	*t*-value	*p*-value	2.5%	97.5%	Remark
H1: Perceived Privacy → Perceived Price Value → Behavioral Intentions	0.006	0.014	0.420	0.675	-0.019	0.039	Not supported
H2: Perceived Security → Perceived Price Value → Behavioral Intentions	-0.051	0.021	2.427	0.015	-0.105	-0.019	Not supported
H3: Perceived Environmental Initiatives → Perceived Price Value → Behavioral Intentions	0.042	0.021	2.037	0.042	0.010	0.092	Supported
H4: Convenience → Perceived Price Value → Behavioral Intentions	0.068	0.027	2.565	0.010	0.026	0.132	Supported
H5: IT Resources → Perceived Price Value → Behavioral Intentions	0.066	0.023	2.830	0.005	0.028	0.119	Supported

Source(s): Authors' own illustration

DISCUSSION

The first research question aims to assess the impact of Sacrifice components in VAM such as perceived privacy and perceived security on the intention to purchase a home in smart cities, mediated by perceived price value. Firstly, the findings indicate that the mediation effects were insignificant for the relationship between perceived privacy and behavioral intention (H1), as well as perceived security and behavioral intention (H2). This may be due to participants prioritizing other factors over perceived price value when considering purchasing a home in a smart city and smart city living may be associated with various perceived benefits beyond price considerations such as environmental, social and technological benefits (Chang and Smith, 2023; Lim et al., 2019). When perceived privacy is highly considered, individuals may be willing to overlook the price, rendering the mediation effect of price value insignificant. Perceived privacy may also be closely tied to the overall quality of life offered by smart cities (Haque et al., 2021; Ismagilova et al., 2019). Even if the cost is high, individuals may be willing to invest in a home that offers enhanced privacy features and amenities, regardless of the price.

Similarly, perceived security may be considered a fundamental aspect of home purchase decisions, often outweighing considerations of price value. In situations where security concerns are paramount, individuals may be less influenced by price fluctuations, resulting in an insignificant mediating effect. Another possible explanation could be due to individuals prioritizing safety and peace of mind over financial savings and individuals might justify higher costs for security features as part of long-term investments (Braun et al., 2018; Yang and Lam, 2021), reducing the relevance of price value as a mediator.

The second research question examines the influence of the Benefit components in VAM, such as perceived environmental initiatives, convenience, and IT resources on homebuyers' inclination towards acquiring residential properties within smart urban environments, mediated by perceived price value. A significant result was reported for H3, in which perceived price value mediates the relationship between perceived environmental initiatives and behavioral intention. This is because individuals may value both environmental sustainability and financial implications of such initiatives (Malik et al., 2022; Puron-Cid and Gil-García, 2022).). The perceived benefits of living in an environmentally friendly smart city may be tempered by concerns about affordability (Mohamad et al., 2019). As a result, perceived price value may serve as a crucial mediator in translating perceived environmental initiatives into actual behavioral intentions. Perceived environmental initiatives in smart cities may often be associated with the adoption of sustainable technologies and practices, which could potentially increase the cost of housing (Chu et al., 2021). In other words, individuals may perceive homes in environmentally friendly smart cities offer various value propositions but at the same time need the assurance that the price they pay aligns with the perceived benefits.

As hypothesized in H4, the finding confirmed that perceived price value mediates the relationship between convenience and behavioral intention regarding home purchase in smart cities. While convenience is a desirable attribute for homebuyers in smart cities, affordability remains a critical consideration. Individuals may perceive homes in smart cities desirable as it offers conveniences but their willingness to purchase may ultimately depend on whether the perceived value aligns with the price they are willing to pay (Zhao and Chen, 2021; Rihar et al., 2015). Individuals will still weigh the perceived convenience benefits against the associated costs (Cole et al., 2018) especially when convenience features such as proximity to amenities, transportation hubs, and employment centers may signal a premium in the housing market.

Finally, H5 is also supported in which perceived price value acts as a mediator for the relationship between IT resources and behavioral intention to purchase homes in smart cities. IT resources offer amenities such as high-speed internet access, smart home features, and digital infrastructure. Individuals may value access to advanced technologies, but they might still be concerned about the affordability, especially those with limited financial resources (Daimon, 2022). Consequently, it reflects individuals' assessments of the affordability of homes equipped with such amenities and the financial implications of accessing these technologies. Perceived price value mediates the relationship by helping individuals evaluate the cost-effectiveness of IT resources in relation to their perceived long-term benefits, influencing their behavioral intentions regarding home purchase.

Implications for Researchers

Research on smart city adoption has generally been concentrated in developed regions, focusing on factors such as technology readiness, environmental sustainability, and digital infrastructure. However, studies examining residential purchase intentions in multicultural, emerging economies such as Malaysia remain scarce. Malaysia's diverse demographic composition and ongoing national initiatives toward digital transformation make it an important setting for testing adoption frameworks, yet empirical work in this area is still limited. This chapter contributes to bridging that gap by contextualizing consumer behavior in a society where smart city development is still in its formative stage and public awareness of such living environments is gradually evolving.

From a theoretical perspective, this chapter advances the literature by not only examining key determinants of purchase intention but also by incorporating perceived price value as a mediating mechanism. Previous studies often treated factors such as privacy, security, convenience, and environmental initiatives as direct predictors of behavioral intentions. By contrast, this study highlights how individuals interpret these factors through the lens of affordability and value-for-money, offering a deeper understanding of the decision-making process. This extension of the VAM demonstrates that perceived value is not merely an outcome of benefits and sacrifices but an active mediator shaping the path to behavioral intention.

The findings also validate the applicability of VAM in a Malaysian smart city context, reinforcing its utility beyond traditional technology adoption research. While prior applications of VAM have largely been in digital service or mobile technology studies, this work broadens its scope by applying it to urban housing decisions—an area where the stakes are higher and the perceived trade-offs more complex. This novel application of VAM provides future scholars with a tested framework for investigating how economic perceptions interact with social, technological, and environmental considerations in shaping consumer intentions.

Moreover, the results open opportunities for comparative studies across different cultural and economic contexts. Researchers can build on this framework to explore whether similar mediating effects of price value are evident in other developing nations, or whether cultural and institutional differences shift the balance of benefits and sacrifices. Such cross-national comparisons could enrich global debates on smart city adoption and provide a more nuanced picture of how residents from varied backgrounds perceive the value of smart living.

Implications for Practitioners

For practitioners, the findings of this chapter underscore the importance of aligning smart city initiatives with the real concerns and priorities of potential homeowners. One of the strongest drivers identified is the perception of environmental responsibility. Developers and policymakers must therefore ensure that sustainability features are not treated as add-ons but as core components of housing projects. Clear communication of how these features—such as energy-efficient systems, green building practices, and waste management solutions—translate into long-term financial savings and environmental benefits can significantly increase consumer confidence. Highlighting tangible contributions to sustainability also positions smart housing as both a lifestyle choice and a socially responsible investment.

Convenience emerged as another pivotal factor shaping purchase intentions. Practitioners should focus on integrating smart technologies seamlessly into daily life by emphasizing their role in saving time, reducing effort, and enhancing comfort. Features such as automated home systems, smart mobility options, and proximity to essential services must be presented in a way that resonates with everyday experiences rather than abstract technological promises. Overstating or misrepresenting such capabilities risks creating a gap between consumer expectations and actual experience, which could undermine trust in both developers and policymakers.

Access to robust IT infrastructure is equally critical. The availability of high-speed connectivity, reliable digital platforms, and well-maintained technological systems plays a direct role in shaping the perceived value of smart homes. Developers should ensure that IT systems are not only installed but also supported through long-term service arrangements. Collaboration with telecommunications providers, technology companies, and local governments can help establish the strong digital backbone required for smart city living.

In addition, practitioners need to recognize that user confidence and literacy are essential for adoption. Advanced technologies can appear to be intimidating potential homeowners who may be unfamiliar with their operation or skeptical about their reliability. Organizing workshops, demonstration sessions, and awareness campaigns in partnership with IT experts can help bridge this gap. These initiatives empower residents by showing them how to operate smart systems effectively, protect their data, and maximize the benefits of smart home technologies.

Finally, practitioners should adopt a transparent and customer-centric marketing approach. By focusing on real-life applications, long-term benefits, and cost-effectiveness rather than purely technical features, developers and marketers can build stronger emotional and financial connections with buyers. Transparency in pricing, sustainability claims, and technology performance is essential to avoid skepticism and foster trust in smart city developments.

Research Limitations

Despite the contributions of this chapter, several limitations should be acknowledged to provide context for the findings and to guide future inquiry. First, the study was conducted exclusively in Malaysia, which limits the ability to generalize the results to other cultural or socio-economic settings. While Malaysia provides a valuable case due to its active pursuit of smart city initiatives, the diversity of global urban development contexts means that factors influencing home purchase intentions may differ significantly across countries. For instance, cultural perceptions of privacy, security concerns, or attitudes toward environmental initiatives might vary in societies with different governance models, economic structures, or technological readiness. Extending the research to include comparative studies across regions would therefore provide a broader perspective and enhance the external validity of the results.

Another limitation lies in the study's reliance on purchase intention as the primary behavioral outcome. While intention is a widely accepted predictor of actual behavior, it does not always translate into final decision-making due to external constraints such as financial capacity, market conditions, or unforeseen personal circumstances. Future studies could strengthen this research stream by examining not only intentions but also actual purchase behaviors or longitudinal data that tracks how intentions evolve into concrete actions over time.

Additionally, the analytical framework employed in this study was centered on VAM with perceived price value as the mediating variable. Although this approach offered useful insights, it did not account for other potential mediators or moderators that may influence consumer decisions. Variables such as trust in government, perceived technological literacy, cultural values, or even demographic attributes (e.g., age, education level, and income) could significantly shape behavioral outcomes. Including such factors in future research could provide a more comprehensive understanding of the complexity of decision-making in the context of smart city housing.

Another limitation relates to the data collection method. The study used a survey-based approach with self-reported data, which is susceptible to biases such as social desirability bias and common method variance. Respondents may have provided answers they considered favorable rather than reflecting their genuine perspectives. Employing a mixed-methods approach in future studies—combining surveys with interviews, focus groups, or even experimental designs—could help triangulate findings and minimize these limitations.

Lastly, the study was conducted at a particular moment in time, capturing consumer perceptions during that period. However, perceptions of smart city living are likely to evolve as technology advances, housing affordability changes, and government policies adapt. A cross-sectional design may therefore overlook dynamic changes in consumer attitudes. Future research could benefit from adopting longitudinal designs to capture shifts in perceptions and intentions over time, providing a richer understanding of how the adoption of smart city housing unfolds in practice.

CONCLUSION

This chapter examined the intention to purchase homes in smart cities using the VAM, with a focus on the mediating role of perceived price value. The findings reveal that while privacy and security are important considerations, they do not significantly influence purchase intentions through price value. Instead, environmental initiatives, convenience, and IT resources emerged as stronger predictors, with affordability shaping how these benefits translate into behavioral intentions. The results highlight that successful smart city adoption depends on balancing technological innovation with financial accessibility. For policymakers and developers, emphasizing environmental sustainability, everyday convenience, and robust IT infrastructure—while ensuring affordability—can enhance buyer confidence and support sustainable urban development. These insights provide a foundation for future research across different cultural and socio-economic contexts, contributing to the broader goal of creating resilient and inclusive smart cities.

ACKNOWLEDGEMENT

This research was supported by Taylor's University through the Centre for Industrial Revolution and Innovation (CIRI) Seed Funding (CIRI-SEED/1/2022/SOMM/002).

REFERENCES

Ahad, M. A., Paiva, S., Tripathi, G., & Feroz, N. (2020). Enabling technologies and sustainable smart cities. *Sustainable Cities and Society*, *61*, 102301.

Balfaqih, M., & Alharbi, S. A. (2022). Associated information and communication technologies challenges of smart city development. *Sustainability*, *14*(23), 16240.

Balfaqih, M., Balfagih, Z., Lytras, M. D., Alfawaz, K. M., Alshdadi, A. A., & Alsolami, E. (2023). A blockchain-enabled IoT logistics system for efficient tracking and management of high-price shipments: A resilient, scalable and sustainable approach to smart cities. *Sustainability*, *15*(18), 13971.

Baudier, P., Ammi, C., & Deboeuf-Rouchon, M. (2020). Smart home: Highly-educated students' acceptance. *Technological Forecasting and Social Change*, *153*, 119355.

Braun, T., Fung, B., Iqbal, F., & Shah, B. (2018). Security and privacy challenges in smart cities. *Sustainable Cities and Society*, *39*, 499–507.

Chang, S., & Smith, M. (2023). Residents' Quality of Life in Smart Cities: A Systematic Literature Review. *Land (Basel)*.

Chatterjee, S., & Kar, A. K. (2018). Effects of successful adoption of information technology enabled services in proposed smart cities of India: From user experience perspective. *Journal of Science and Technology Policy Management*, *9*(2), 189–209.

Chu, Z., Cheng, M., & Yu, N. (2021). A smart city is a less polluted city. *Technological Forecasting and Social Change*, *172*, 121037.

Cole, J., McDonald, J., Wen, X., & Kramer, R. (2018). Marketing energy efficiency: Perceived benefits and barriers to home energy efficiency. *Energy Efficiency*, *11*, 1811–1824.

Daimon, H. (2022). A Fundamental Study on the Influences of Smart City on Residential Location Choice. *Toshi Keikaku Rombunshu*.

Dogra, N., & Adil, M. (2024). Should we or should we not? Examining travelers' perceived privacy, perceived security and actual behavior in online travel purchases. *Journal of Vacation Marketing*, *30*(1), 123–142.

Giffinger, R., & Gudrun, H. (2010). Smart cities ranking: an effective instrument for the positioning of the cities? *ACE: Architecture. City and Environment*, *4*(12), 7–26.

Habib, A., Alsmadi, D., & Prybutok, V. R. (2020). Factors that determine residents' acceptance of smart city technologies. *Behaviour & Information Technology*, *39*(6), 610–623.

Hair, J. F., Risher, J. J., Sarstedt, M., & Ringle, C. M. (2019). When to use and how to report the results of PLS-SEM. *European Business Review*, *31*(1), 2–24.

Han, L., & Han, X. (2023). The influence of price value on purchase intention among patients with chronic diseases in medical e-commerce during the COVID-19 pandemic in China. *Frontiers in Public Health*, *11*, 1081196. PMID: 36817906

Haque, A., Bhushan, B., & Dhiman, G. (2021). Conceptualizing smart city applications: Requirements, architecture, security issues, and emerging trends. *Expert Systems: International Journal of Knowledge Engineering and Neural Networks*, 39.

Hayat, P. (2016). Smart cities: A global perspective. *India Quarterly*, *72*(2), 177–191.

Henseler, J., Ringle, C. M., & Sarstedt, M. (2015). A new criterion for assessing discriminant validity in variance-based structural equation modeling. *Journal of the Academy of Marketing Science*, *43*, 115–135.

Herath, H. M. K. K. M. B., & Mittal, M. (2022). Adoption of artificial intelligence in smart cities: A comprehensive review. *International Journal of Information Management Data Insights*, *2*(1), 100076.

Ismagilova, E., Hughes, D., Dwivedi, Y., & Raman, K. (2019). Smart cities: Advances in research - An information systems perspective. *International Journal of Information Management*, *47*, 88–100.

Jabbar, W. A., Kian, T. K., Ramli, R. M., Zubir, S. N., Zamrizaman, N. S., Balfaqih, M., & Alharbi, S. (2019). Design and fabrication of smart home with internet of things enabled automation system. *IEEE Access : Practical Innovations, Open Solutions*, *7*, 144059–144074.

Jiang, L., Jun, M., & Yang, Z. (2016). Customer-perceived value and loyalty: How do key service quality dimensions matter in the context of B2C e-commerce? *Service Business*, *10*(2), 301–317.

Kim, H. W., Chan, H. C., & Gupta, S. (2007). Value-based adoption of mobile internet: An empirical investigation. *Decision Support Systems*, *43*(1), 111–126.

Kock, N. (2015). Common method bias in PLS-SEM: A full collinearity assessment approach. [ijec]. *International Journal of e-Collaboration*, *11*(4), 1–10.

Kumar, T. V., & Dahiya, B. (2017). Smart economy in smart cities. In *Smart Economy in Smart Cities* (pp. 3–76). Springer.

Leow, N. X., & Krishnaswamy, J. (2021). Smart cities need environmental consciousness and more social responsibilities as an outcome of COVID-19 – reflections from urban road commuters. *Foresight, ahead-of-print*(ahead-of-print). https://doi .org/DOI: 10.1108/fs-02-2021-0035

Lim, Y., Edelenbos, J., & Gianoli, A. (2019). Identifying the results of smart city development: Findings from systematic literature review. *Cities (London, England)*, *95*, 102397.

Lin, C., Zhao, G., Yu, C., & Wu, Y. J. (2019). Smart city development and residents' well-being. *Sustainability*, *11*(3), 676.

Malik, H., Yong, H., & Mohamad, Z., CharlesRamendranS.P., R., Yip, C., Fong, C., Rehman, M., & Zailani, S. (2022). Sustainable Smart Homes and Community Happiness in the Malaysian Context. *International Journal of Asian Business and Information Management*, *13*, 1–18.

Merhi, M., Hone, K., & Tarhini, A. (2019). A cross-cultural study of the intention to use mobile banking between Lebanese and British consumers: Extending UTAUT2 with security, privacy and trust. *Technology in Society*, *59*, 101–151.

Mohamad, Z., Nee, A., Yang, F., Rehman, M., & Yin, Y. (2019). *Achieving Community Happiness through Affordable Eco-Friendly Smart Houses*. KnE Social Sciences.

Nitzl, C., Roldan, J. L., & Cepeda, G. (2016). Mediation analysis in partial least squares path modeling: Helping researchers discuss more sophisticated models. *Industrial Management & Data Systems*, *116*(9), 1849–1864.

Ogiemwonyi, O., Alam, M. N., Alshareef, R., Alsolamy, M., Azizan, N. A., & Mat, N. (2023). Environmental factors affecting green purchase behaviors of the consumers: Mediating role of environmental attitude. *Cleaner Environmental Systems*, *10*, 100130.

Pratama, A. B., & Imawan, S. A. (2019). A scale for measuring perceived bureaucratic readiness for smart cities in Indonesia. *Public Administration and Policy*, *22*(1), 25–39.

Puron-Cid, G., & Gil-García, J. (2022). Are Smart Cities Too Expensive in the Long Term? Analyzing the Effects of ICT Infrastructure on Municipal Financial Sustainability. *Sustainability*.

Rihar, M., Hrovatin, N., & Zorič, J. (2015). Household valuation of smart-home functionalities in Slovenia. *Utilities Policy*, *33*, 42–53.

Shamsuzzoha, A., Nieminen, J., Piya, S., & Rutledge, K. (2021). Smart city for sustainable environment: A comparison of participatory strategies from Helsinki, Singapore and London. *Cities (London, England)*, *114*, 103194.

Venkatesh, V., Thong, J. Y., & Xu, X. (2012). Consumer acceptance and use of information technology: Extending the unified theory of acceptance and use of technology. *Management Information Systems Quarterly*, •••, 157–178.

Vimalkumar, M., Sharma, S. K., Singh, J. B., & Dwivedi, Y. K. (2021). 'Okay google, what about my privacy?': User's privacy perceptions and acceptance of voice based digital assistants. *Computers in Human Behavior*, *120*, 106763.

Wirth, J., Maier, C., Laumer, S., Oehlhorn, C. E., & Weitzel, T. (2020, June). Benefits in privacy research: a literature review, status quo and future research directions. In *Proceedings of the 2020 on Computers and People Research Conference* (pp. 135-141).

Yang, W., & Lam, P. (2021). Evaluating non-market costs of ICT involving data transmission in smart cities. *Building Research and Information*, *49*, 715.

Zhao, S., & Chen, L. (2021). Exploring Residents' Purchase Intention of Green Housings in China: An Extended Perspective of Perceived Value. *International Journal of Environmental Research and Public Health*, 18. PMID: 33924311

Chapter 10
Role of Next Generation Advanced Wireless Techniques for Disaster Management in Smart Cities

Pankaj Kumar
Manipal Institute of Technology, India

Sagnik Bhattacharyya
https://orcid.org/0000-0003-4315-8281
Indian Institute of Technology, Ropar, India

Sam Darshi
https://orcid.org/0000-0001-6254-0270
Indian Institute of Technology, Ropar, India

ABSTRACT

The evolution of smart cities, driven by 5G and B5G, enhances quality of life and sustainability but faces challenges in disaster management. This chapter explores solutions using advanced communication technologies to ensure operational continuity post-disaster. Device-to-device (D2D) communication facilitates crucial message exchange between victims and relief teams without relying on base stations. Drone-assisted D2D cooperative communication (DA-DD-CC) improves reliability by incorporating drones as relays, enabling direct, drone-assisted, or combined links. Rate-splitting multi-user cooperation by drone (RSMUCD) integrates rate-splitting multiple access (RSMA) with multi-antenna drones to mitigate interference,

DOI: 10.4018/979-8-3373-0154-9.ch010

optimize bandwidth, and boost reliability. This framework examines multi-cluster DA-DD-CC scenarios and evaluates RSMUCD against non-orthogonal multiple access (NCSD) and network-coded cooperation (NCC) schemes using metrics like outage probability, data rate, and throughput. Findings demonstrate significantly enhanced performance for disaster management in smart cities.

INTRODUCTION

In recent years, the fifth generation (5G) and beyond 5G (B5G) technologies have seen major advancements which have been aimed at increasing the data rates and network capacity for traditional cellular networks. Systems like Multiple Input Single Output (MISO), Multiple Input Multiple Output (MIMO), Multi-User Multiple Input Single Output (MU-MISO) are employed to achieve these goals through several techniques including network traffic offloading (Choi et al., 2017; Phunchongharn et al., 2013; Vahidi and Saberinia, 2019). These systems require multiple antennas at the end nodes. However, adding multiple antennas requires additional hardware, which increases cost and size. As a result, these methods are generally unsuitable for battery-operated tinny devices. To improve cellular network performance, traffic offloading techniques are also discussed in the literature. Traffic offloading is primarily achieved through two methods: (i) *deploying small cells* (micro, pico, or femto) within macro-cells, and (ii) *utilizing device-to-device (D2D) communication* (Asadi et al., 2014). The first method may be inconvenient due to limited flexibility caused by fixed access points, whereas the second approach allows for dynamic offloading with or without support from an infrastructure network. In D2D communication, two nearby devices set up a direct link for data transmission. This communication can occur in two modes: in-band and out-band. In-band and out-band modes are each further divided into two sub-modes: in-band includes underlay and overlay, while out-band includes controlled and autonomous modes (Tehrani et al., 2014). Among these four sub modes, underlay, overlay, and controlled modes require base station assistance as shown in Figure 1(a).

Figure 1. D2D communication assisted with base station, (b) D2D communication without the assistance of a base station and (c) D2D communication through a drone/UAV

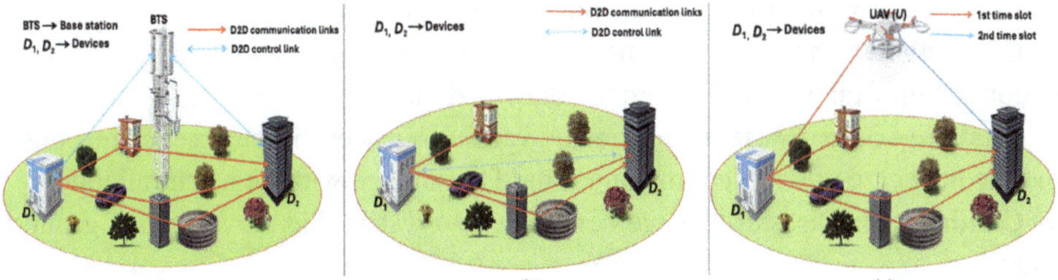

Whereas autonomous mode can operate independently without a base station (Fodor et al., 2014) as shown in Figure 1(b). Our proposal leverages this unique characteristic of the D2D technique. Communication systems often fail when the base station or eNB (evolved Node B) is disrupted following natural or human-made disasters. In such situations, the autonomous mode of the D2D technique can be utilized to establish communication links between nearby users. This type of setup can be beneficial in search and rescue missions during post disaster management operations. It facilitates seamless communication between rescue teams and victims, ultimately reducing the recovery process time. While numerous studies on device-to-device (D2D) communication exist, as discussed in (Schreck et al., 2018), the available studies have not fully explored the potential of autonomous D2D communication techniques.

D2D communication faces limitations such as device size, battery constraints, and channel randomness. To improve reliability and reduce deep fades, using UAVs or drones with cooperative communication (CC) techniques in D2D links is proposed. This approach aims to improve communication quality and resilience in the face of challenging conditions. This introduces an additional path for communication between devices, enhancing network performance and improving the accuracy of critical decisions through increased diversity. Therefore, to combine CC with D2D links in an autonomous mode for enhancing the reliability of smart cities in disaster scenarios a network is proposed as shown in Figure 1(c). UAVs and drones are emerging as pivotal use cases for next-generation wireless networks. By strategically adjusting the position of the drone, the likelihood of achieving a Line of Sight (LoS) can be significantly enhanced.

D2D communication aided by base stations can be essential in disaster management situations in smart cities to maintain connectivity in the event of partial or complete disruption of infrastructure. In case of infrastructure failure, a few of the primary problems during post disaster scenario are as follows:

Challenge 1: Reliability may decrease if the central node (base station) is damaged, causing ground users to lose connectivity with others.

Challenge 2: The damage to the central node may overload neighbouring base stations as they take on users from the damaged one.

Challenge 3: In disaster management, various services like emergency response and medical care are needed, making it hard to manage without a central node.

Challenge 4: Disaster management may also restrict the movement of ground users.

Challenge 5: Due to limited battery size, mobile users' devices have short battery life, and recharging may not be possible in disasters, requiring immediate rescue services.

Drone-assisted D2D communication can effectively address challenges in disaster management within smart cities by providing temporary infrastructure, improving connectivity, and enabling adaptability when conventional infrastructure fails.

The integration of next generation multiple access techniques such as Rate Splitting Multiple Access (RSMA) and non-orthogonal multiple access techniques (NOMA) in drone-assisted networks brings more crucial advantages. In this chapter, we propose the use of RSMA in disaster management environments. RSMA, as a foundation of 6G technology, will offer unique advantages in post-disaster scenarios where network conditions are highly dynamic and unpredictable. It enables the system to adapt to rapidly changing channel conditions and user densities that are typical in disaster-affected areas by intelligently splitting messages into common and private streams. This flexibility becomes crucial when emergency responders, survivors, and rescue teams all need reliable and fast connectivity despite varying signal strengths and interference levels.

In disaster scenarios, where power sources may be limited or unreliable, the ability of RSMA to dynamically allocate power between common and private streams ensures an extended operation time for emergency communication systems. This power efficiency, along with the inherent flexibility of RSMA in managing multiple users, makes it an ideal technology for maintaining sustainable communication networks during extended disaster recovery periods.

Furthermore, in post-disaster scenarios, where network infrastructure is often damaged or overloaded, the novel interference management technique employed by RSMA proves to be highly beneficial. When integrated with drone-based CC systems, RSMA enables efficient service to multiple vulnerable ground users simultaneously, even under challenging channel conditions due to debris, collapsed structures, or adverse weather. This capability makes RSMA particularly well-suited for coordinating rescue operations and ensuring reliable communication in areas where conventional networks are unavailable or compromised.

Now, when communication infrastructures like base stations are damaged in emergency situations like disaster management, earthquakes, and floods, an autonomous D2D communication mode becomes essential. This mode makes it possible to establish communication linkages without a base station, which makes disaster management procedures run more smoothly (Mukherjee et al., 2023). The motivation behind using drone as an aerial relay in D2D communications are listed as: 1) In such a situation the drone may play an important role for providing rapid response among the ground users by temporary deploying as an aerial base station. 2) By serving as an aerial relay, a drone can assist in establishing or restoring communication between affected communities and emergency agencies (Gohari et al., 2022). 3) Using drone as an aerial node guarantees the smooth real-time flow of vital information, including survivor locations and resource needs. 4) Drones can work in unreachable places that are hard or impossible for people to get to, like sites that have flooded, landslide, or collapsed (Kurt et al., 2021). 5) During and after disasters, drones can gather a wide variety of data, such as thermal images, topographical maps, and environmental conditions. 6) Drones can improve operational efficiency and situational awareness in emergency situations by integrating with the current smart city infrastructure, including IoT devices and data analytics systems (Balfaqih et al.,2023; Qu et al., 2023).

Overall, the integration of drones as aerial relays in smart city disaster management systems significantly enhances response capabilities, improves safety, and optimizes resource utilization during emergencies. Their versatility and adaptability make them indispensable tools in modern disaster management strategies. Based on these advantages we aim to use drone along with few modern technologies which would be supremely beneficial for disaster management.

This chapter focuses on overcoming the challenges associated with disaster management by utilizing next-generation communication technologies. To tackle these challenges effectively, it presents a range of potential solutions designed to maintain operational functionality and ensure the seamless continuation of post-disaster activities. The ***proposal includes the following key components***:

1. **Device-to-Device (D2D) communication:** This ensures crucial message exchange between victims and relief teams in scenarios where base station is unavailable.
2. **Drone-Assisted Device-to-Device Cooperative Communication (DA-DD-CC):** This approach utilizes the autonomous nature of D2D communication to establish links without relying on a central node (such as a base station). To enhance system reliability, drones facilitate CC within the D2D session.

The post-disaster services for victim users are delivered through three possible scenarios:

(a) Direct communication between victims and rescue teams without drone assistance.

(b) Communication facilitated solely by the drone between victims and rescue teams.

(c) A combined effect of drone assistance and direct signals from the rescue team to the victims.

3. **Rate Splitting Aided Multi-User Cooperation by Drone (RSMUCD) System:** This approach introduces the use of Rate Splitting Multiple Access (RSMA), a 6G technology, integrated with a multiantenna drone. The RSMUCD system provides multi-user cooperation to improve reliability and ensure robust post-disaster communication.

In the first case (case 2(a)), the communication links between ground-to-ground (G2G) users are strong enough for them to communicate with each other and with the rescue team.

In the second case (case 2(b)), the communication links among G2G users are sufficient, but for increased reliability and accuracy of services, the rescue team may deploy a drone acting as relay. The drone facilitates CC, requiring two time slots to receive two copies of the signal at the ground unit (GU). In the first time slot, the transmitter sends the signal to the affected user, which is also received by the drone. In the second time slot, the drone employs amplify-and-forward (AF) or decode-and-forward (DF) and broadcasts the processed signal to the GUs, which combine both copies to evaluate performance.

In the third case (case 2(c)), the direct communication links among the GUs may experience deep fading. Therefore, the GUs receive the necessary rescue information via the drone only.

In the fourth case (case 3), the focus is on further enhancing reliability by addressing the limitations of previously existing techniques and using a multi-antenna drone. Over the years, CC has evolved as a reliable technology to provide reliable data to users. This is evident through the development of PHY and MAC protocols for various CC scenarios (Malik et al., 2020; Sami et al., 2016). However, as the number of devices keeps increasing, CC faces challenges in meeting the requirements of next-generation systems. This is due to the reason that when extended for *N* source-destination pairs, a total of 2*N* timeslots are required for each pair to be served. This affects the throughput and becomes spectrally inefficient which is a crucial factor, especially in disaster/ emergency scenarios. This led to the development of network-coded cooperation

(NCC) which incorporates network coding (NC) to address the throughput concerns of CC while providing the same diversity. In an NCC-based system, the source nodes first transmit their data in the first *N* orthogonal consecutive time slots (i.e. following TDMA. It can also be transmitted following carrier sense multiple access/ collision avoidance (CSMA/ CA)) as depicted in Figure 2.

Figure 2. Multi-users-D2D communication scenario assisted by a UAV acting as the relay while employing NCC. At first, a TDMA scenario is illustrated (represented as solid arrows) following which the relay transmits superposed data following either AF, DF or hybrid relaying (represented by dashed arrows)

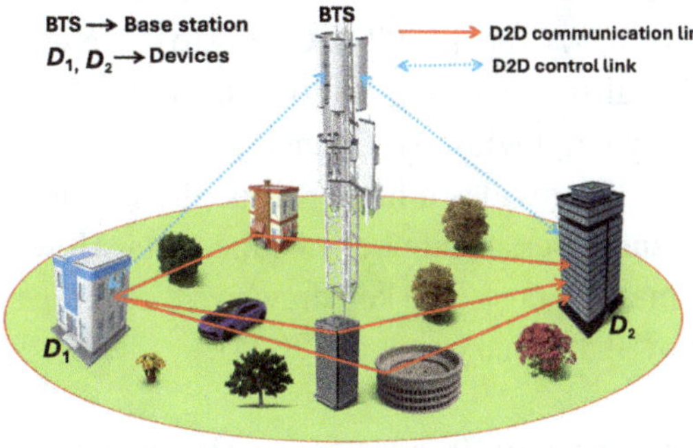

During this transmission, each destination and relay nodes work in promiscuous mode (i.e. overhears the data being transmitted by other source nodes even if it is not meant for that particular node). Once all the data from the source nodes have been transmitted, the relay transmits the superposed data in the final timeslot following any of the prevalent relaying schemes like AF, DF or a hybrid scheme (Bhattacharyya et al., 2021, 2023, 2024). Thus, compared to the 2*N* time-slots required in the case of CC, NCC requires a total of *N*+1 time-slots which improves the throughput to quite an extent. Owing to this benefit, considerable strides have been made in the domain of domain of NCC (Bhattacharyya et al., 2023, 2021a,b, 2024; Kumar et al., 2022). Since TDMA does not always provide a proper quality-of-service (QoS) as it transmits irrespective of the channel conditions, authors in (Bhattacharyya et al., 2023) proposed a cross-layer medium access control (MAC) protocol to exploit multi-user diversity (MUD) gain by letting the source nodes contend and transmit only when the channel condition is better between the source and the relay. This enhanced the outage performance however, this enhancement comes at the cost of a higher latency owing to the overhead and collisions due to contentions. To enhance the rate, a full duplex (FD) NCC system was proposed in (Bhattacharyya et al., 2021). However, the FD system is highly vulnerable to the decoding errors. Furthermore, all the previous works used to work on a fixed relaying scheme (i.e. sticking to either AF or DF). Now, AF enjoys a low system complexity however, comes at the cost of higher NC noise which is experienced at the end user. On the other hand, DF experiences a lower NC noise however, it is a devil in disguise. This is because improper decoding of data at the relay can lead to heavy error propagation. Thus, sticking to AF or DF blindly might not be beneficial always. This led to the development of a hybrid-combine-and-forward (HCF) relaying scheme in (Bhattacharyya et al., 2021) based on the SNR of the received signal at the relay. This led to the higher number of packets being successfully decoded at the destination nodes when compared to both AF or DF. The performance of the HCF scheme was further enhanced by the proposal of a MAC scheme which introduced re-transmissions by the source nodes (Bhattacharyya et al., 2024). Owing to these several benefits along with its simplicity, NCC also saw its applications in a drone-based scenario where drone was used as a relay employing AF (Kumar et al., 2022). It is clear from the above discussion that NCC provides a higher throughput than CC based systems owing to the requirement of lesser number of time-slots. However, the enhancement in throughput comes at a cost. The key drawbacks of these approaches can be summarized as follows (Bhattacharyya et al., 2024):

1. **Unreliable Communication:** Broadcast scenarios are unreliable due to lack of acknowledgments, worsened by the broadcast storm problem causing collisions and congestion. Adding acknowledgments would also hurt latency, critical in disaster scenarios.
2. **Rate Limitation:** During or after a disaster, both connectivity and data rate are critical to saving lives. In broadcast scenarios, channel conditions between the source and each destination node limit performance. To ensure all nodes receive data, the transmission rate is restricted to that of the weakest node, reducing overall efficiency.
3. **Dependence on Overheard Data:** In NCC, nodes use promiscuous mode to overhear data for a second copy. However, errors in this data can propagate, worsening outage performance and endangering lives in disaster scenarios.
4. **Error Propagation:** Imperfect Channel State Information at the Receiver (CSIR) can cause error propagation, reducing reliability and performance, especially in dynamic or adverse conditions.

These raise significant concerns about the reliability offered by NCC in meeting the stringent QoS demands of a disaster-struck area. As saving human lives is of utmost importance in such cases, ensuring the seamless exchange of safety messages at high data rates with enhanced reliability becomes crucial in these challenging conditions which NCC cannot meet. Consequently, it becomes imperative to incorporate next-generation multiple access technologies (NG-MATs) to effectively save lives in such dynamic scenarios. The incorporation of NG-MATs like space division multiple access (SDMA) (Wu and Dai, 2023), non-orthogonal multiple access (NOMA) (Liu et al., 2022) and rate splitting multiple access (RSMA) (Mao et al., 2022; Naser et al., 2020) would be highly beneficial in such cases. To do so a novel system, rate splitting aided multi-user cooperation by drone (RSMUCD) has been proposed and discussed further in the next section.

To provide solutions for post-disaster management, we thus present a robust framework aimed at supporting sustainable growth in smart city infrastructure. This framework delivers a comprehensive analysis that is crucial for ensuring the future success of smart cities. Additionally, a multi-cluster DA-DD-CC scenario is meticulously examined to validate the effectiveness of the proposed approach. However, they face certain disadvantages as discussed above. Building on this, we evaluate the performance of the proposed RSMUCD system, comparing it against other prevalent schemes, such as a Non-Orthogonal Multiple Access-based Cooperation Scheme aided by drone (NCSD) and Network-Coded Cooperation (NCC) employing two different relaying strategies. Key performance metrics, including outage probability, data rate, and throughput, are thoroughly analysed. The findings from this investigation not only justify the proposed work but also provide actionable insights for the optimal selection of critical system parameters, ensuring significantly enhanced performance during post-disaster scenarios.

Chapter Organization

In Section 2 the system model is described. Initially the DA-DD-CC system is proposed with AF and DF relaying at the drone, respectively in consequent sub-sections. In the following sub-section, the RSMUCD system is proposed and discussed in depth following which the transmitter and receiver architectures are provided along with the respective SNR/ SINR equations. The performance metrics considered for the work are discussed in details in Section 3. Finally, the results obtained are discussed in Section 4 while Section 5 concludes the chapter.

SYSTEM MODEL

This section explores communication among devices, specifically ground users (GUs), within smart cities for disaster management scenarios. In such critical situations, traditional communication infrastructure may be partially or entirely disrupted due to disasters such as earthquakes, floods, or storms. To address this, drones or unmanned aerial vehicles (UAVs) have emerged as pivotal tools for establishing and maintaining communication links between GUs and rescue teams, ensuring uninterrupted flow of information and coordination. Beyond communication, drones can serve multiple roles, such as delivering food packets, medical supplies, and other essential resources to ground users based on their specific needs. This dual role of drones not only enhances their utility in disaster-stricken areas but also significantly contributes to improving the overall effectiveness of disaster response operations. While direct communication links among ground users may suffice in certain cases, the presence of obstacles, interference, or long distances can degrade the reliability of these links. To overcome these challenges and ensure the QoS, rescue teams may deploy a drone or UAV to function as an aerial relay. Acting as a communication intermediary, the drone facilitates CC among devices, thereby enhancing the reliability and robustness of the network. In this CC setup, two time slots are typically required for the ground users to receive two copies of the desired signal. In the first time slot, the transmitter GU sends the desired signal, x to the affected disaster GU. This signal is also received by the drone during the first time slot. After receiving the signal, the drone can use either amplify-and-forward (AF) or decode-and-forward (DF) for processing. In the second time slot, the processed signal is broadcast by the drone to the GUs. Once the GUs receive both the copies, one directly from the transmitter GU and one from the drone. The receiver GU combine these copies to evaluate performance.

Analysis of DA-DD-CC System Using AF Relaying at Drone

Here, the receiver GU received the desired copy sent by the transmitting GU in two time slots. In the first time slot, both drone and receiver GU receive data from transmitting GU. After that, the drone performs AF operation over the signal received in the first time slot from the GU. In the second time slot, drone broadcast the amplified data towards the receiver GUs. For the sake of mathematical formulation, we denote transmitter GU, drone, and receiver GU as S, R, and D, respectively. The information obtained during the first time slot at the R and any receiver node D can be expressed as

$$Y_{SR} = \sqrt{P_S d_{SR}^{-\alpha}}\, h_{SR} x + \eta_R \tag{1}$$

$$Y_{SD} = \sqrt{P_S d_{SD}^{-\alpha}}\, h_{SD} x + \eta_D \tag{2}$$

where Y_{ab} represents the signal transmitted from node a and received at node b. Power transmitted by the S can be denoted by P_S, h_{ab} represents the channel coefficients between a and b nodes. Considering zero mean and equal variance (σ^2), the background noise at the node R and D are denoted as η_R and η_D, respectively. The variable α represents the path loss exponent.

Using AF Relaying at Drone

The data received from node S is amplified by node R and sent to node D in the second time slot, which can be expressed as

$$Y_{RD} = Q\sqrt{d_{RD}^{-\alpha}}\, h_{RD}\, Y_{SR} + \eta_D \tag{3}$$

where Q indicates the amplification factor given as (Kumar et al., 2021),

$$Q = \sqrt{\frac{P_R}{P_S d_{SR}^{-\alpha} h_{SR}^2 + \sigma_R^2}} \tag{4}$$

where P_R represents the power transmitted by the node R and σ_R^2 denotes variance of background noise at node R. Substituting Y_{SR} from (1) to (3), and rewritten Y_{RD} as given below

$$Y_{RD} = Q\sqrt{d_{RD}^{-\alpha}}\left(\sqrt{P_S d_{SR}^{-\alpha}}\, h_{SR} x + \eta_R\right) h_{RD} + \eta_D \tag{5}$$

After simplification, (5) can be further written as

$$Y_{RD}^{AF} = \underbrace{Q\sqrt{d_{RD}^{-\alpha}}\left(\sqrt{P_S d_{SR}^{-\alpha}}\, h_{SR} x\right) h_{RD}}_{\text{1st term}} + \underbrace{Q\sqrt{d_{RD}^{-\alpha}}\, \eta_R h_{RD} + \eta_D}_{\text{2nd term}} \tag{6}$$

In (6), the first term denotes the signal of the second copy of the desired signal (first copy being the one received directly from the source as shown in (2)) and second term represents the combined effect of both AF-noise (amplified by the relay while amplifying the signal and is a part of AF relaying) and background noise at node D, denoted by ξ_{AF} and can be rewritten as,

$$\xi_{AF} = Q\sqrt{d_{RD}^{-\alpha}}\, \eta_R h_{RD} + \eta_D \tag{7}$$

Using DF Relaying at Drone

This is the second approach used at the drone for processing the received signal from the transmitting GU in the first time slot. In DF, the signal received at the drone is decoded, re-encoded, and then re-transmitted to the node D.

At the R and D nodes, the received data in the first time-slot are defined as:

$$Y_{SR} = \sqrt{P_S d_{SR}^{-\alpha}}\, h_{SR} x + \eta_R \tag{8}$$

$$Y_{SD} = \sqrt{P_S d_{SD}^{-\alpha}}\, h_{SD} x + \eta_D \tag{9}$$

In second time-slot, as per the DF scheme, R decodes the received signal, re-encodes it and forward the encoded signal towards the node D and can be written as:

$$Y_{RD}^{DF} = \sqrt{P_R d_{RD}^{-\alpha}}\, h_{RD} x + \eta_D \tag{10}$$

Analysis of Rate Splitting Aided Multi-User Cooperation by Drone

Over the years, various MATs have been developed for different systems. Systems like SDMA have shown promising results which uses precoding in multi-antenna systems like MIMO, MU-MISO etc to enhance the system performance. In under-loaded conditions (i.e. the number of transmit antennas at the transmitter is greater than the number of receiving antennas) when perfect channel state information at the transmitter (CSIT) is available, SDMA can successfully suppress or eliminate multi-user interference if the channels not correlated (Mao et al., 2022). However, as discussed in (Mao et al., 2022), SDMA is highly sensitive to network load and is suitable for underloaded systems. For overloaded systems (i.e. for a MU-MISO case, the number of single antenna receivers are larger than the number of transmit antennas), the performance of SDMA drops significantly. Furthermore, SDMA is sensitive to user deployment angles, channel correlation and CSIT inaccuracy. This imposes several stricter requirements which is a headache for network developers. Hence, SDMA works well in low interference scenarios with perfect CSIT. However, as the number of devices keep rising, overloaded systems are often seen in practical scenarios where SDMA will not perform well.

On the other hand, technologies like NOMA have been developed for single-input single-output (SISO) broadcast channel (BC) systems. The primary advantage of NOMA lies in its ability to enhance spectral efficiency (SE) in highly overloaded scenarios by simultaneously serving users with differing channel strengths within the same time-frequency resources. Motivated by the advantages of NOMA over orthogonal multiple access (OMA) schemes in SISO networks, significant efforts have been dedicated in recent years to integrating NOMA with multi-antenna networks like MU-MISO, MIMO (Shim et al., 2019; Ding et al., 2016; Liu et al., 2022). However, as the number of users increases, the performance of NOMA starts taking a hit. For example, considering a N user SISO BC system, the user with the strongest channel among N users has to undergo N-1 successive interference cancellations (SIC) to decode and remove interference from the previous N-1 messages. Now, in such cases, SIC error propagation is highly aggravated. Furthermore, multi-antenna NOMA systems like MISO BC NOMA may suffer from a loss of degree of freedom (DoF) due to the inefficient usage of SIC (Mao et al., 2022). Along with this, multi-antenna NOMA can introduce substantial computational complexity at both the transmitter and receivers which will be unhelpful in disaster-based scenarios as communication link are to be established for the vulnerable humans as long as possible. Another matter of concern is the CSI inaccuracy which would be detrimental in multi-antenna NOMA cases. Hence, NOMA might show a better result in extremely high interference scenarios while its performance drop in medium or low interference scenarios.

The limitations of existing multiple access (MA) schemes as discussed before, stem primarily from their approach to managing interference. According to the authors in (Etkin et al., 2008), treating interference purely as noise is effective only when multi-user interference is weak, while fully decoding interference is recommended only when it is sufficiently strong. In scenarios with moderate interference (i.e. neither too strong nor too weak), partially decoding the interference provides significant performance gains. Therefore, it may be advantageous to dynamically adjust the proportion of interference that can be decoded with respect to the portion that is treated as noise. This approach allows for efficient interference management and enhances the system performance significantly. This adaptive strategy as discussed in (Etkin et al., 2008) represents a key insight, offering valuable guidance for the design and optimization of future MA schemes. Now, since SDMA and NOMA are optimally suited for handling weak and strong interference levels, respectively, to enable adaptive interference management, one straightforward approach is to dynamically alternate between SDMA and NOMA, as suggested in (Chen et al., 2016). However, this method restricts the system to supporting either SDMA or NOMA at a given time, making it ineffective in scenarios with medium interference levels. To handle such scenarios, a new MA scheme has been developed and has been attracting the attention of researchers. Rate-Splitting Multiple Access (RSMA) has recently emerged as a leading candidate for beyond 5G (B5G) and 6G networks, drawing attention for its potential to deliver higher data rates and enhanced resilience, especially in scenarios with imperfect Channel State Information (CSI) (Clerckx et al., 2023; Mao et al., 2018). RSMA builds on concepts from SDMA and NOMA, creating a versatile communication framework that balances interference management and decoding complexity. By employing both interference cancellation and noise-tolerant methods, RSMA adapts to a broad spectrum of interference levels, effectively managing inter-user interference while optimizing data throughput.

The backbone of RSMA lies in its approach to data splitting. In the 1-layer RSMA framework, each user's data is split into two sub-messages: a *common* and a *private*. The common sub-messages of all users are concatenated together into a single common message that must be decoded by all the users, while the private sub-messages are maintained as separate, individual messages designated for specific users. This division allows RSMA to control interference more flexibly. Now, power allocation is central to the effectiveness of RSMA. Typically, a larger share of power is assigned to the common stream to ensure reliable primary decoding across all the users, while the private streams remain active as background interference during this stage. Once the common stream is decoded, the system applies SIC, which partially removes interference by decoding stream the common stream. Each user then decodes their own private stream, treating any residual interference from other users' private streams as noise.

This sophisticated decoding strategy enables RSMA to dynamically adapt to varying interference levels, thereby optimizing system performance in diverse environments. By blending interference management and data separation, RSMA positions itself as a robust solution that meets the high throughput, and flexible interference handling requirements of disaster-stricken scenarios. This motivates us to incorporate RSMA as a relaying scheme which would be employed by the drone and thus, remove the dependency of overheard signals as can be seen in Figure 3.

Figure 3. Illustration of the proposed RSMUCD system where N S_n-D_n pairs are aided by a multi-antenna drone/ UAV. The source nodes transmit following TDMA protocol (represented through solid arrows) while in the final time-slot, the relay transmits RSMA data (represented by dashed arrow)

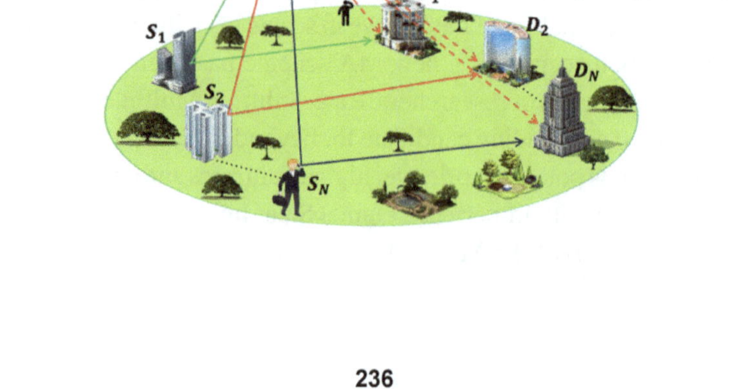

Figure 4. Internal architecture of a drone acting as the relay and employing RSMA. The N scheduled data were obtained over N time-slots from the respective source nodes following which it decoded the signals received

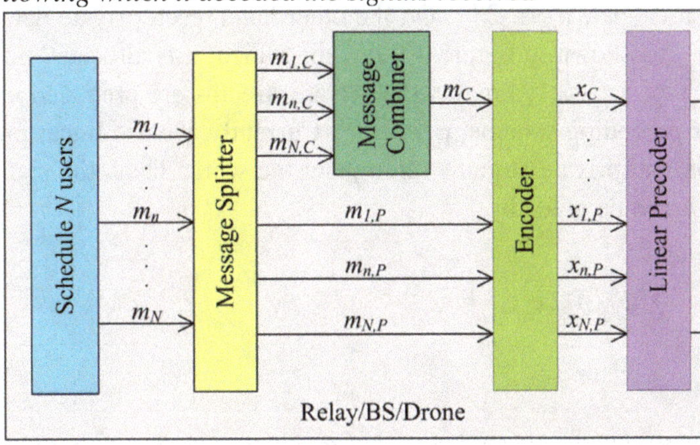

Introduction and Analytical Framework of RSMUCD

In the aftermath of/ during a disaster, effective communication is the lifeline of relief efforts, and organizing nodes into source-destination pairs proves to be a transformative approach. By mimicking the real-world urgency of connecting rescue teams to command centers or aid suppliers to distribution hubs, this strategy streamlines the network, reduces interference, and ensures that critical messages are delivered with precision when they are needed most. Let us consider N single antenna source-destination pairs (S_n-$D_n \forall n \in \{1,...,N\}$) aided by relay, R equipped with M transmitting antennas as shown in Figure 3. Each S_n has message m_n (modulated to x_n) to be transmitted by user-n. All the users follow TDMA protocol and transmit their data which is received by the respective destination, D_n and R. Thus, the signals received at D_n and R can be given as:

$$Y_{S_nD_n} = \sqrt{P_{S_n} d_{S_nD_n}^{-\alpha}}\, h_{S_nD_n} x_n + \eta_{D_n} \tag{13}$$

$$Y_{S_nR} = \sqrt{P_{S_n} d_{S_nR}^{-\alpha}}\, h_{S_nR} x_n + \eta_R \tag{14}$$

where η_{Dn}, η_R represents the additive white gaussian noise (AWGN) with 0 mean and variance $\sigma_{D_n}^2$ and σ_R^2, respectively at D_n and R. On receiving the signal, R decodes the messages to obtain $\widehat{m}_n$. Assuming perfect decoding, $\widehat{m}_n = m_n$. Message m_n is then split into the common and the private part given as, $m_{n,C}$ and $m_{n,P}$, respectively. Each common part, $m_{n,C}$ is then concatenated with each other to form the overall common data, $m_C=[m_{1,C},...,m_{N,C}]$ as shown in Figure 4. Following this, m_C is modulated to form the common stream x_C, allocated appropriate power ϕ_C and precoded using an appropriate precoder, $\mathbf{p}_C \in \mathbb{C}^{M\times 1}$. On the other hand, each private data, $m_{n,P}$ is modulated to $x_{n,P}$ as shown in Figure 4 following which, it is allocated power $\phi_{n,P}$ such that $\phi_C+ \sum_{n=1}^{N}\phi_{n,P} \leq 1$. Finally, the private streams are precoded using appropriate linear precoding scheme, $\mathbf{p}_n \in \mathbb{C}^{M\times 1}$. Currently, we use linear precoding for the common and private streams throughout the work. Thus, the signal to be transmitted by R can be given by,

$$X_R = \phi_C \mathbf{p}_C x_C + \sum_{n=1}^{N} \phi_{n,P} \mathbf{p}_{n,P} x_{n,P} \tag{15}$$

Figure 5. Internal architecture of a user decoding the required data in a RSMUCD system when the drone transmits RSMA data

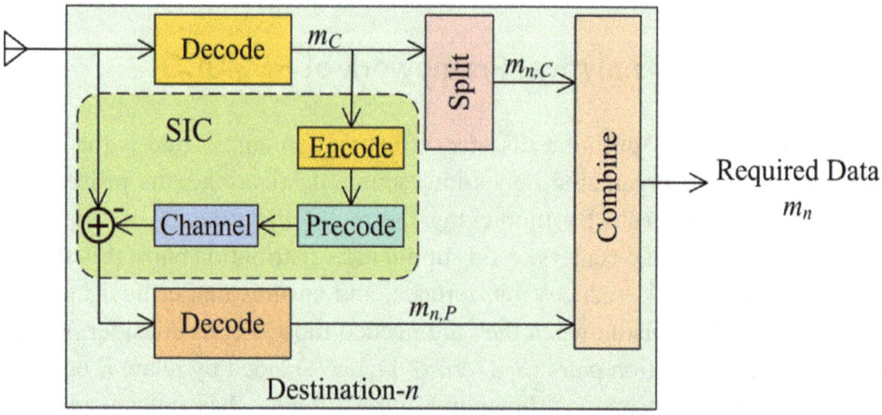

Finally, considering $\mathbf{H}_{RD}=[\mathbf{h}_{RD1}, \mathbf{h}_{RD2},..., \mathbf{h}_{RDn},..., \mathbf{h}_{RDN}]^H$ where $\mathbf{H}_{RD} \in \mathbb{C}^{N\times M}$ as the multi-user multiple input single output (MU-MISO) channel matrix between the multi-antenna drone and the single antenna destinations, the signal received at D_n can be given as,

$$Y_{RD_n} = \sqrt{P_R d_{RD_n}^{-\alpha}}\, \mathbf{h}_{RD_n} X_R + \eta_{D_n} = \sqrt{P_R d_{RD_n}^{-\alpha}}\, \mathbf{h}_{RD_n} \left(\phi_C p_C x_C + \sum_{n=1}^{N} \phi_{n,P} \mathbf{p}_{n,P} x_{n,P} \right) + \eta_{D_n} \tag{16}$$

where $\mathbf{h}_{RDn} = [h_{n1},...,h_{nM}] \in \mathbb{C}^{1\times M}$ is the channel vector between nodes R and D_n and h_{nM} is the individual channel between user D_n and M^{th} antenna of R. On receiving the signal, it is decoded. On decoding the common stream, m_C is obtained as it is allocated a higher power. From m_C the required part of the common stream can be obtained which is $m_{n,C}$ in this case. Following this, the common data is re-modulated, re-precoded and re-allocated power and SIC is performed and the required private data is then decoded as shown in Figure 5. On obtaining the required private data, $m_{n,P}$ it is then concatenated with $m_{n,C}$ to form m_n. Hence, using (16) the signal-to-interference noise ratio (SINR) of the common and private streams at D_n can be respectively given as:

$$SINR_C = \frac{P_R \phi_C d_{RD_n}^{-\alpha} \left| \boldsymbol{h}_{RD_n} \boldsymbol{p}_C \right|^2}{P_R (1 - \phi_C) d_{RD_n}^{-\alpha} \sum_{i=1}^{N} \left| \boldsymbol{h}_{RD_n} \boldsymbol{p}_i \right|^2 + \sigma_{D_n}^2} \tag{17}$$

$$SINR_{n,P} = \frac{P_R \phi_{n,P} d_{RD_n}^{-\alpha} \left| \boldsymbol{h}_{RD_n} \boldsymbol{p}_n \right|^2}{P_R d_{RD_n}^{-\alpha} \sum_{i=1, i \neq 1}^{N} \phi_{i,P} \left| \boldsymbol{h}_{RD_n} \boldsymbol{p}_i \right|^2 + \sigma_{D_n}^2} \tag{18}$$

Now, for the work we have considered zero forcing (ZF) for the private streams while singular value decomposition (SVD) is considered for the common stream. Now, utilising the concepts of SVD, the channel matrix can be rewritten as, $\mathbf{H}_{RD}$= [$\mathbf{h}_{RD1}$,$\mathbf{h}_{RD2}$,...,$\mathbf{h}_{RDn}$,...,$\mathbf{h}_{RDN}$]^{H}=$\mathbf{U}\Sigma V^{H}$. From this, we obtain $\mathbf{p}_C$=$\mathbf{U}$(:,1) where $\mathbf{U}$(:,1) is the largest left singular vector of the channel matrix, $\mathbf{H}_{RD}$. Such a precoding for the common stream has been done in (Mao et al., 2018). Furthermore, since ZF is being employed, the common stream will experience interference due to its own private stream while the interference due to the other private streams have been nullified. This is because ZF typically decouples the channels into multiple independent single-user channels and thus, nullifies the interference (Fatema et al., 2018). Thus, $\mathbf{h}_{RDn}\mathbf{p}_i=0 \ \forall \ i \neq n$ where $\mathbf{p}_i$ is the ith column of the private stream precoding matrix, $\mathbf{P}$ formed by the pseudo-inverse of $\mathbf{H}_{RD}$ which can be given as $\mathbf{P}$ (Bhattacharyya et al., 2025). Based on these discussions, (17) and (18) can be re-written as:

$$SINR_C = \frac{P_R \phi_C d_{RD_n}^{-\alpha} \left| \boldsymbol{h}_{RD_n} \boldsymbol{p}_C \right|^2}{P_R \phi_{n,P} d_{RD_n}^{-\alpha} \left| \boldsymbol{h}_{RD_n} \boldsymbol{p}_n \right|^2 + \sigma_{D_n}^2} \tag{19}$$

$$SINR_{n,P} = \frac{P_R \phi_{n,P} d_{RD_n}^{-\alpha} \left| \boldsymbol{h}_{RD_n} \boldsymbol{p}_n \right|^2}{\sigma_{D_n}^2} \tag{20}$$

Thus, it can be observed here that two copies of the same data are being received here and the second copy of the received data is not dependent on any overheard signal unlike NCC as shown in Figure 3. Finally, using the value of (19) and (20) along with the SNR of the copy directly received at D_n when source S_n transmitted, $\mathrm{SNR}_{Sn,Dn}$, the best of the two copies is selected. Hence, the signal providing higher mutual information can be selected thus, enhancing the performance of the system and providing the vulnerable users affected from disaster a reliable data in the time of need.

In the following section we discuss the evaluation metrics considered for the proposed DA-DD-CC and RSMUCD systems.

PERFORMANCE EVALUATION METRICS AT THE GROUND USERS

The proposed scenarios can be evaluated using various performance metrics like outage probability, data rate, spectral efficiency, throughput, etc. Out of these metrics, outage probability, data rate and throughput are used in this discussion for evaluating the performance for measuring the quality-of-services (QoS) at the ground users. First we illustrate the same for the proposed DA-DD-CC system following which we discuss the formulas used for obtaining results of the proposed RSMUCD system.

Performance Evaluation Metrics for DA-DD-CC

Data Rate

Another metric used to measure QoS is the data rate. It is a function of SN-R_{Received} along with the available bandwidth and can be mathematically defined as (Bhattacharyya et al., 2023; Kumar and Majhi, 2024) below:

$$\mathscr{R} = B \times \log_2\left(1 + SNR_{Received}\right) \tag{21}$$

where B denotes the system bandwidth associated with wireless network.

Outage Probability

It is used for measuring the QoS at the ground users. It is defined as probability that the received SNR ($SNR_{Received}$) is less than the pre-defined SNR threshold ($SNR_{Pre\text{-}defined}$) at the receiving nodes and can be given as (Bhattacharyya et al., 2024):

$$P_{out} = P\left(SNR_{Received} \leq SNR_{Pre-defined}\right) \tag{22}$$

where the value of the received signal-to-noise ratio ($SNR_{Received}$) varies depending on the specific scenario considered, as well as the relaying scheme employed. For instance, when AF scheme is used by R, the value of $SNR_{Received}$ can be determined using (6) if only the relay path is available. On the other hand, when DF scheme is employed and is the only path available, the $SNR_{Received}$ can be obtained using equation (10). Furthermore, as in the first scenario, if only the direct path is available, $SNR_{Received}$ can be found using (2) or (9). Furthermore, if both the direct and relayed copies of the signal are available like the one discussed before, the overall $SNR_{Received}$ will be determined by the diversity combining technique applied, which combines the two copies to enhance the reliability of the received signal.

Throughput

Throughput is another important metric using which the QoS of a system can be measured. It can be given as:

$$T = \mathscr{R} \times (1 - P_{out})$$

Performance Evaluation Metrics for RSMUCD

The performance evaluation of RSMUCD using various metrics is of prime importance specially when it comes to the safety of human life in the aftermath of a disaster/ during a disaster. In such a case, metrics like data rate, outage probability and throughput are of supreme importance to evaluate the QoS being provided by the system.

Data Rate

Using (19) and (20), the instantaneous rates of the common and private streams at D_n can be given as:

$$\mathscr{R}_{C,n} = B \times \log_2(1 + SINR_C) \tag{23}$$

$$\mathscr{R}_{P,n} = B \times \log_2(1 + SINR_{P,n}) \tag{24}$$

Furthermore, since the common stream is to be decoded at all the users, the achievable rate is upper limited due to channel constrains and channel capacity. Thus, the achievable rate of the common stream can be represented as,

$$\mathscr{R}_C = min\{\mathscr{R}_{C,1}, \ldots, \mathscr{R}_{C,n}, \ldots, \mathscr{R}_{C,N}\} \tag{25}$$

out of which only a certain part is required by D_n. Hence, $\mathscr{R}_C$ can be further given as $\mathscr{R}_C = \sum_{i=1}^{N} \mathscr{C}_i$, where $\mathscr{C}_{i=n}$ is the part of rate R_C allocated for D_n. On the other hand, the instantaneous rate of the private stream is the achievable rate as the private streams are to be decoded at the individual users only. Thus, the achievable rate of a nth vulnerable user in danger through the relay can be given as:

$$\mathscr{R}_{R,D_n} = \mathscr{C}_n + \mathscr{R}_{P,n}$$

Furt e more, using (13) the rate of the signal received directly by D_n can be given is: (26)

$$\mathscr{R}_{S_n,D_n} = B \times \log_2\left(1 + SINR_{S_n,D_n}\right) \tag{27}$$

where $SNR_{S_n,D_n} = \frac{P_{S_n} d_{S_n D_n}^{-\alpha} \left|h_{S_n D_n}\right|^2}{\sigma_{D_n}^2}$. Now, out of the available copies shown using (24) and (25), the copy with the higher mutual information is selected. Thus, the overall achievable rate of D_n can be given as:

$$\mathscr{R}_{D_n} = \frac{1}{N+1} \max\left(\mathscr{R}_{S_n,D_n}, \mathscr{R}_{R,D_n}\right) \tag{28}$$

Outage Probability

As discussed earlier, outage probability is the probability with which the SNR of a signal is below the decoding threshold. Thus, the outage probability can be calculated as:

$$P_{out} = \mathrm{P}\left(\mathscr{R}_{D_n} \leq \mathscr{R}_{Th}\right) = \mathrm{P}\left(\frac{1}{N+1} \max\left(\mathscr{R}_{S_n,D_n}, \mathscr{R}_{R,D_n}\right) \leq \mathscr{R}_{Th}\right) \tag{29}$$

where $\mathscr{R}_{Th}$ is the threshold rate for correctly decoding a signal. The factor N+1 comes from the fact that a total of N+1 time-slots are required for the entire process which includes transmission by N source nodes followed by the transmission of RSMA data by the relay.

Throughput

Apart from data rate and outage probability, throughput is another important metric which can be used to illustrate the QoS of a system. The throughput is the amount of data/ number of bits received correctly and can be calculated as a function of the rate and the outage probability. Thus, the throughput can be calculated as follows:

$$T = \mathscr{R}_{D_n} \times (1 - P_{out}) \quad (30)$$

where $(1 - P_{out})$ shows the probability with which the signal is not in outage.

RESULTS AND DISCUSSION

This section provides an overview of the two systems proposed in this chapter, namely DA-CC-D2D and RSMUCD, and their respective performance evaluations. The discussion is divided into two subsections for clarity. The first subsection focuses on the performance analysis of the DA-CC-D2D system, while the second subsection examines the performance of the RSMUCD system in detail.

Figure 6. Outage probability when AF is used as the relaying scheme at drone

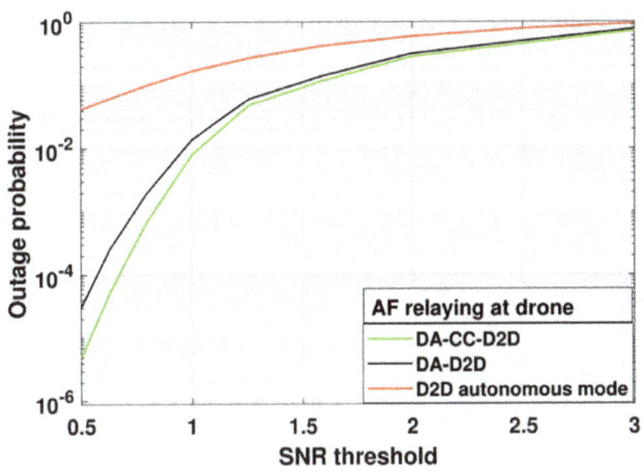

Performance Analysis of DA-CC-D2D

In this section, we compare the outage performance of the DA-CC-D2D system with the DA-D2D and autonomous D2D modes under two relaying mechanisms employed by the drone namely AF and DF. This comparison highlights the relative strengths and weaknesses of each approach in disaster management scenarios.

Figure 6 illustrates the outage probability for different scenarios when AF is employed as the relaying scheme at the drone. As discussed earlier, this chapter considers three distinct cases for disaster situations in smart cities. The first case involves providing services to GUs in autonomous mode, where communication links are established directly between the GUS without relying on any external infrastructure. In the second case, communication is established between the GUs whenever base stations are not functional, with the drone serving as a relay node to restore connectivity. The reliability in this case is based on the signal reaching the GUs solely via the drone. The third case further enhances the reliability by combining the signals received from both the rescue team and the drone, providing a more robust communication link for the GUs. During the analysis of these three cases, Nakagami-*m* fading is assumed for the A2G links, while Rayleigh fading is considered for the ground-to-ground links. From the results presented in Figure 6, it is evident that the third case offers the best reliability for the GUs, as it combines the effects of both the direct rescue team signal and the relay signal from the drone. In contrast, the first case, which only considers the non-line-of-sight paths, provides the worst reliability. The second case lies in between, as it relies solely on the signal from the drone, improving reliability over the first case but not as much as the third case. This outcome highlights the advantage of combining multiple signal sources for enhanced communication reliability in disaster situations.

Figure 7. Outage probability when DF is used as the relaying scheme at drone

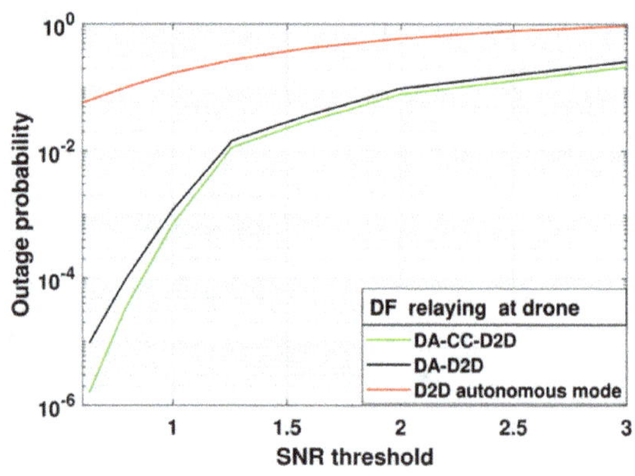

Figure 7 presents the outage probability for various scenarios where DF is employed as the relaying scheme by the drone. The analytical methodology remains consistent with the approach explained for Figure 6, with the key distinction being the relaying scheme applied at the drone. The results indicate that the reliability at the GUs is slightly enhanced when DF is utilized compared to AF. This improvement arises because, in the DF scheme, the signal received by the drone during the first time slot is decoded, re-encoded, and re-transmitted towards the GUs. In contrast, AF simply amplifies and forwards the received signal, which inadvertently amplifies the noise along with the signal. While AF is computationally simpler, the noise amplification inherent to the scheme compromises its reliability, making DF a more robust choice despite its added complexity. However, DF is more prone to error propagation in cases where the data is not decoded correctly by the relay.

Performance Analysis of RSMUCD

In this section, we compare the performance of RSMUCD with NOMA based Cooperation Scheme aided by drone (NCSD), NCC (DF), and NCC (AF). We first compute the respective rate for each scheme using a similar equation like (26) using which the outage probability is analysed. After the analysis of the data rate and outage probability, we focus on throughput which is obtained using (28).

Figure 8. Comparison of outage probability (Y_1) and achievable rate (bps/Hz) (per user) (Y_2) concerning transmit power (dBm) (P_{S_n}, P_R) for $d_{S_n}R \leq$ 100m and M= 2

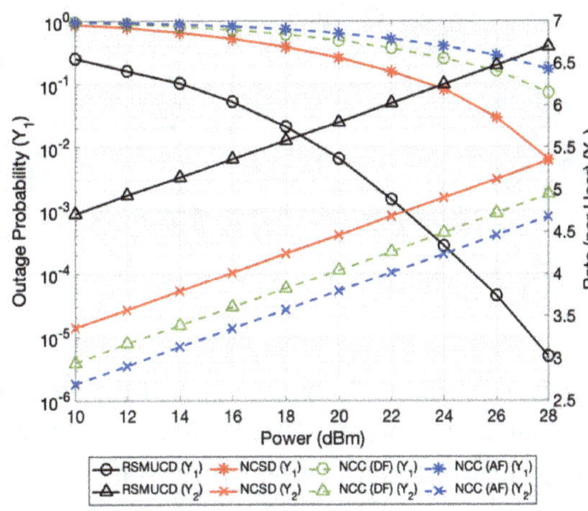

Figure 8 illustrates the comparison of outage probability (Y_1) and per-user achievable rate (Y_2) for the aforementioned schemes with varying transmit power. It is evident from the figure that RSMUCD outperforms the other schemes in both outage probability and achievable rate. The superior performance of RSMUCD can be attributed to its ability to handle interference effectively by allocating different power levels to the common and private streams and applying precoding to the data streams. This approach enables SIC, allowing some interference to be decoded while treating the rest as noise. In contrast, NCSD decodes interference strictly, resulting in NCSD being second in line. The performance of NCC (DF) further deteriorates due to the presence of digital network coding (DNC) noise. NCC (AF) performs the poorest among all the schemes because of analog network coding (ANC) noise, which has a more significant impact compared to DNC noise. This is a disadvantage of the AF-based relaying scenarios as the noise gets amplified however, DF-based relaying can also be harmful and may cause error propagations if the data are not decoded correctly at the relay.

Figure 9. Comparison of outage probability (Y_1) and achievable rate (per-user) (Y_2) with the distance between source nodes and relay. The transmit power considered is (P_{Sn}, P_R)=15 dBm and M=2

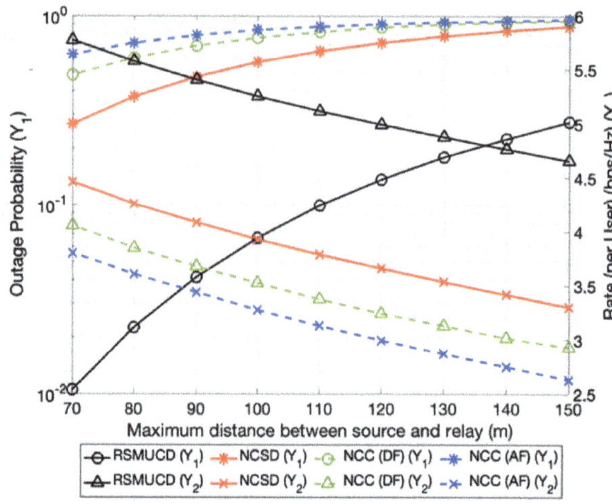

In Figure 9, the outage probability and rate (per user) are compared with the distance between the source and the relay. Here, an upper threshold acting as the maximum value of d_{SnR} and during simulation we generate random values between 10m and the varying threshold. As anticipated, the rate exhibits a downward trend for all the schemes, primarily due to the detrimental effects of large-scale fading. This decline in rate corresponds with a worsening of the outage probability. Among the schemes considered, RSMUCD demonstrates the best performance, owing to the rate enhancement it achieved through the employment of RSMA. Furthermore, it can handle interference in a dynamic manner due to which it sees an enhancement in performance. This is followed by NCSD, NCC (DF), and NCC (AF). As discussed previously, NCC (AF) shows the poorest performance, mainly because of the negative impact of the ANC noise, while NCC (DF) performs slightly better than NCC (AF) due to its fact that the effects of DNC noise are lesser in comparison to the effects of ANC noise. NCSD, which benefits from decoding the interference through the implementation of SIC, outperforms NCC (DF) and NC (AF). However, RSMUCD stands out as the most effective scheme, surpassing all the others, thanks to its superior interference management capabilities.

Figure 10. Throughput in contrast with transmit power (dBm) (P_{Sn}, P_R) for $d_{Sn}R_{Sn}$ ≤ 100m and M= 2

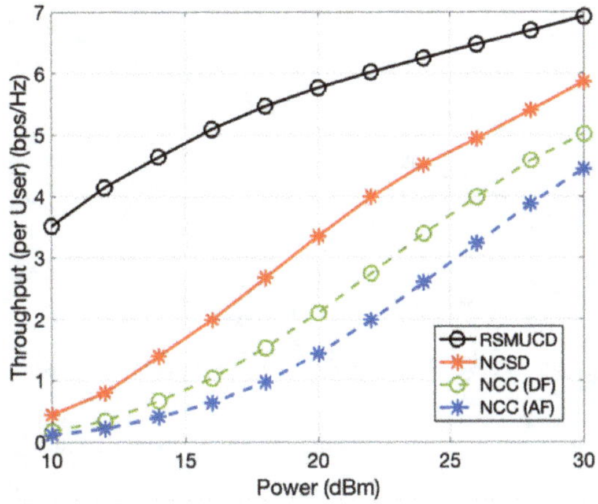

Figure 10 compares the per-user throughput of all the schemes as a function of transmit power. It can be observed that the throughput improves for every scheme with an increase in power. In low-power scenarios, NCC (AF) and NCC (DF) exhibit similar performance; however, the distinct effects of ANC and DNC noise become evident as NCC (DF) outperforms NCC (AF) at higher power levels since ANC noise effects the signals more in comparison to the DNC noise. RSMUCD demonstrates the best performance due to its effective interference mitigation and precoding capabilities, while NCSD achieves the second-best performance among the compared schemes as it depends on SIC for its performance.

Figure 11. Comparison of the sum rate with respect to transmit power (dBm) for RSMUCD, NCSD, NCC (DF) and NCC (AF) for M= 1 and 2, respectively

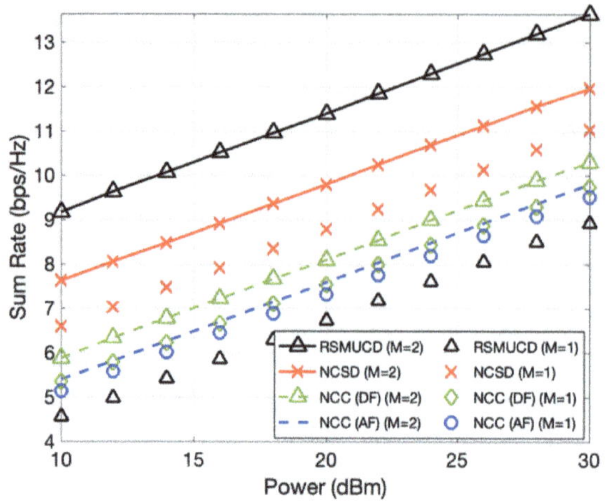

Figure 11 compares the sum rate of all the schemes as a function of transmit power for various value of M. This is done to illustrate the effectiveness of RSMA based systems for multi-antenna drones while underlying the fact that splitting should not be done blindly. To illustrate this, we consider two cases of M=2 (balanced scenario) and $M = 1$ (overloaded scenario). When $M = 2$, the sum rate follows the expected trend, with RSMUCD achieving the best performance, followed by NCSD, while NCC (DF) and NCC (AF) remain at the bottom due to the effects of DNC and ANC, respectively. However, for $M = 1$, the performance dynamics change significantly as RSMUCD experiences a sharp decline in this overloaded scenario. This drop is attributed to the characteristics of ZF precoding, which typically decouples a multi-user channel into independent single-user channels—a process that becomes ineffective in overloaded scenarios, leading to high interference. In such high-interference situations, decoding interference proves advantageous. NCSD, which employs this approach, outperforms the other schemes in this scenario, although its performance is slightly worse compared to $M = 2$ due to the loss of spatial diversity gain. Similarly, the performance of NCC (AF) and NCC (DF) also declines compared to their respective results for $M = 2$, with ANC and DNC continuing to affect them as before. It is also worth noting that RSMUCD leverages RSMA, a generalized multiple access framework that bridges the extremes of SDMA and NOMA, with both being special cases of RSMA. In high-interference scenarios, RSMA can adapt by allocating power accordingly, effectively making the scheme similar to NOMA. As a result, RSMUCD can demonstrate performance comparable to that of NCSD under such conditions.

CONCLUSION

In this chapter, we explore two distinct systems for next-generation advanced wireless technologies aimed at enhancing disaster management in smart cities and evaluate their performance metrics. A drone with a single antenna is used as a relay in three operational scenarios for post-disaster communication. It employs AF or DF relaying, and we derive outage probability and data rate models to evaluate performance for ground victim users. These metrics show how well the system delivers communication after a disaster. The DA-CC-D2D system outperforms DA-D2D and standalone D2D methods, proving effective for disaster-hit smart cities. It offers reliable, high-quality communication even when traditional networks are down. The systems in the first case suit single-antenna setups where minimizing outage matters more than maximizing throughput. They aim for reliable communication in critical scenarios like disaster recovery. Though techniques like NCC can boost throughput, they have drawbacks. NCC depends on overheard signals for diversity, which can be unreliable in dynamic environments. This raises QoS concerns, as such signals aren't always available or consistent. So, while NCC may help with throughput, its limits must be weighed carefully, especially when QoS is crucial.

Due to CC and NCC limitations, a rate-splitting multi-user cooperation system with a drone (RSMUCD) was proposed. A multi-antenna drone acts as a relay using RSMA. RSMUCD is compared to NCC (AF), NCC (DF), and NCSD. Results show RSMUCD achieves better rate, sum rate, outage, and throughput due to its antennas and interference-cancelling precoding. While the scheme offers improved performance, it requires more processing power. Channel acquisition overhead remains a challenge for future work. Our analysis shows the scheme is less effective in overloaded scenarios, where interference mitigation weakens. In such cases, decoding interference gives NCSD an edge. NCSD outperforms AF and DF in these scenarios due to the lack of network coding noise. Exploring cases with more users, where power allocation becomes crucial, would be valuable. Interference may have a more severe impact than NC noise in such cases. Investigating the effects of imperfect and outdated CSI on the RSMUC system is also a promising direction for future work. Additionally, we aim to derive analytical outage expressions for the proposed RSMUCD system to support reliable communication in post-disaster scenarios.

REFERENCES

Asadi, A., Wang, Q., & Mancuso, V. (2014). A survey on device-to-device communication in cellular networks. *IEEE Communications Surveys and Tutorials*, *16*(4), 1801–1819.

Balfaqih, M., Balfagih, Z., Lytras, M. D., Alfawaz, K. M., Alshdadi, A. A., & Alsolami, E. (2023). A blockchain- enabled iot logistics system for efficient tracking and management of high-price shipments: A resilient, scalable and sustainable approach to smart cities. *Sustainability*, *15*(18), 13971.

Bhattacharyya, S., Darshi, S., Kumbhani, B., Pan, J., & Almohammedi, A. A. (2024). RSMUC: An alternative to coded cooperation for QoS enhancements in next generation vehicular network. In 2024 *IEEE Middle East Conference on Communications and Networking (MECOM) (IEEE MECOM 2024)*, pp. (1–6), Abu Dhabi, United Arab Emirates.

Bhattacharyya, S., Darshi, S., & Lin, Z. (2025). Outage analysis of downlink rsma in vehicular cluster: A copula- based approach. *IEEE Wireless Communications Letters*, 1–5.

Bhattacharyya, S., Kumar, P., Darshi, S., Agarwal, S., & Shailendra, S. (2023). Cross layer MAC protocol for a peer conscious opportunistic network coded cooperation system. *IEEE Transactions on Mobile Computing*, *22*(12), 7014–7026.

Bhattacharyya, S., Kumar, P., Darshi, S., Burnwal, O., & Singhal, K. (2021, December). Full duplex relay assisted coded cooperation for next generation wireless networks. In *2021 IEEE 18th India Council International Conference (INDICON)* (pp. 1-6). IEEE.

Bhattacharyya, S., Kumar, P., Darshi, S., Majhi, S., Kumbhani, B., & Almohammedi, A. A. (2024). Hybrid relaying based cross layer MAC protocol using variable beacon for cooperative vehicles. *IEEE Transactions on Vehicular Technology*.

Bhattacharyya, S., Kumar, P., Sharma, S., Darshi, S., & Almohammedi, A. A. (2021, December). A hybrid-combine-and-forward relaying scheme for network coded cooperative systems. In *2021 IEEE International Conference on Advanced Networks and Telecommunications Systems (ANTS)* (pp. 426-431). IEEE.

Chen, Z., Ding, Z., Dai, X., & Karagiannidis, G. K. (2016). On the application of quasi-degradation to MISO-NOMA downlink. *IEEE Transactions on Signal Processing*, *64*(23), 6174–6189.

Choi, J., Park, J., & Evans, B. L. (2017). Spectral efficiency bounds for interference-limited SVD-MIMO cellular communication systems. *IEEE Wireless Communications Letters*, *6*(1), 46–49.

Clerckx, B., Mao, Y., Jorswieck, E. A., Yuan, J., Love, D. J., Erkip, E., & Niyato, D. (2023). A primer on rate-splitting multiple access: Tutorial, myths, and frequently asked questions. *IEEE Journal on Selected Areas in Communications*, *41*(5), 1265–1308.

Ding, Z., Adachi, F., & Poor, H. V. (2015). The application of MIMO to non-orthogonal multiple access. *IEEE Transactions on Wireless Communications*, *15*(1), 537–552.

Etkin, R. H., David, N. C., & Wang, H. (2008). Gaussian interference channel capacity to within one bit. *IEEE Transactions on Information Theory*, *54*(12), 5534–5562.

Fatema, N., Hua, G., Xiang, Y., Peng, D., & Natgunanathan, I. (2018). Massive MIMO linear precoding: A survey. *IEEE Systems Journal*, *12*(4), 3920–3931.

Fodor, G., Parkvall, S., Sorrentino, S., Wallentin, P., Lu, Q., & Brahmi, N. (2014). Device-to-device communications for national security and public safety. *IEEE Access : Practical Innovations, Open Solutions*, *2*, 1510–1520.

Gohari, A., Ahmad, A. B., Rahim, R. B. A., Supa'at, A. S. M., Abd Razak, S., & Gismalla, M. S. M. (2022). Involvement of surveillance drones in smart cities: A systematic review. *IEEE Access : Practical Innovations, Open Solutions*, *10*, 56611–56628.

Kumar, P., Bhattacharyya, S., Darshi, S., Sharma, A., Almohammedi, A. A., Shepelev, V., & Shailendra, S. (2022). Drone assisted network coded cooperation with energy harvesting: Strengthening the lifespan of the wireless networks. *IEEE Access : Practical Innovations, Open Solutions*, *10*, 43055–43070.

Kumar, P., Darshi, S., & Shailendra, S. (2021). Drone assisted device to device cooperative communication for critical environments. *IET Communications*, *15*(7), 957–972.

Kumar, P., & Majhi, S. (2024). Uav-assisted network coded cooperation by using height-dependency shaping parameters in nakagami-m faded channel. *IEEE Access : Practical Innovations, Open Solutions*.

Kurt, A., Saputro, N., Akkaya, K., & Uluagac, A. S. (2021). Distributed connectivity maintenance in swarm of drones during post-disaster transportation applications. *IEEE Transactions on Intelligent Transportation Systems*, *22*(9), 6061–6073.

Liu, Y., Zhang, S., Mu, X., Ding, Z., Schober, R., Al-Dhahir, N., & Shen, X. (2022). Evolution of NOMA toward next generation multiple access (NGMA) for 6G. *IEEE Journal on Selected Areas in Communications*, *40*(4), 1037–1071.

Malik, P. K., Wadhwa, D. S., & Khinda, J. S. (2020). A survey of device to device and cooperative communication for the future cellular networks. *International Journal of Wireless Information Networks*, *27*(3), 411–432.

Mao, Y., Clerckx, B., & Li, V. O. (2018). Rate-splitting multiple access for downlink communication systems: Bridging, generalizing, and outperforming SDMA and NOMA. *EURASIP Journal on Wireless Communications and Networking*, *2018*, 1–54. PMID: 30996723

Mao, Y., Dizdar, O., Clerckx, B., Schober, R., Popovski, P., & Poor, H. V. (2022). Rate-splitting multiple access: Fundamentals, survey, and future research trends. *IEEE Communications Surveys and Tutorials*, *24*(4), 2073–2126.

Mukherjee, A., De, D., Dey, N., Crespo, R. G., & Herrera-Viedma, E. (2023). DisastDrone: A disaster-aware consumer Internet of Drone Things system in ultra-low latent 6G network. *IEEE Transactions on Consumer Electronics*, *69*(1), 38–48.

Naser, S., Sofotasios, P. C., Bariah, L., Jaafar, W., Muhaidat, S., Al-Qutayri, M., & Dobre, O. A. (2020). Rate-splitting multiple access: Unifying NOMA and SDMA in MISO VLC channels. *IEEE Open Journal of Vehicular Technology*, *1*, 393–413.

Phunchongharn, P., Hossain, E., & Kim, D. I. (2013). Resource allocation for device-to-device communications underlaying LTE-advanced networks. *IEEE Wireless Communications*, *20*(4), 91–100.

Qu, C., Sorbelli, F. B., Singh, R., Calyam, P., & Das, S. K. (2023). Environmentally-aware and energy-efficient multi- drone coordination and networking for disaster response. *IEEE Transactions on Network and Service Management*, *20*(2), 1093–1109.

Sami, M., Noordin, N. K., Khabazian, M., Hashim, F., & Subramaniam, S. (2016). A survey and taxonomy on medium access control strategies for cooperative communication in wireless networks: Research issues and challenges. *IEEE Communications Surveys and Tutorials*, *18*(4), 2493–2521.

Schreck, J., Jung, P., & Stańczak, S. (2018). Compressive rate estimation with applications to device-to-device communications. *IEEE Transactions on Wireless Communications*, *17*(10), 7001–7012.

Shim, K., Nguyen, T. V., & An, B. (2019). Exploiting opportunistic scheduling schemes to improve physical-layer security in MU-MISO NOMA systems. *IEEE Access : Practical Innovations, Open Solutions*, *7*, 180867–180886.

Tehrani, M. N., Uysal, M., & Yanikomeroglu, H. (2014). Device-to-device communication in 5G cellular networks: Challenges, solutions, and future directions. *IEEE Communications Magazine*, *52*(5), 86–92.

Vahidi, V., & Saberinia, E. (2019). MIMO–OFDM communication systems for traffic data transmission in 5G drone small cells. *IET Communications*, *13*(20), 3565–3574.

Wu, Z., & Dai, L. (2023). Multiple access for near-field communications: SDMA or LDMA? *IEEE Journal on Selected Areas in Communications*, *41*(6), 1918–1935.

Chapter 11
Geospatial Data Complexities and Technology Challenges in Smart Cities

Sangeetha Radhakrishnan
https://orcid.org/0000-0001-6830-9483
Vellore Institute of Technology, India

Vijaya Kittu Manda
https://orcid.org/0000-0002-1680-8210
PBMEIT, India

ABSTRACT

Smart Cities must integrate geospatial and computing technologies to enhance urban service delivery. These technologies help city administrators make superior decisions and incorporate sustainability practices. This chapter identifies and analyzes various data complexities and technology challenges encountered. Attention is drawn to Geospatial AI, an emerging technology that combines Geospatial technologies and Artificial Intelligence (AI). The challenges are categorized into data quality and management, scalability and performance, data privacy and security, interoperability, workforce gaps, legal and ethical issues, and emerging concerns such as federated learning. The lack of skilled professionals and the outdated state of educational programs are highlighted. This chapter highlights the barriers and offers actionable insights for researchers, developers, policymakers, and urban administrators seeking to build robust, future-ready systems.

DOI: 10.4018/979-8-3373-0154-9.ch011

INTRODUCTION

Geospatial AI (GeoAI) is a new branch of study that emerged by integrating Geospatial Technologies with Artificial Intelligence (AI) (Jitta et al., 2024). GeoAI also incorporates other computing and communication technologies, such as the Internet of Things (IoT), Big Data (Bhardwaj et al., 2024), Cloud Computing, and Digital Twins (Silvestri et al., 2025). The bundled technologies improve the intelligence levels of the city (Zhao et al., 2022) and offer significantly newer opportunities and applications for improving Smart City (SC) services for its citizens (Manda et al., 2024). Decision-makers and city administrators can use these technologies to improve resource efficiency and make the cities sustainable. Such cities tend to improve the economic prosperity of their citizens. However, the development and deployment journey involves numerous data- and technology-related challenges and hurdles.

Geospatial AI research requires interdisciplinary methods (Fu et al., 2025). Hence, researchers need to study the challenges of integrating geospatial technologies with computing technologies for several reasons (Al-Yadumi et al., 2021). Technologies are already playing a critical role in diverse fields of SC operations. Figure 1 shows some components of Smart cities that have recently witnessed significant improvements. Because of location data, privacy and ethical concerns must be addressed. Researchers can develop frameworks and protocols to protect individual privacy and ensure data security. There is a need to establish ethical guidelines for using geospatial information responsibly to achieve this.

Figure 1. Components of smart cities (Created by the Authors)

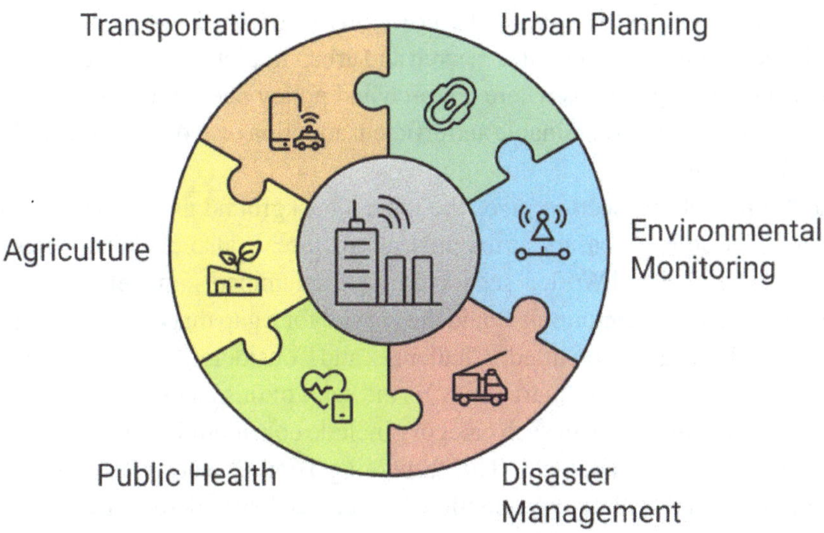

Researchers must overcome data timeliness, accuracy, and infrastructure limitations. These are common expectations from the geospatial application because improving response times can save lives and reduce disaster-related costs. It can only accrue benefits when these technologies fit the disaster management cycle well (Hierink et al., 2024). Researchers can prescribe managing large-scale geospatial data more effectively. Finding areas where optimizing algorithms is necessary and where cloud-based solutions need improvement is to be done. Researchers need to explore how these technologies interact, address compatibility challenges, and understand their convergence's potential risks and benefits. Research can help ensure that the data used for decision-making is reliable, accurate, and up-to-date. Data privacy and security are critical concerns that require attention. They are mentioned here due to their significance and the growing concerns surrounding them in the future. The chapter also intends to discuss the technical and operational barriers. The chapter identifies the importance of skilled professionals in GIS, remote sensing, AI, and data governance. It encourages educational advancements and training programs to fill gaps in technical expertise necessary for implementing and maintaining geospatial technologies in SCs.

This chapter identifies six key challenges in implementing geospatial technologies within Smart Cities. These challenges are selected based on their recurring prominence in literature and their practical impact on system performance and governance. They are also critical for ensuring scalability, security, and interoperability. These challenges represent bottlenecks that consistently restrict real-world deployments. The evidence from both academic research and urban implementation supports this. The aim is to highlight areas where research and policy need immediate attention. Addressing it facilitates sustainable and efficient adoption of geospatial technologies in urban environments.

The "Introduction" section gives the topic's background and explains why researcher attention is necessary at the present juncture. It also explains the chapter objectives. The "Related Works" section critically examined some related literature and highlighted their limitations. It laid the ground for a gap that the chapter intends to fill. The following section titled "Challenges and Considerations in implementing Geospatial technologies in Smart Cities" is the heart of this chapter. It discusses six challenges. Sub-categories in each category include considerations that help understand and address the challenges. The "Emerging Topics" section discusses future trends in Geospatial Technology adoption. It discussed key Federated Learning (FL) considerations in Geospatial AI implementations and other potential challenges and opportunities. The character ends with conclusion remarks.

RELATED WORKS

Integrating geospatial technologies with Artificial Intelligence (AI) and other digital infrastructure has attracted significant academic and practical interest. The development and deployment of these technologies in Smart City (SC) accelerated the need. Existing literature has acknowledged the transformative potential of these technologies. However, it often lacks a critical examination of the data and infrastructural challenges.

For example, Jitta et al. (2024) highlight the convergence of GeoAI with technologies like IoT, Big Data, and Digital Twins. The authors argue that these technologies improve urban services. However, such studies largely adopt a techno-optimistic stance. They tend to understate persistent issues related to interoperability, data heterogeneity, and lack of scalable deployment models. Similarly, Manda et al. (2024) explore the benefits of GeoAI in SCs. However, they offered a limited analysis of real-world deployment barriers. They miss topics like semantic integration and standardization across diverse city systems.

Bhardwaj et al. (2024) delve into the role of AI, IoT, and Big Data in territorial management. However, they insufficiently address the spatial complexity and performance limitations when these technologies are applied in high-density urban environments. The work also falls short in offering mitigation strategies for data governance issues. Governance issues arise when deploying decentralized systems across multiple administrative jurisdictions.

Several studies have emphasized the role of supporting infrastructure in SCs. Silvestri et al. (2025) propose a data platform for subsystem integration, emphasizing modularity. While promising, their model assumes ideal data flow and neglects real-time applications' latency, semantic drift, and versioning issues.

Additional works deserve attention for their relevance and the gaps they expose. For instance, "*Associated Information and Communication Technologies Challenges of Smart City Development*" is a notable work. It identified interoperability, security, and real-time communication limitations (Balfaqih & Alharbi, 2022). Their work aligns closely with the present study. However, the authors overlooked the federated data governance models and their role. Spatial data harmonization's role in mitigating these ICT-related pitfalls was insufficient.

"*Challenges and Factors Influencing the Implementation of Green Logistics: A Case Study of Saudi Arabia*" (Khayyat et al., 2024) addressed barriers such as regulatory inertia, lack of standardized practices, and inadequate infrastructure. Their green logistics perspective offers insights into the operational bottlenecks relevant to smart urban mobility and logistics. However, the analysis made macro-level observations. It did not drill into the spatial and data-infrastructure layers required to support location-aware systems.

Finally, a blockchain-IoT integrated model was proposed to bring transparency and traceability in smart logistics (Balfaqih et al., 2023). While the architecture is conceptually robust, the scalability and real-time geospatial analytics capabilities are underexplored. Furthermore, the study does not address the performance trade-offs introduced by blockchain's inherent computational overhead. These overheads become important when applied in real-time urban logistics.

The related works have underscored the potential of emerging technologies in enabling smart urban ecosystems. However, most fail to address the practical, technical, and ethical integration challenges this chapter foregrounds. These include data standardization and quality, semantic interoperability, and real-time analytics under resource constraints. Further, ethical concerns regarding surveillance and data privacy are missing. A systems-oriented approach is necessary to bridge the gap between theoretical capability and real-world implementation.

CHALLENGES AND CONSIDERATIONS IN IMPLEMENTING GEOSPATIAL TECHNOLOGIES FOR SMART CITIES

A gradual adaptation of processes and emergent forms of urban politics is seen (Masik et al., 2021). The increasing adoption of GeoAI applications increases the priority of data, format, standardization, and improving geospatial interoperability (Arundel et al., 2023). Representing, managing, and analyzing spatially complex and large utility networks is a significant challenge. This is further illustrated by Ji (2020), who conducted a comparative study of various database management tools, finding that NoSQL graph databases significantly improve data reading and network analysis performance. Such performance considerations highlight why data management, integration, and scalability challenges are central in smart city geospatial applications. However, the selection of the six key challenges in this chapter is based not only on performance concerns but also on their recurring prominence in literature, practical impact on system governance, and the critical need for scalability, security, and interoperability in real-world deployments.

Figure 2. Challenges in implementing geospatial technologies for smart cities (Created by the Authors)

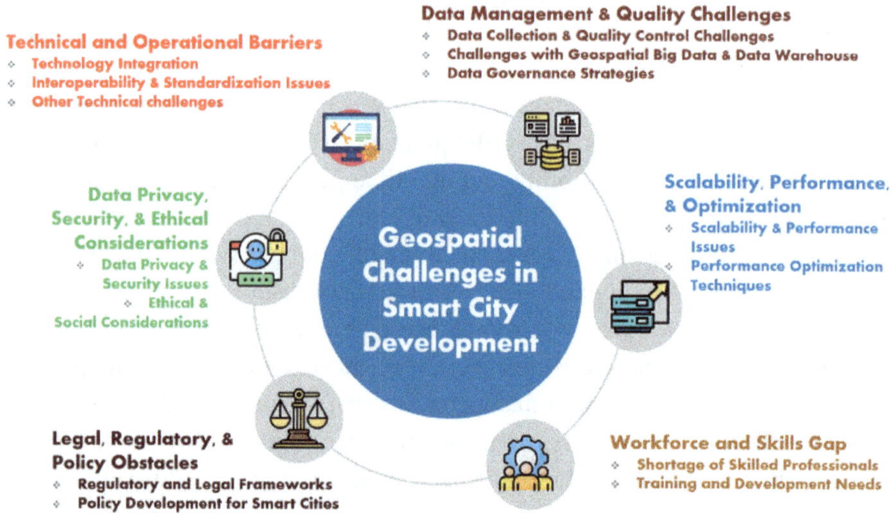

Data Management and Quality

Addressing data-related challenges requires careful data management practices. Data standardization, data validation, and data quality control can help. Before assessing the quality and accuracy of geospatial data, it is necessary to consider the specific requirements of the intended application.

Data Collection & Quality Control

Triangulation methods are used for data collection to help verify data accuracy. This is done by by comparing it with data from other sources. The data is often collected in a modular and scalable manner (Davoli et al., 2024) with numerous subsystems. Integrating these subsystems requires specialized ICT solutions supporting data collection, management, sharing, and integration (Silvestri et al., 2025). Newer forms of data quality and accuracy challenges arise because of geospatial data integration. Further, data is to be processed before use. Each piece of data will have varying levels of precision, completeness, accuracy, and timeliness. Differences in data formats, standards, and temporal resolutions are prevalent. Using data from these sources and forms can make it challenging to ensure the overall reliability of the integrated data. For example, differences in resolution or spatial reference systems can lead to misalignment when combining data from multiple sources. So, the system that will be built requires interoperability at its core, even when dealing with diverse data sources. Only then will the system be accurate and reliably called for in geospatial analysis.

The most common challenges encountered using geospatial data in urban settings include:

Data Quality & Availability. Geospatial data needs to maintain accuracy and quality levels. This is necessary for the data to be fed adequately into analytic systems and for proper analysis. Inaccurate, insufficient, or outdated data can lead to inefficient decision-making. For example, while groundwater data on maps is currently available, data related to water quality levels is limited. Also, limited studies point to the best ML/DL algorithm that works for such data (Razavi-Termeh et al., 2024). If high-quality data is sought after, there is a need for regular updates and validation. Also, newer and advanced data collection techniques will be used in remote sensing and LiDAR. The quality of data fed to AI models should be good at filtering noisy and outlier data to avoid incorrect and inaccurate results (Ramu et al., 2022).

Data Heterogeneity (Sources & Formats). Geospatial data can come from various sources. Some popular sources include satellite imagery, drones, IoT sensors, mobile applications, and government databases. These data sources may use different formats, coordinate systems, resolutions, scales, and detail levels. Figure 3 presents a few geospatial data categories.

Figure 3. Geospatial data categories (Created by the Authors)

Each source generates data with varying levels of detail, temporal resolution, and accuracy. Diverse data formats are used. GeoJSON, KML, shapefiles, and different proprietary formats are a few examples. The presence of multiple formats complicates the integration and analysis of these datasets. For instance, satellite images provide comprehensive spatial coverage. However, they may have lower temporal resolution, while IoT sensors provide real-time data but often cover limited spatial areas. Similarly, technical challenges may arise when integrating real-time IoT sensor data with traditional GIS layers (Davoli et al., 2024). The data interchange often happens in JSON or XML formats. This diversity in data formats requires two types of strong data management systems capabilities:

1 Handling multiple formats and
2 Efficient ETL (Extract, Transform, Load) processes to create unified spatial databases.

Combining the data requires transformation processes involving reformatting, resampling, and often converting between coordinate systems. If not handled properly, these processes can introduce errors such as loss of detail or misalignment of spatial features. Therefore, a strong geospatial data infrastructure with tools for automatic format conversion and real-time processing is essential.

Data Distortions. Dense urban settings/environments are complex and called "urban canyons." They are dense infrastructure, tall buildings, and street corridors formed by closely spaced tall buildings on either side, as well as fluctuating populations. The presence of urban canyons can cause distortions in data collection. A signal loss in GPS or occlusions in imagery are common examples. High population density areas may lead to overcrowding location-based services such as Wi-Fi and cellular networks. Algorithms are used for optimal mobile positioning with signals of opportunity (SOOP). Data transmission delays may affect data quality. Such disturbances lead to inaccurate mapping or poor real-time data. 3D visualizations help visualize thermal images of urban canyons. However, visual clutter makes users lose focus when searching for a 2D representation of a 3D geographic space (K.-C. Wu et al., 2022).

Data Accuracy and Reliability. This is a critical requirement. Even a tiny error can propagate into significant discrepancies in the final results of spatial mapping. Accuracy and reliability depend on several factors. Three key factors are:

- The quality of the source data
- The methods used to collect and process the data
- The scale of the data.

Urban settings have high-density buildings, infrastructure, and other features. This factor makes it challenging to collect accurate geospatial data. For example, inaccuracies in satellite imagery or GPS data can lead to significant mapping and spatial analysis errors. Graph neural networks (GNNs) need sufficient training data to make accurate predictions. While sufficient training data can be sourced and fed to the network, it might not always fit the circumstances. It brings in domain generalization challenges (Sun et al., 2024).

Data Currency. Smart cities are dynamic urban environments. The geospatial data collected could become outdated quickly. Temporal changes, such as environmental changes because of construction, natural disasters, or seasonal variations, can quickly make the data outdated. So, the system must be refreshed with periodic/continuous updates to remain relevant. Necessary mechanisms to handle regular updates and periodic maintenance are to be implemented.

Semantic and Syntactic Heterogeneity. Semantic heterogeneity refers to the differences in the meaning of data between different sources. These generally exist because of varying terminologies, classifications, or data structures representing similar spatial phenomena. For example, one data source may use the term "road" to refer to all paved roads. Another data source may use the term "road" to refer only to major highways. Syntactic heterogeneity refers to the differences in data structure between different sources. Some such differences are variations in attribute names, data types, or file structures. For example, one data source may store data in a tabular format, while another may store data in a hierarchical format. These differences come when datasets use different terminologies, classifications, or coordinate systems. The differences lead to inconsistencies and misinterpretations. In smart cities, this heterogeneity can reduce geospatial data's effective integration and utilization. It requires extensive data harmonization efforts. Developing common ontologies and metadata standards can help.

Data Integration and Interoperability. Data integration with other platforms and software systems is challenging, especially when crowd-sourced platforms are involved. These platforms introduce biases or errors. These challenges complicate urban planning, traffic management, and other Smart city applications. In these circumstances, more advanced methods for data verification, fusion, and real-time accuracy improvements.

Data Gaps and Missing Values. Data gaps and missing values are a common challenge when dealing with data. Smart city applications rely heavily on high-resolution, continuous, and complete geospatial datasets to provide insights. Figure 4 shows four key types of data gaps and missing values to be dealt with in Smart City Applications.

Figure 4. Data gaps and missing values in smart city applications (Created by the Authors)

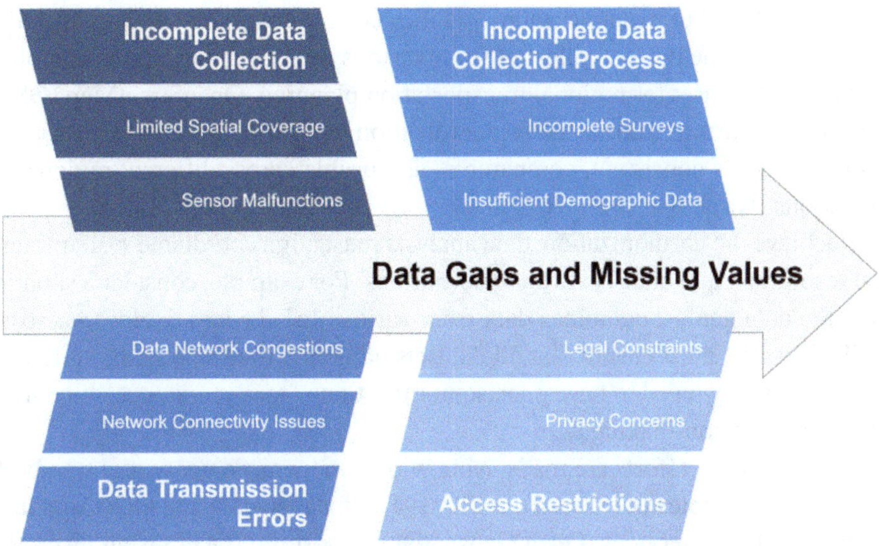

Dealing with data gaps and missing values is a mandatory requirement. Filling data gaps while maintaining spatial consistency or integrating data from additional sources is important. Failure to deal with it will compromise proper analysis, modeling, and decision-making processes. Eventually, stakeholders will lead to less reliable or inaccurate outcomes. Such analysis can even have bias. Technologies are now available to fill the data gaps. However, they require interpolation. Interpolation estimates missing values based on known data (Abdulmanov et al., 2021). Another alternative method is to do data fusion. Data from different sources are combined to fill the gaps to estimate missing values (Priyashani et al., 2023). Sometimes, both these ways can introduce further uncertainty or error. Advanced data imputation techniques are sometimes required to deal with missing values in the geospatial data. One such popular approach is the Machine Learning (ML)-based Missing Value Imputation (MVI) approach. The approach uses Clustering and instance-based algorithms. Percentage of correct prediction (PCP) and root mean square error (RMSE) are the most used evaluation metrics. These evaluation metrics can predict missing values. At the same time, they will preserve the spatial and temporal integrity of the datasets (Fouad et al., 2021; Thomas & Rajabi, 2021).

Data Harmonization and Standardization. Data is captured continuously from traffic sensors, public safety cameras, weather stations, and other devices. Data managers use data harmonization and standardization when using diverse geospatial data. Checks are done for consistency in spatial reference systems, attribute structures, and data models. For example, municipalities collect and maintain data related to road networks. The data is collected from various municipalities and is combined into a unified dataset for regional transportation planning. OpenStreetMap (OSM) relies on data harmonization and standardization to ensure that the crowd-sourced geospatial data is consistent, comparable, and usable across different regions and applications (Nirandjan et al., 2022).

To achieve the harmonization, data analysts must align coordinate systems, temporal resolution, and dataset attribute definitions. For example, consider a situation where the data analyst combines data from various IoT devices deployed across a city. The analyst must ensure that all datasets adhere to the same spatial reference system. Also, they need to have consistent timestamps for accurate real-time monitoring and decision-making.

Standardization efforts are made using widely recognized geospatial standards. Two famous examples of this are the ISO 19115 for metadata and OGC standards for geospatial web services. Open Geospatial Consortium (OGC) is an institution that promotes standardization. It aims to establish standard formats and protocols for geospatial data exchange. However, implementing these standards across diverse urban systems and legacy databases can be challenging. Smart city data is dynamic, and real-time updates keep coming. The data will have varying temporal resolutions. These reasons add up to the complexity of the harmonization process.

Geospatial Big Data & Data Warehouse

Geospatial Big Data (GBD) evolved as a new field owing to the extensive use of Big Data in geospatial applications (J. Wu et al., 2024). Geospatial Data Warehouse (GDW) became logical extensions to traditional data warehouses. This is perhaps because of their ability to scale non-relational databases horizontally (Ferro et al., 2023). Implementing GBD and GDW comes with a few challenges. Figure 5 shows five core challenges that need to be addressed.

Figure 5. Five core challenges with geospatial big data & data warehouse (Created by the Authors)

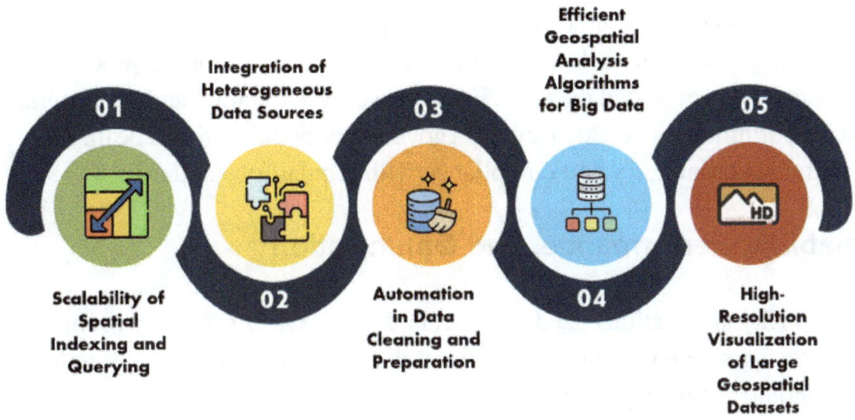

The first of the challenges is the need for specialized data processing and management tools to handle three aspects of data:

1. **Data Volume**: A vast amount of data needs to be processed. Traditional data processing tools cannot handle the volume and complexity of Big Data.
2. **Data Quality**: Structured and unstructured data
3. **Data Sources**: Data comes from diversified sources such as social media, IoT devices, and public records.

As a result, new tools and techniques are required for big data processing and management. These tools and techniques can be used to:

1. Ingest and store large amounts of Big Data
2. Clean and prepare Big Data for analysis
3. Analyze Big Data using a variety of methods
4. Visualize Big Data to identify patterns and trends

Data Governance Strategies

Geospatial analysis has to formulate and put appropriate data governance strategies in place. Depending on the actors (stakeholders) involved, four popular data governance models are famous – Data Sharing Pools (DSPs), Data Cooperatives (DCs), Public Data Trusts (PDTs), and Personal Data Sovereignty (PDS) (Micheli et al., 2020). A data governance strategy ensures data quality, security, and compliance with privacy regulations. Automated data cleansing, validation, and conflation techniques help maintain data integrity since multiple sources capture Smart City data. Furthermore, distributed computing helps in robust analysis, and cloud-based storage brings scalability solutions. These are important for storing, processing, and sharing massive geospatial datasets. By doing so, collaboration and interoperability for SC opportunities arise. Stringent data governance policies should be implemented to maintain data integrity and compliance with regulatory standards.

Scalability, Performance, and Optimization

Smart City applications need to be scalable. The system needs to accommodate the increasing size and nature of captured data. At the same time, system performance should not be degraded.

Scalability & Performance Issues

Scalability is the ability of a system, network, or process to handle growth. Growth can come in the form of increased data volume, user demand, or complexity. At the same time, there should not be any compromise in functionality or efficiency. For example, as the city expands, a geospatial platform must be scalable enough to support additional users. More city planners, emergency responders, and citizens joining the Smart City will mean a proportionate increase in data. The performance of a system refers to its speed, responsiveness, and reliability in processing and delivering geospatial data and services. Table 1 explains how the scalability and performance of geospatial technologies become critical for a Smart City.

Table 1. Scalability and performance challenges of geospatial technologies

Challenge	Description
Distributed computing	Implementing efficient distributed computing (serverless) frameworks for geospatial data processing is challenging due to the spatial nature of the data and the need to maintain spatial relationships during parallel processing (Bebortta et al., 2020).
Spatial indexing	Developing and maintaining efficient spatial indexing structures that can handle large volumes of dynamic geospatial data is crucial for query performance but presents significant technical challenges (Ghosh & Eldawy, 2022).
Cloud-native geospatial processing	This involves adapting geospatial algorithms and workflows to cloud-native architectures to take advantage of scalable computing resources. In the process, optimizing data transfer and storage costs can be a challenge (Bachiega et al., 2021).

continued on following page

Table 1. Continued

Challenge	Description
Edge computing for geospatial applications	Implementing edge computing solutions for geospatial data processing in Smart City environments requires careful optimization of algorithms to run on resource-constrained devices while maintaining accuracy and timeliness (Das et al., 2021).
Handling temporal dynamics	Many Smart City applications require processing and analyzing time-series geospatial data. Developing scalable solutions that efficiently handle large datasets will be a challenge. Also, treating temporal dimensions is computationally expensive (Pereira et al., 2022).
Optimizing data structures	Developing and implementing efficient data structures for storing and querying large volumes of geospatial data. Space-filling curves or hierarchical spatial data structures are crucial for performance. They require careful optimization (Semeniuk, 2024).

Performance Optimization Techniques

Performance optimization techniques are a collection of methods, tools, and strategies to improve systems or applications' efficiency, speed, and resource utilization (Akintayo et al., 2024). Optimization is often a collective effort of various stakeholders. Some famous performance optimizations of geospatial systems are:

Spatial Indexing. These methods involve efficient indexing structures like R-trees, K-D trees, Quad-trees, or hexagonal grids (such as H3) (Kim et al., 2024). The methods provide faster access to spatial data, improving query performance by organizing data to reduce search space and retrieval time (Liu et al., 2021). Implementing hierarchical spatial indexing structures reduces query execution time by approximately 40-60% compared to traditional B-tree indexes.

Distributed and Parallel Processing. Several distributed computing frameworks are now available thanks to the enhancements in multi-core processors and GPUs (Akintayo et al., 2024). Some popular names include MapReduce, Hadoop, and Apache Spark. Also, unlike in the past, parallelization is now feasible. Parallelization helps split tasks and distribute them across multiple machines. Consequently, the overall processing process speeds up. This is useful in scenarios where large geospatial datasets are used. It is also a consideration when the application requires real-time processing, as in the case of Smart Cities. Parallel processing architectures also can use spatial partitioning schemes. They distribute computational workloads. GPU clusters are used to implement map-matching algorithms. The clusters show a 15-20x speedup compared to CPU-based processing for large-scale trajectory data analysis. When spatial decomposition techniques are used, load balancing across distributed computing nodes can be achieved. Even quad-tree and grid-based partitioning are used in such environments.

Data Partitioning. This process involves dividing data into manageable chunks. For example, by using Geohashing and Hilbert curve-based partitioning, optimized storage and retrieval can be achieved. This data organization and processing becomes efficient in distributed or cloud-based environments.

Data Privacy, Security, and Ethical Considerations

A Smart City must ensure data privacy, security, and ethical use of the data captured. Data issues can come because of personal location in data, infrastructure details, and real-time sensor inputs. Protecting this data from unauthorized access, misuse, or breaches is critical. These aspects are discussed in the following sub-sections.

Data Privacy and Security Issues

By its inherent nature and applications, geospatial data involves collecting, aggregating, analyzing, and storing vast quantities of data. This geospatial and operational data about various parts of the city and people is highly sensitive. Generally speaking, geospatial technologies are not designed with privacy in mind. The topic became vital because of an apparent rise in significant privacy and security concerns. More so because of geoprivacy and location privacy (Zurbaran et al., 2020). The data may expose vulnerabilities that potentially lead to unauthorized access or misuse. Stringent privacy-preserving mechanisms are to be taken up. Efforts are needed to disallow unauthorized access and data breaches. Privacy Enhancement Technologies (PET) can be used in SCs in a layered manner. PET can include technical, governance, and community mechanisms. Homomorphic Encryption, Differential Privacy, and Federated Learning (FL) are some popular PETs.

Studies showed the integration of digital twins with federated learning to ensure privacy preservation and trustworthiness. Apart from PET, sophisticated geospatial privacy techniques, such as k-anonymity and spatial generalization, are available. They can be employed to minimize the exposure of identifiable location data without compromising the utility of the data for urban analytics. Two popular techniques to address this are:

1. Implementing data minimization strategies through anonymization
2. Reducing precision, such as spatial blurring

These techniques ensure individual privacy protection while enabling geospatial data in urban management. Geospatial privacy methods (such as removing unnecessary details and spatial blurring) should be used when storing location data related to citizens. Since Smart city projects involve several third parties and vendors, comprehensive oversight mechanisms must be implemented. Smart cities implementing PET will win residents' trust and recognize their digital rights (R. Gupta & Sharma, 2024).

Ethical and Social Considerations

Individual data is collected from CCTV cameras, mobile phone data, and GPS in Smart Cities. Geospatial technologies will involve mass surveillance and the erosion of privacy. Ethics considerations regarding surveillance and public data usage are essential, especially considering that AI is used in urban settings. Integrating high-resolution satellite imagery, LiDAR data, and IoT sensor networks further creates new unanticipated capabilities for urban monitoring. This makes it inevitable for Smart cities to:

1. Establish ethical guidelines regarding the use of geospatial AI in urban settings.
2. Addressing issues such as algorithmic bias in spatial analysis
3. The right to locational privacy
4. Equitable access to geospatial services across diverse socio-economic groups.

Ethical considerations question the benefits of Smart city technologies on one side and individual rights and freedoms on the other. Public engagement and transparent governance are essential to maintaining trust and upholding ethical standards.

Explainable AI (XAI) is a trending topic of interest for researchers and practitioners, especially those interested in AI's ethical use. XAI is a system that is capable of providing explanations for its judgments or predictions. XAI methods help classify Smart cities based on various metrics and explain how the results arrived. However, some challenges come from XAI integration into Smart city applications. The list includes (Algarni & Mishra, 2024):

1. Security risks due to extensive data and communication networks
2. Real-time data processing demands
3. Vulnerabilities in IoT devices
4. Ensuring the accountability of AI system operators, and
5. There is a need for secure AI systems to protect sensitive information.

More research in XAI is necessary. The existing approaches are limited across four critical dimensions: accessibility, cultural reflexivity, situatedness, and visibility into internal representations (S. Gupta & Loukissas, 2023).

Technical and Operational Barriers

Overcoming significant technical and operational barriers is important in Smart Cities. The challenges can come from three routes:

1. Complexities of technology integration
2. The need for interoperability and standardization
3. The need to address computational and algorithmic limitations.

These aspects are discussed in the following sub-sections.

Technology Integration

Integrating geospatial technologies with other advanced technologies will immensely benefit Smart Cities. However, it can be complex because of operational complexities. Most web mapping tools and libraries currently available are one-off solutions. They may not be universally applicable and often have a disconnect between research and day-to-day practice (González et al., 2020). Integration of the technologies with various GIS platforms and diverse data sources will be a nightmare. User interfaces of the tool may not appeal to end decision-makers like city planners. Personnel who use the applications need additional training to handle the new technologies (Ioanid & Andrei, 2024). Improper training could lead to the technology deployments becoming useless. Most tools might perform well with static data and poorly handle dynamic and constantly evolving data, such as in a SC. Advanced spatial statistical methods and network analysis are necessary to make the tools even more powerful(Bruno, 2024).

Further, technology integration will be a costly activity requiring considerable financial commitments. Technologies keep advancing at a fast pace, and hence, constant efforts must be made to modify or upgrade it. The updates would be in the form of advancements in sensor networks, AI models, and data processing algorithms. Overall, it will be a continuous cycle of investment and adaptation. It can put pressure on thin city and municipal budgets and resource allocation strategies.

Interoperability and Standardization Issues

Geospatial projects for smart cities are large and complex systems. It involves collaboration and parallel development of technologies by diverse stakeholders ranging from governmental agencies to private firms and citizens. This can create significant interoperability challenges.

Standardization efforts must focus on establishing standard systems/repositories, data formats, APIs, and system architectures that can seamlessly interoperate. Without standardization, bottlenecks in integrating various data sources will be created. The impacts will be high in GIS, remote sensing data, and IoT networks. Standards include adhering to Open Geospatial Consortium (OGC) standards for geospatial data exchange and implementing ISO/TC 211 geographic information standards. ITU-T Focus Group on Smart Sustainable Cities is a key protocol. Harmonizing these standards with Smart city-specific protocols is vital for seamless integration. The recent joining of the International Cartographic Association (ICA) as an associate member of OGC furthers the promotion of standardization of cartography and geospatial semantics and ontologies (Coetzee et al., 2023). Because IoT devices come from different manufacturers and follow various formats, geospatial web standards are to be followed to allow for interoperability. OGC's Sensor Observation Service and SensorThings AP are popular standards (Marsh-Hunn et al., 2021).

Other Technical Challenges

Computational Complexity and Computational Resources. Federated Learning with remote sensing data presents challenges due to the high computational demands on each device. Such situations would lead to clients participating infrequently in training and using outdated model parameters (Moreno-Álvarez et al., 2024).

Algorithm Development and Optimization. Developing GeoAI algorithms for Smart cities comes with several challenges. Table 2 lists the challenges specific to the developing and optimizing phases of the AI algorithms for Smart Cities.

Table 2. Challenges associated with developing and optimizing AI algorithms for smart cities

Challenge	Description
Spatial autocorrelation	Geospatial data often show spatial autocorrelation. This means that the nearby features are more related than the distant ones. So, it will be challenging to develop algorithms that accurately account for this phenomenon while maintaining computational efficiency.
Non-stationarity	Many geospatial processes are non-stationary. So, their statistical properties vary across space. Algorithms must be adaptable to handle these spatial variations. Consequently, model development will become complicated. There will be an increase in computational requirements, too.
Multi-scale analysis	Smart City applications often require analysis at multiple spatial and temporal scales. Developing algorithms that efficiently handle multi-scale data and provide meaningful insights across different resolutions is a significant challenge.
Integration of heterogeneous data sources	Smart cities deal with diverse data sources. Data is obtained from satellite imagery, LiDAR, and IoT sensors and sometimes is crowd-sourced. Developing algorithms that can effectively integrate and analyze these heterogeneous data types while accounting for their varying spatial and temporal resolutions is complex.
Real-time processing	Many Smart City applications require real-time or near-real-time data processing and analysis. So, algorithms are to be optimized to meet these stringent time constraints. At the same time, accuracy must also be maintained. This will be a persistent challenge.

Workforce and Skills Gap

Building smart cities requires skilled professionals to plan, develop, and implement systems and applications. However, there is a shortage of experts with the right skills and training in these advanced technologies. As more cities aspire to become smart cities, the demand for human resources keeps increasing. These following sub-sections discuss these aspects.

Shortage of Skilled Professionals

Professionals with different expertise and skill levels must work together as a team when building Smart City applications. Some roles of the professions for this are:

Geospatial Data Scientists. These professionals combine spatial analysis with advanced data science techniques, including AI and ML. They create models to predict natural disasters such as landslides (Pawar & Sharma, 2025). Models also help analyze urban growth, optimize delivery routes in a city, or even help in environmental conservation. They can track changes in forest cover.

Remote Sensing Specialists. It is evident that the volume and variety of Earth observation data is increasing. There is a growing need for experts who can process and interpret multispectral and hyperspectral imagery. LiDAR and Synthetic Aperture Radar (SAR) data need higher precision and processing (Pawar & Sharma, 2025).

Geospatial Software Developers. Programmers with specialized knowledge of spatial algorithms and geodatabases are called Geospatial Software Developers. They develop custom geospatial applications, web mapping services, and location-based mobile apps that require.

Geospatial Systems Architects are professionals who can design and implement complex spatial data infrastructures. They decided to build distributed geospatial computing environments and real-time sensor networks, which are crucial for Smart city implementation.

Geospatial Data Quality Analysts ensure data accuracy, completeness, and consistency across various spatial datasets. They are necessary because of the increasing dependency on geospatial data for decision-making.

There is a shortage of qualified professionals with expertise in geospatial technologies for smart cities. The reasons for this situation are many:

1. Insufficient talent pool
2. The gap between demand and expertise
3. Deficit in the skilled workforce
4. Lack of expertise in advanced geospatial technologies
5. Shortage of cross-disciplinary expertise (geography, computer science, and data ethics)
6. Limited availability of highly trained specialists
7. Underdeveloped workforce for emerging geospatial applications.

Training and Development Needs

Upgrading the traditional geography class with geospatial technologies in the school curriculum is an excellent place to start. More importance should be on applying technology rather than solely on theoretical understanding (Ridha & Kamil, 2021). The first level of the curriculum can deal with core aspects such as geospatial web services and fieldwork based on GIS. Later, emerging technologies such as AI, ML, and Blockchain can introduced as an advanced layer (Lee, 2023).

Developing specialized training and education programs can address the shortage of skilled professionals in geospatial technologies. Traditional geospatial curricula give a solid foundation in GIS, cartography, and remote sensing. The traditional curricula complement the advanced skill sets necessary for modern Smart city applications. Further, industry partnerships and collaborations with academic institutions help. They accelerate the development of certifications, workshops, and practical learning experiences tailored to the needs of smart cities. These programs should stress hands-on experience with:

1. Real geospatial datasets and cloud infrastructures like Google Earth Engine or ArcGIS Online are helpful.
2. AI-powered platforms can help develop proficiency in handling large-scale urban geospatial challenges.

Moreover, continuous learning opportunities for current geospatial professionals are necessary. Geospatial education programs should offer flexible learning formats to reach a wider audience and accommodate working professionals. Some formats are online courses, MOOCs (Massive Open Online Courses), and blended learning approaches. With this, learners can stay in touch with technological advancements. They can improvise in federated learning, machine learning models, and spatial data mining.

Location analytics gives valuable insights for decision-making across various industries (Ferro-Diez et al., 2020). However, academic Information Systems (IS) research poorly integrates them. Education on the topic remains limited despite its practical usage. Experts believe that spatial methodologies must be part of IS research and curricula. They say there can be a shift towards more spatially informed approaches to enhance research rigor. Updated curricula also prepare students for the growing field of location-based analytics (Erskine et al., 2024).

Legal, Regulatory, and Policy Obstacles

Using AI and geospatial technologies in smart cities raises regulatory and legal concerns. Regulators are generally skeptical when new technologies are used because of the uncertainty of the results that the technologies bring. Some examples of this are data ownership, privacy, and accountability. These areas require coordinated policies and ethical frameworks to ensure innovation while protecting public safety and individual rights. These aspects are discussed in the following sub-sections.

Regulatory and Legal Frameworks

Artificial Intelligence (AI) (Zhou & Kankanhalli, 2021) and geospatial technologies (Merodio Gómez et al., 2022) in Smart Cities can attract regulatory questioning. Regulatory checks are almost always standard when emerging technologies are involved. Autonomous decision-making systems generally trigger legal data ownership, accountability, and liability concerns. Regulatory frameworks help address these concerns. They do not compromise or disturb the innovation aspect. Policies are designed to focus on the ethical implications of geospatial data collection and use. They help in balancing public safety and individual privacy. There is a need for coordination between national, regional, and local governance. The trio can formulate and establish coherent regulations that govern geospatial technologies in Smart Cities. Legal frameworks should address the unique challenges posed by these technologies. Two key focus areas are data ownership rights and liability in automated decision-making processes. Compliance with evolving data protection regulations such as GDPR in the context of geospatial data collection and processing is required (Eskhita et al., 2021).

The United Nations Committee of Experts on Global Geospatial Information Management (UN-GGIM) collaborated with the World Bank. The duo has put efforts into developing the Integrated Geospatial Information Framework (IGIF) (UN-IGIF, 2024). It acts as a blueprint and a practical guide for geospatial infrastructure. IGIF connects to regulations, standards, and policies at national and international levels worldwide. The framework, however, suffers from criticism in that it does not address ethics-related dimensions. Consequently, ethics assessment lists were developed (Calzati & Van Loenen, 2023).

Policy Development for Smart Cities

Policy development for Smart Cities involves various stakeholders, as Figure 6 points out. Policymakers generally welcome frameworks in their decision-making. The frameworks give them a structured way of managing risks, ensuring compliance, and making effective decisions. This is true in projects that involve geospatial technologies to enhance urban planning and management. An effective policy mandates using Geographic Information Systems (GIS) for spatial data collection, analysis, and visualization.

Figure 6. Stakeholders in policy development for smart cities (Created by the Authors)

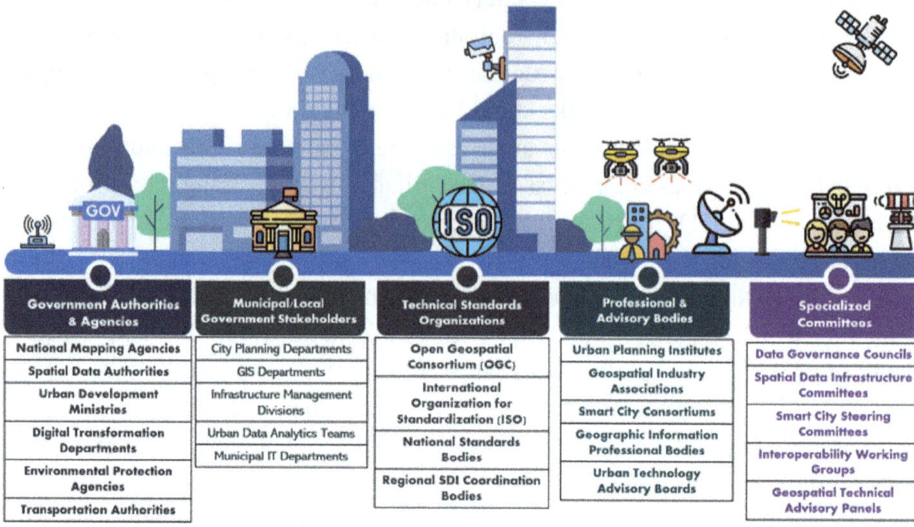

The policy ensures accurate mapping of urban infrastructure, land use, and environmental resources. The policies are also expected to address data interoperability and standardization. Critical policy considerations are:

1. Spatial reference system standardization
2. Topology rules enforcement
3. Metadata schema specifications that align with ISO 19115 and INSPIRE directives.

These are required for seamless geospatial data integration between diverse sources and various technologies. Privacy and security protocols are put in place to protect sensitive geospatial data.

Similarly, ethical guidelines give fair and equitable access and use of geospatial information to its citizens. Collaborative governance models are welcome because they promote public-private partnerships (Benfeldt et al., 2020). Such models can drive innovation and resource sharing and promote sustainable urban development. Finally, a timely and continuous policy evaluation and adaptation must be mandated because of technological advancements and evolving urban challenges. Policies should consider the need for a skilled workforce and their training needs. Support from the workforce is necessary to drive the adoption and implementation of geospatial technologies in smart cities. It includes providing training and development opportunities for GIS, data science, and urban planning professionals. It is more important because GeoAI and deep learning are emerging as key environmental sustainability tools (Fu et al., 2025). Since technologies keep evolving faster, there may be some cushion for the policies. The policies need to be dynamic and accommodative. They should have space for advancements, such as AI-driven spatial analysis and federated learning.

EMERGING TOPICS IN GEOSPATIAL TECHNOLOGY ADOPTION

3

Challenges in Federated Learning for Geospatial Data

Federated Learning (FL) for Geospatial data allows collaborative learning from decentralized data sources (Cao et al., 2024). The universal feeling is that centralizing data is unnecessary because of potentially sensitive location information. Preserving data privacy and sensitivity has become increasingly important because of the vast amounts of personal and sensitive data generated and collected. It led to a new area called privacy-preserving data analysis (PPDA) (David Shamoo Excel, 2024). While FL promises many benefits, it comes with some challenges too:

Model Synchronization and Scalability

In FL within Smart Cities, aggregating models combine insights or learnings from multiple decentralized sources. Eventually, they form a cohesive, centralized model or decision framework. The data sources can be edge devices, sensors, and local servers. These models send continuous updates from multiple devices. So, ensuring synchronization can be challenging, especially when a large-scale Smart city environment with numerous nodes is involved. Network designers should consider scalability and network latency issues that can potentially arise.

Data Imbalance and Variability

Urban data management involves data collection from different sources. They vary in quality, volume, and patterns. It can lead to biased models. Federated Learning needs to account for these disparities and adjust model training accordingly (Wang et al., 2021).

Communication and Computation Overheads

FL reduces the need for raw data transfer. However, frequent communication of model updates between nodes and the central server can still cause overheads. Further, Smart cities use fog and edge devices. Some examples of such deployments are smart traffic lights, smart meters, utility sensors, and environment sensors. These devices have minimal computational power and bandwidth capacities. Some issues that future research should address are (Almanifi et al., 2023):

- Large-scale implementations
- Security-efficiency trade-off
- Unified framework and
- Standardized benchmarking

Security and Trustworthiness

While FL offers enhanced privacy, model updates could still be susceptible to attacks (e.g., model poisoning or adversarial attacks). Ensuring the integrity of the models and safeguarding against malicious updates are critical. Future research should address challenges such as overcoming the challenge of data memorization, developing hybrid approaches, and explainability for fairness (Zhang et al., 2024).

Potential Future Challenges and Opportunities

Storage and retrieval of large quantities of data is important. Increased penetration of geospatial technologies and deployment of more IoT devices necessitate this. Researchers will have to focus on developing efficient data processing and storage solutions. Geospatial data will meet data from emerging sources such as social media and sensors. Techniques for merging such data to create a unified view will be necessary. AI-powered geospatial analysis can help identify patterns and trends in urban data. Advanced analytics and ML techniques are required to get helpful information from large datasets that city administrators can use. However, these methods require significant computational resources and expertise. Though computing power and staffing resources have become commodities that can be purchased online, they come at a considerable cost. Blockchain will ensure geospatial data integrity and security. The use of edge computing and fog computing will also become increasingly crucial. Developing new geospatial standards and protocols will ensure interoperability and seamless geospatial data integration from diverse sources.

CONCLUSION

Integrating geospatial technologies with computing technologies can revolutionize the application development of Smart Cities. Several benefits and new opportunities accrue for city planners and administrators. Implementing the technologies will improve urban services and enhance citizens' quality of life. However, various challenges are faced in the integration process. This chapter draws the attention of practitioners and researchers to the fundamental complexities and technological challenges of implementing geospatial technologies in Smart Cities. The continuous evolution of technology necessitates ongoing adaptation and investment. The future of Smart Cities depends on smoothly connecting and combining different technologies and modern innovations. Policymakers must encourage a data-driven urban management and governance approach. Practitioners encounter several data management and quality challenges. These include scalability and performance issues, data privacy and security concerns, and technical and operational barriers. Developing robust data governance strategies, investing in performance optimization techniques, and establishing clear regulatory and policy frameworks to overcome challenges are essential. There is a need for a skilled workforce with expertise in geospatial technologies. Continous training and development opportunities are necessary to drive the adoption and implementation of these technologies in Smart Cities. The future evolution of technologies will likely address the challenges and help Smart Cities stay sustainable and resilient.

REFERENCES

Abdulmanov, R., Miftakhov, I., Ishbulatov, M., Galeev, E., & Shafeeva, E. (2021). Comparison of the effectiveness of GIS-based interpolation methods for estimating the spatial distribution of agrochemical soil properties. *Environmental Technology & Innovation*, *24*, 101970. DOI: 10.1016/j.eti.2021.101970

Akintayo, T. A., Olobo, N. A., & Atinuke, A. T., & AbdulKareem, I. O. (2024). Computing Performance Optimization Through Parallelization: Techniques and Evaluation. *International Journal of Education, Management, and Technology*, *2*(3), 418–439. DOI: 10.58578/ijemt.v2i3.4210

Al-Yadumi, S., Xion, T. E., Wei, S. G. W., & Boursier, P. (2021). Review on Integrating Geospatial Big Datasets and Open Research Issues. *IEEE Access : Practical Innovations, Open Solutions*, *9*, 10604–10620. DOI: 10.1109/ACCESS.2021.3051084

Algarni, M., & Mishra, S. (2024). A Secure and Reliable Framework for Explainable Artificial Intelligence (XAI) in Smart City Applications. *Engineering, Technology &. Applied Scientific Research*, *14*(4), 15291–15296. DOI: 10.48084/etasr.7676

Almanifi, O. R. A., Chow, C.-O., Tham, M.-L., Chuah, J. H., & Kanesan, J. (2023). Communication and computation efficiency in Federated Learning: A survey. *Internet of Things : Engineering Cyber Physical Human Systems*, *22*, 100742. DOI: 10.1016/j.iot.2023.100742

Arundel, S. T., Li, W., & Campbell, B. B. (2023). Reimagining standardization and geospatial interoperability in today's GeoAI culture. *Proceedings of the 6th ACM SIGSPATIAL International Workshop on AI for Geographic Knowledge Discovery*, 83–84. https://doi.org/DOI: 10.1145/3615886.3627744

Bachiega, J., Holanda, M., & Araujo, A. P. F. (2021). A proposal to minimize the cost of processing big geospatial data in public cloud providers. *Transactions in GIS*, *25*(3), 1599–1624. DOI: 10.1111/tgis.12754

Balfaqih, M., & Alharbi, S. A. (2022). Associated Information and Communication Technologies Challenges of Smart City Development. *Sustainability*, *14*(23), 16240. DOI: 10.3390/su142316240

Balfaqih, M., Balfagih, Z., Lytras, M. D., Alfawaz, K. M., Alshdadi, A. A., & Alsolami, E. (2023). A blockchain-enabled IoT logistics system for efficient tracking and management of high-price shipments: A resilient, scalable and sustainable approach to smart cities. *Sustainability*, *15*(18), 13971.

Bebortta, S., Das, S. K., Kandpal, M., Barik, R. K., & Dubey, H. (2020). Geospatial Serverless Computing: Architectures, Tools and Future Directions. *ISPRS International Journal of Geo-Information, 9*(5), 311. DOI: 10.3390/ijgi9050311

Benfeldt, O., Persson, J. S., & Madsen, S. (2020). Data Governance as a Collective Action Problem. *Information Systems Frontiers, 22*(2), 299–313. DOI: 10.1007/s10796-019-09923-z

Bhardwaj, A., Bachu, S., Rajendran, S. P., Rastogi, S., Ramakrishna, K., Balalakshmi, C., & Sharma, P. (2024). The Future of Territorial Management: AI, IoT, and Big Data Synergies. In Alla, L., Alj, B., & Bentalha, B. (Eds.), (pp. 239–272). Advances in Electronic Government, Digital Divide, and Regional Development. IGI Global., DOI: 10.4018/979-8-3693-6854-1.ch008

Bruno, S. da M. (2024). *Processing Geospatial Data from Public Datasets to Assess Cycling Infrastructure in Smart Cities* [Master in Electrical and Computer Engineering Thesis, Faculdade De Engenharia, Universidade Do Porto]. https://repositorio-aberto.up.pt/bitstream/10216/162384/2/694027.pdf

Calzati, S., & Van Loenen, B. (2023). An ethics assessment list for geoinformation ecosystems: Revisiting the integrated geospatial information framework of the United Nations. *International Journal of Digital Earth, 16*(1), 1418–1438. DOI: 10.1080/17538947.2023.2200041

Cao, F., Liu, B., He, J., Xu, J., & Xiao, Y. (2024). Privacy preservation-based federated learning with uncertain data. *Information Sciences, 678*, 121024. DOI: 10.1016/j.ins.2024.121024

Coetzee, S., Camboim, S. P., Griffin, A. L., Kubicek, P., Harvey, F., & Behr, F.-J. (2023). “Cartography and Open Standards”: A Proposal for an Open Geospatial Consortium Technical Paper. *Abstracts of the ICA, 6*, 1–2. https://doi.org/DOI: 10.5194/ica-abs-6-39-2023

Das, J., Ghosh, S. K., & Buyya, R. (2021). Geospatial Edge-Fog Computing: A Systematic Review, Taxonomy, and Future Directions. In Mukherjee, A., De, D., Ghosh, S. K., & Buyya, R. (Eds.), *Mobile Edge Computing* (pp. 47–69). Springer International Publishing., DOI: 10.1007/978-3-030-69893-5_3

Davoli, L., Belli, L., Dall’Olio, A., Di Nocera, F., Adorni, P., Cantelli, A., & Ferrari, G. (2024). Data Integration in a Smart City: A Real Case. In Menozzi, R. (Ed.), *Information and Communications Technologies for Smart Cities and Societies* (Vol. 5, pp. 11–24). Springer Nature Switzerland., DOI: 10.1007/978-3-031-39446-1_2

Erskine, M. A., Pick, J. B., Satpathy, A., Díaz López, A., Sarkar, A., & Shin, N. (2024). Location Analytics in Information Systems: Opportunities for Research and Teaching. *Communications of the Association for Information Systems*, *55*. https://aisel.aisnet.org/cais/vol55/iss1/21

Eskhita, R., Manda, V. K., & Hlali, A. (2021). Dubai and Barcelona as Smart Cities: Some Reflections on Data Protection Law and Privacy. *Environmental Policy and Law*, *51*(6), 403–407. DOI: 10.3233/EPL-210023

Excel, D. S. (2024). Privacy-preserving data analysis. *World Journal of Advanced Research and Reviews*, *23*(3), 597–609. DOI: 10.30574/wjarr.2024.23.3.2724

Ferro, M., Silva, E., & Fidalgo, R. (2023). AStar: A modeling language for document-oriented geospatial data warehouses. *Data & Knowledge Engineering*, *145*, 102174. DOI: 10.1016/j.datak.2023.102174

Ferro-Diez, L. E., Villegas, N. M., & Diaz-Cely, J. (2020). Location Data Analytics in the Business Value Chain: A Systematic Literature Review. *IEEE Access : Practical Innovations, Open Solutions*, *8*, 204639–204659. DOI: 10.1109/ACCESS.2020.3036835

Fouad, K. M., Ismail, M. M., Azar, A. T., & Arafa, M. M. (2021). Advanced methods for missing values imputation based on similarity learning. *PeerJ. Computer Science*, *7*, e619. DOI: 10.7717/peerj-cs.619 PMID: 34395861

Fu, B., Zhang, J., Wu, X., & Meadows, M. E. (2025). Geography's hotspots and frontiers: Diverse, systematic, and intelligent trends. *Geography and Sustainability*, *100285*. Advance online publication. DOI: 10.1016/j.geosus.2025.100285

Ghosh, S., & Eldawy, A. (2022). AID*: A Spatial Index for Visual Exploration of Geo-Spatial Data. *IEEE Transactions on Knowledge and Data Engineering*, *34*(8), 3569–3582. DOI: 10.1109/TKDE.2020.3026657

González, A., Kelly, C., & Rymszewicz, A. (2020). Advancements in web-mapping tools for land use and marine spatial planning. *Transactions in GIS*, *24*(2), 253–267. DOI: 10.1111/tgis.12603

Gupta, R., & Sharma, P. (2024). Protecting Privacy in the Smart City: Novel Approaches for Urban Informatics and Geospatial Data Management. *Emerging Trends in Machine Intelligence and Big Data*, 17.

Gupta, S., & Loukissas, Y. A. (2023). Making Smart Cities Explainable: What XAI Can Learn from the "Ghost Map." *Extended Abstracts of the 2023 CHI Conference on Human Factors in Computing Systems*, 1–8. https://doi.org/DOI: 10.1145/3544549.3585847

Hierink, F., Yaghmaei, N., Bakker, M. I., Ray, N., & Van Den Homberg, M. (2024). Geospatial tools and data for health service delivery during disasters: Opportunities and challenges across the management cycle. *Geospatial Health, 19*(2). Advance online publication. DOI: 10.4081/gh.2024.1284 PMID: 39470294

Ioanid, A., & Andrei, N. (2024). *Artificial Intelligence and Geospatial Technologies for Sustainable Maritime Logistics. Case Study: Port Of Constanta, Romania.* 15. https://marlog.aast.edu/files/marlog13/MARLOG13_paper_112.pdf

Jitta, A. S., Manda, V. K., & Tarnanidis, T. (2024). Geospatial Technologies for Smart Cities. In Darwish, D., & Chemingui, H. (Eds.), (pp. 103–136). Advances in Geospatial Technologies. IGI Global., DOI: 10.4018/979-8-3693-8054-3.ch004

Khayyat, M., Balfaqih, M., Balfaqih, H., & Ismail, M. (2024). Challenges and Factors Influencing the Implementation of Green Logistics: A Case Study of Saudi Arabia. *Sustainability, 16*(13), 5617. DOI: 10.3390/su16135617

Kim, J., Hong, S., Jeong, S., Park, S., & Yu, K. (2024). SGIR-Tree: Integrating R-Tree Spatial Indexing as Subgraphs in Graph Database Management Systems. *ISPRS International Journal of Geo-Information, 13*(10), 346. DOI: 10.3390/ijgi13100346

Lee, J. (2023). Beyond Geospatial Inquiry—How Can We Integrate the Latest Technological Advances into Geography Education? *Education Sciences, 13*(11), 1128. DOI: 10.3390/educsci13111128

Liu, Y., Hao, T., Gong, X., Kong, D., & Wang, J. (2021). Research on Hybrid Index Based on 3D Multi-Level Adaptive Grid and R+ Tree. *IEEE Access : Practical Innovations, Open Solutions, 9*, 146010–146022. DOI: 10.1109/ACCESS.2021.3115510

Manda, V. K., Christy, V., & Hlali, A. (2024). Current Trends, Opportunities, and Futures Research Directions in Geospatial Technologies for Smart Cities. In Darwish, D., & Chemingui, H. (Eds.), (pp. 239–270). Advances in Geospatial Technologies. IGI Global., DOI: 10.4018/979-8-3693-8054-3.ch009

Marsh-Hunn, D., Trilles, S., Gonzalez-Perez, A., Torres-Sospedra, J., & Ramos, F. (2021). A Comparative Study in the Standardization of IoT Devices Using Geospatial Web Standards. *IEEE Sensors Journal, 21*(4), 5512–5528. DOI: 10.1109/JSEN.2020.3031315

Masik, G., Sagan, I., & Scott, J. W. (2021). Smart City strategies and new urban development policies in the Polish context. *Cities (London, England), 108*, 102970. DOI: 10.1016/j.cities.2020.102970

Merodio Gómez, P., Ramírez Santiago, A., García Seco, G., Casanova, R., MacKenzie, D., & Tucker, C. (2022). Ethics in the use of geospatial information in the Americas. *Technology in Society*, *69*, 101964. DOI: 10.1016/j.techsoc.2022.101964

Micheli, M., Ponti, M., Craglia, M., & Berti Suman, A. (2020). Emerging models of data governance in the age of datafication. *Big Data & Society*, *7*(2), 2053951720948087. DOI: 10.1177/2053951720948087

Moreno-Álvarez, S., Paoletti, M. E., Sanchez-Fernandez, A. J., Rico-Gallego, J. A., Han, L., & Haut, J. M. (2024). Federated learning meets remote sensing. *Expert Systems with Applications*, *255*, 124583. DOI: 10.1016/j.eswa.2024.124583

Nirandjan, S., Koks, E. E., Ward, P. J., & Aerts, J. C. J. H. (2022). A Spatially Explicit Harmonized Global Dataset of Critical Infrastructure. *Scientific Data*, *9*(1), 150. DOI: 10.1038/s41597-022-01218-4 PMID: 35365664

Pawar, N. S., & Sharma, K. V. (2025). Comprehensive review of remote sensing integration with deep learning in landslide forecasting and future directions. *Natural Hazards*. Advance online publication. DOI: 10.1007/s11069-025-07171-w

Pereira, P. J., Costa, N., Barros, M., Cortez, P., Durães, D., Silva, A., & Machado, J. (2022). A Comparison of Automated Time Series Forecasting Tools for Smart Cities. In Marreiros, G., Martins, B., Paiva, A., Ribeiro, B., & Sardinha, A. (Eds.), *Progress in Artificial Intelligence* (Vol. 13566, pp. 551–562). Springer International Publishing., DOI: 10.1007/978-3-031-16474-3_45

Priyashani, N., Kankanamge, N., & Yigitcanlar, T. (2023). Multisource Open Geospatial Big Data Fusion: Application of the Method to Demarcate Urban Agglomeration Footprints. *Land (Basel)*, *12*(2), 407. DOI: 10.3390/land12020407

Ramu, S. P., Boopalan, P., Pham, Q.-V., Maddikunta, P. K. R., Huynh-The, T., Alazab, M., Nguyen, T. T., & Gadekallu, T. R. (2022). Federated learning enabled digital twins for smart cities: Concepts, recent advances, and future directions. *Sustainable Cities and Society*, *79*, 103663. DOI: 10.1016/j.scs.2021.103663

Razavi-Termeh, S. V., Sadeghi-Niaraki, A., Abba, S. I., Ali, F., & Choi, S.-M. (2024). Enhancing spatial prediction of groundwater-prone areas through optimization of a boosting algorithm with bio-inspired metaheuristic algorithms. *Applied Water Science*, *14*(11), 244. DOI: 10.1007/s13201-024-02301-4

Ridha, S., & Kamil, P. A. (2021). The Problems of Teaching Geospatial Technology in Developing Countries: Concepts, Curriculum, and Implementation in Indonesia. *The Journal of Geography*, *120*(2), 72–82. DOI: 10.1080/00221341.2021.1872681

Semeniuk, V. (2024). Development of Optimization Algorithms for Geospatial Search in Large Volumes of Data. *Наука і Техніка Сьогодні, 5(33)*. https://doi.org/DOI: 10.52058/2786-6025-2024-5(33)-1083-1094

Silvestri, S., Tricomi, G., Damiano, E., Sicuranza, M., & Ciampi, M. (2025). A Data Platform for the Integration of Smart City Subsystems. In L. Barolli (Ed.), *Advances on P2P, Parallel, Grid, Cloud and Internet Computing* (Vol. 232, pp. 187–198). Springer Nature Switzerland. https://doi.org/DOI: 10.1007/978-3-031-76462-2_17

Sun, Y., Chen, C., Xu, Y., Xie, S., Blum, R. S., & Venkitasubramaniam, P. (2024). On the generalization discrepancy of spatiotemporal dynamics-informed graph convolutional networks. *Frontiers of Mechanical Engineering, 10*, 1397131. DOI: 10.3389/fmech.2024.1397131

Thomas, T., & Rajabi, E. (2021). A systematic review of machine learning-based missing value imputation techniques. *Data Technologies and Applications, 55*(4), 558–585. DOI: 10.1108/DTA-12-2020-0298

UN-IGIF. (2024). *United Nations Integrated Geospatial Information Framework (UN-IGIF)*. https://ggim.un.org/UN-IGIF/

Wang, L., Xu, S., Wang, X., & Zhu, Q. (2021). Addressing Class Imbalance in Federated Learning. *Proceedings of the AAAI Conference on Artificial Intelligence, 35*(11), 10165–10173. DOI: 10.1609/aaai.v35i11.17219

Wu, J., Gan, W., Chao, H.-C., & Yu, P. S. (2024). Geospatial Big Data: Survey and Challenges. *IEEE Journal of Selected Topics in Applied Earth Observations and Remote Sensing, 17*, 17007–17020. DOI: 10.1109/JSTARS.2024.3438376

Wu, K.-C., Chen, C.-H., & Hsu, C.-H. (2022). Three-dimensional visualization of thermal environments in urban canyons. *Geocarto International, 37*(27), 16261–16282. DOI: 10.1080/10106049.2022.2107713

Zhang, Y., Zeng, D., Luo, J., Fu, X., Chen, G., Xu, Z., & King, I. (2024). A Survey of Trustworthy Federated Learning: Issues, Solutions, and Challenges. *ACM Transactions on Intelligent Systems and Technology, 15*(6), 1–47. DOI: 10.1145/3678181

Zhao, Y., Li, S., Chen, H., & Xu, Y. (2022). Application of Smart City Construction in a New Data Environment. *Frontiers in Energy Research, 10*, 908338. DOI: 10.3389/fenrg.2022.908338

Zhou, Y., & Kankanhalli, A. (2021). AI Regulation for Smart Cities: Challenges and Principles. In Estevez, E., Pardo, T. A., & Scholl, H. J. (Eds.), *Smart Cities and Smart Governance* (Vol. 37, pp. 101–118). Springer International Publishing., DOI: 10.1007/978-3-030-61033-3_5

Zurbaran, M. A., Salazar, A., Brovelli, M. A., & Wightman, P. M. (2020). An Evaluation Framework for Assessing the Impact of Location Privacy on Geospatial Analysis. *IEEE Access : Practical Innovations, Open Solutions*, *8*, 158224–158236. DOI: 10.1109/ACCESS.2020.3019631

KEY TERMS AND DEFINITIONS

AI Bias: Artificial Intelligence algorithms can be biased. Such algorithms would give unfair, discriminatory, or unexpected outcomes. It is a particular concern for geospatial technologies because AI is used to make decisions regarding allocating resources and services to its citizens.

Distributed Computing: It involves breaking down complex geospatial computations into smaller tasks. Smaller tasks can be executed in parallel across multiple nodes or machines. Using this helps reduce processing time and increase scalability. Spatial data partitioning is a technique used to divide large geospatial datasets into smaller, manageable pieces. Each of the pieces can then be processed or analyzed independently. This partitioning is particularly valuable in distributed computing environments. Data can be spread across multiple nodes or servers to speed up processing and reduce the complexity of handling large datasets.

Explainable AI (XAI): Explainable AI is a collection of various aspects of AI systems. They deliver clear and transparent justifications for their outputs (judgments/results). For Smart Cities, XAI ensures that stakeholders understand AI-driven geospatial insights. It addresses accountability, fairness, and ethics in urban planning and decision-making. City administrators and decision-makers can understand the result of an AI system and how the result is arrived.

Federated Learning (FL): Federated Learning is a training mechanism. It is a decentralized system that teaches machine learning (ML) models across several devices. It does not transfer raw data to a central server (central location). In Smart Cities, FL enables collaborative geospatial analytics and preserves data privacy. FL can effectively distribute data across edge devices such as smartphones and IoT devices. It is suitable for applications that require data from multiple private sources.

Geospatial AI: Geospatial AI or GeoAI combines Artificial Intelligence with geospatial data. It analyzes and interprets spatial information and enhances decision-making in various aspects of Smart Cities. It uses AI techniques, such as machine learning and neural networks. They process and analyze spatial components (such as location-based information). They derive spatial insights and patterns and provide insights into complex urban phenomena.

Urban Canyons: Urban Canyns are dense, closely spaced urban environments with tall buildings on either side of the street. They generally have a high population density. They hinder the data collection process and lead to signal loss in GPS. They cause occlusions in imagery, compromise the geospatial data collection process, and lead to inaccurate analysis.

Public Data: Smart cities collect vast amounts of data about their citizens. This data is analyzed using analytic tools that provide insights. The insights appear on the dashboards of city administrators who can use them for effective decisions. The data can help improve city services but raises concerns about data privacy and security.

Spatial Indexing: Spatial indexing is the creation of data structures that allow efficient access and querying of spatial data. Spatial indexing methods such as R-trees, K-D trees, or Quad-trees are used. They help geospatial systems to organize spatial data hierarchically. These techniques optimize search and retrieval and improve the performance and scalability of large urban datasets.

Surveillance: Surveillance in Smart Cities involves using geospatial technologies to collect and examine data. Data can be about the movement of objects (people, vehicles, amongst others), often in real-time across urban spaces. These technologies collect vast amounts of location data. They enhance public safety, traffic management, and emergency response by providing insights into patterns and potential risks. This is done using Location-based services (LBS). LBS include GPS, Wi-Fi, cellular network data, CCTV cameras, IoT sensors, and Satellite imaging. Surveillance using geospatial technologies raises several concerns related to privacy, misuse, bias, and security.

Chapter 12

Global Perspectives on Smart Cities:

Strategies, Technologies, and Lessons Learned

Mohammed Balfaqih
https://orcid.org/0000-0002-6498-2408
University of Jeddah, Saudi Arabia

Ehab Ali
https://orcid.org/0009-0007-7737-8481
University of Benghazi, Libya

ABSTRACT

This chapter explores the evolution of smart cities and their role in tackling global urban challenges through advanced technologies and sustainability-driven frameworks. It examines key components—technological, economic, social, and governance—and provides a comparative analysis of five leading smart cities: Zurich, Oslo, Canberra, Geneva, and Singapore. It further aligns smart city initiatives with select UN Sustainable Development Goals (SDGs), highlighting their impact on healthcare, energy, mobility, inclusion, and environmental resilience. The chapter concludes with key lessons on policy, ethics, and collaboration for guiding future sustainable urban transformation. This multidimensional perspective offers valuable insights for researchers, policymakers, and urban planners committed to shaping intelligent, equitable, and resilient cities.

DOI: 10.4018/979-8-3373-0154-9.ch012

INTRODUCTION

The concept of the smart city first gained prominence in the mid-1990s, emerging as a response to the integration of advanced information and communication technologies (ICT) in urban environments. Initially, it was associated with city branding efforts, digital governance initiatives, and the attraction of technology-driven investment to spur regional economic development. Over time, the term "smart" has evolved to represent intelligent, automated systems designed to enhance the efficiency and effectiveness of various urban functions (Ahad et al., 2020). At its core, the smart city vision encompasses the application of innovation across all aspects of urban life—ranging from infrastructure, governance, and mobility to sectors such as healthcare, education, energy, industry, agriculture, and the environment (Guelzim et al., 2016).

Smart cities serve as platforms for applied research and innovation that align urban development with sustainability goals. They are defined by the integration of cutting-edge technologies like the Internet of Things (IoT), 5G networks, robotics, and automation, all of which enhance service delivery and improve day-to-day urban living. These digital solutions facilitate more efficient housing systems, communication networks, and city services, while also addressing pressing global challenges such as climate change and aging populations. One of the key ambitions in smart city development is the realization of carbon-neutral cities, supported by renewable energy technologies, energy storage systems, and sustainable resource management. Equally important are the social dimensions of smart urbanism, which emphasize inclusivity, community well-being, and equitable access to opportunities.

Today, smart city development has become a global movement, closely aligned with the United Nations' 2030 Agenda for Sustainable Development and its 17 Sustainable Development Goals (SDGs). Among these, the goal of creating inclusive, safe, resilient, and sustainable cities holds particular significance. Smart city initiatives often serve as the operational framework for achieving these long-term objectives, utilizing ICT-enabled innovations to foster more sustainable, efficient, and livable urban environments (Rodrigues & Franco, 2019). Governments around the world are leveraging smart technologies not only to respond to urban challenges—such as rapid population growth, resource depletion, and environmental degradation—but also to meet their national and international sustainability commitments.

This chapter delves into the key elements that define smart city development and offers a comprehensive examination of their implementation across diverse global contexts. The discussion begins with an in-depth overview of smart city components and initiatives, exploring the technological foundations, human and social capital, business ecosystems, and digital infrastructure essential to their success. The chapter then presents a comparative analysis of five leading smart cities—Zurich,

Oslo, Canberra, Geneva, and Singapore—each exemplifying unique approaches to governance, innovation, and sustainable urban planning. Building on this, the chapter explores the relationship between smart city strategies and selected SDGs, illustrating how technological advancement can support global development goals. Finally, the chapter synthesizes key lessons learned from pioneering smart cities, highlighting effective policy frameworks, stakeholder engagement strategies, cross-city collaboration, and best practices for future urban transformation. Together, these sections provide a holistic understanding of how smart city initiatives contribute to the creation of intelligent, inclusive, and sustainable urban futures.

SMART CITY COMPONENTS AND INITIATIVES

Smart city initiatives have demonstrated diverse approaches, heavily reliant on robust qualitative insights to effectively guide urban transformation. Comprehensive data encompassing social, economic, technological, and environmental aspects provides researchers and policymakers with valuable information, enabling informed decisions to foster sustainable urban growth. Businesses investing in environmentally sustainable urban developments also rely on this multifaceted data for strategic planning. Conventional smart city frameworks typically consist of six essential dimensions: a smart economy fostering business-friendly environments, smart people promoting collaborative networks, smart governance enhancing social and cultural interactions, smart mobility integrating advanced technology in transport systems, smart environment emphasizing sustainable infrastructure and buildings, and smart living, characterized by digitalized urban spaces.

In addition, foundational infrastructure elements critical to the success of smart city projects include efficient waste management systems, digitally integrated housing solutions, reliable urban mobility networks, comprehensive safety and security measures, innovative materials and technologies, cleanliness and effective waste disposal, accessible urban transportation, affordable housing options, robust IT infrastructure and digital connectivity, and transparent governance mechanisms incorporating e-governance platforms and active citizen participation. The potential components and strategic priorities for smart city initiatives, reflective of varied governmental decisions across different regions, are comprehensively illustrated in Figure 1.

Figure 1. Smart city initiatives and features

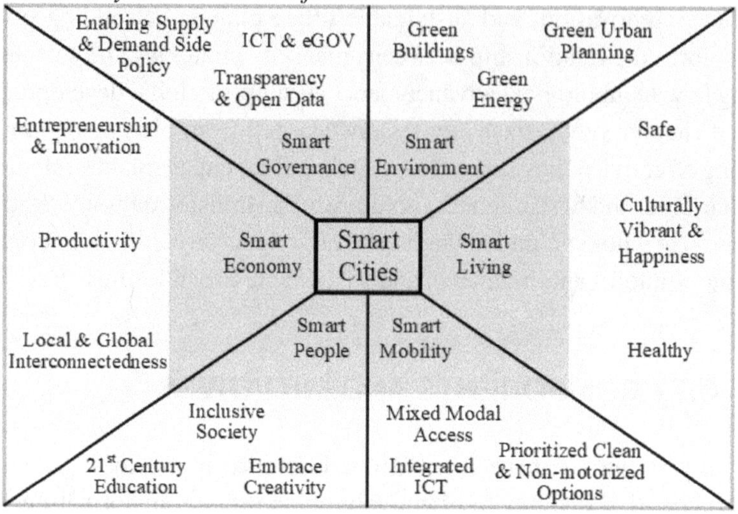

Several core components underpin the development and operation of smart urban environments, each playing a significant role in achieving comprehensive and sustainable city development. The primary urban-centric components can be outlined as follows:

Technology Integration in Smart Environments

In a smart urban environment, technology plays a crucial role by efficiently collecting, organizing, and distributing vast amounts of data to residents and city officials. Technological advancements and improved accessibility facilitate interconnected sensing systems and IoT infrastructure, significantly enhancing urban functionality. By streamlining access to critical information and services—such as smart transportation, effective communication channels, and utility monitoring—technology substantially improves residents' daily experiences. This ease of access also promotes informed decision-making and innovative practices, directly contributing to urban sustainability and enhancing municipal services (Angelidou, 2014; Allwinkle & Cruickshank, 2011; Komninos, 2011; Schaffers et al., 2011; Tranos & Gertner, 2012).

Development of Human and Social Capital

Smart cities actively foster human capital and social innovation by nurturing environments conducive to knowledge exchange, increased citizen engagement, and digital literacy enhancement. Collaborative platforms facilitate the collective generation and sharing of knowledge, creating an inclusive intellectual ecosystem. Smart cities attract highly skilled individuals who creatively leverage technology, becoming primary drivers of urban innovation. It is widely recognized that creative and skilled talent significantly accelerates urban development through novel ideas and strategies generated within digital communities, thus enriching the overall urban ecosystem (Komninos, 2008; Nam & Pardo, 2020; Gillanders & Whelan, 2014).

Enhancement of Smart Commerce and Business Environment

Advancing the business sector is a fundamental objective within smart city initiatives, aiming to stimulate entrepreneurial dynamism and innovation. Urban areas adopting smart city models benefit from intelligent enterprise solutions and digital market strategies, promoting balanced economic development through socio-digital innovations. Smart governance frameworks support business-led urban transformations by creating environments favorable for entrepreneurship. Such initiatives involve cutting-edge mobility solutions, digital banking, and organizational tools designed for affordability and accessibility, thereby facilitating effective civic engagement, business sustainability, and strengthened social connectivity (Mahmood, 2018; Savithramma, Sumathi, & Sudhira, 2022; Turetken et al., 2019; Zavratnik et al., 2020).

Advanced Smart Networking

Contemporary networking primarily leverages digital platforms, fostering transnational economic and international collaboration, especially within smart city contexts. Smart city analyses often emphasize intra- and inter-city networking to enhance public image, disseminate effective practices, diversify economic activities, and achieve economies of scale. Urban administrations increasingly prioritize innovative and high-quality services, leveraging partnerships to meet growing citizen expectations in the digital era. Such strategic collaborations and networks facilitate resource sharing, knowledge exchange, and reinforce each city's unique identity. Consequently, contemporary marketing and communication strategies are now integral components of urban development planning, prominently within the smart city framework.

COMPARATIVE ANALYSIS OF SMART CITY INITIATIVES WORLDWIDE

Many leading global cities have actively embraced the transformative impacts of smart technology, recognizing its capacity to significantly enhance urban life. The positive outcomes brought about by these technological advancements have notably improved residents' experiences and quality of life. In this context, Zurich, Oslo, Canberra, Geneva, and Singapore have emerged as exemplary cities, highlighted by their positions in the IMD Smart City Index (2024). The IMD Smart City Index offers a comprehensive assessment of 142 cities worldwide, evaluating their effectiveness in utilizing smart technologies and infrastructures. It employs detailed survey data reflecting residents' perceptions of urban performance and quality of life. The index, distinguished by its stability and methodological maturity, leverages moving averages calculated over three consecutive editions, ensuring reliable and insightful comparisons. Consequently, the IMD Smart City Index remains a critical resource, empowering city leaders to effectively navigate emerging urban challenges and bridge digital inequalities.

Zurich

Zurich has implemented a range of sustainability strategies aimed at reducing its environmental footprint while accommodating urban growth. Central to its approach is urban densification combined with advanced ICT integration, enabling the city to balance compact urban development with ecological responsibility (Schrotter & Hürzeler, 2020). Research institutions also contribute significantly by fostering innovation in service-oriented sectors, further supporting Zurich's green agenda (Duygan et al., 2022).

In the transport sector, Zurich prioritizes sustainable mobility by expanding public transport and cycling infrastructure, significantly reducing emissions and improving air quality (Menendez & Ambühl, 2022). Green infrastructure projects, especially in high-density areas like District 9, are designed to enhance biodiversity, provide ecosystem services, and support urban ecological health (Perić et al., 2023).

Collaborative planning plays a key role in Zurich's urban sustainability, with redevelopment projects such as the Koch Area exemplifying the benefits and challenges of multi-stakeholder engagement, including high costs and the need for stronger public participation (Peric Momcilovic et al., 2022).

Technologically, Zurich leverages digital twins and smart platforms to improve land-use governance, resource efficiency, and municipal service delivery (Schrotter & Hürzeler, 2020; Duygan et al., 2022). The city's ambitious "2000-watt Society" initiative, targeting major reductions in energy use and emissions, positions Zurich as a leader in sustainable urban mobility and energy transition (Menendez & Ambühl, 2022).

Integrating green spaces into city planning has also contributed to passive cooling, reduced dependence on air conditioning, and enhanced overall air quality (Perić et al., 2023). Despite limited quantitative data, Zurich's consistent top rankings in global smart city indices attest to the effectiveness of its environmental governance (Toh, 2022).

Waste management is another strong area, with smart technologies and strategic planning improving recycling and collection systems. Community-driven planning further strengthens these efforts, though empirical validation is still needed (Duygan et al., 2022; Toh, 2022; Peric Momcilovic et al., 2022).

Finally, Zurich's advancements in air quality result from a combination of alternative transportation, robust green planning, and tech-enabled solutions—showcasing a comprehensive model of smart and sustainable urban development (Duygan et al., 2022; Menendez & Ambühl, 2022; Perić et al., 2023).

Oslo

Oslo has emerged as a leading example of urban sustainability through the implementation of comprehensive strategies to combat climate change. At the core of its efforts is a climate and energy plan that emphasizes cross-sectoral collaboration to reduce emissions and foster resilience (Hofstad & Torfing, 2017). The city effectively utilizes mobility data and promotes public participation, underscoring the importance of civic engagement and digital innovation in sustainable urban governance (Julsrud & Krogstad, 2020).

A key pillar of Oslo's strategy is its investment in electric vehicles (EVs) and cycling infrastructure. Incentives such as toll exemptions and free parking have helped make EV adoption widespread, positioning Oslo as a global leader in electric mobility (Ruggieri et al., 2021). The city has also pioneered biogas-powered public transport and initiated a car-free city center to encourage non-motorized transport and improve air quality (Nikolov, 2024). Projects like Hovinbyen aim to establish self-sustaining districts that merge economic growth with ecological goals (Senior et al., 2021).

Oslo's ambitious climate targets include a 50% reduction in greenhouse gas emissions by 2020 and carbon neutrality by 2050. These efforts have already achieved tangible results, with annual emissions reduced by approximately 4,500 metric tonnes (Nikolov, 2024). However, challenges such as the high cost of zero-emission construction persist (Andersen & Skrede, 2017).

In waste management, Oslo employs big data analytics to optimize operations and improve service delivery. This digital approach enhances efficiency and supports broader climate resilience strategies (Julsrud & Krogstad, 2020; Nikolov, 2024). Hofstad and Torfing (2017) emphasize that such integrated planning reflects Oslo's commitment to sustainable urban development.

Transport initiatives also play a vital role in improving air quality. The city's efforts to reduce fossil fuel use and expand public transport have contributed to a 22% decrease in PM2.5 and NO2 levels since 2016 (Ruggieri et al., 2021). These findings are supported by Senior et al. (2021), who highlight the success of climate-smart mobility policies in fostering healthier urban environments.

Canberra

Canberra ranks as Australia's top smart city, excelling across key domains such as People, Mobility, Livability, and Governance. It holds second place nationally in economic performance, trailing only Sydney. Canberra leads in postgraduate education attainment, cultural participation, organized sports, and managerial employment, reflecting its top ranking in the "People" category. Notably, cultural engagement rates—3.4% for library visits and 2.7% for museum/art gallery visits—exceed national averages, while 34.7% of residents participate in organized sports compared to 24.4% nationwide (ACT Planning Strategy, 2024). Other contributing factors include high income levels, strong digital connectivity, civic engagement, and environmental initiatives (Tariq et al., 2020). However, environmental performance in water and solar initiatives remains an area for improvement, with results comparable to less-performing regions like Greater Darwin and Hobart (Tariq et al., 2020).

Canberra's vision for sustainability centers on achieving net-zero emissions by 2045 and enhancing resilience to climate change. City planning integrates sustainable design, efficient resource use, and ecosystem protection. The city has committed to 100% renewable electricity since 2020 and follows a detailed Climate Adaptation Strategy (ACT Planning Strategy, 2024). Emission reduction milestones are set at 50–60% by 2025, 65–75% by 2030, and 90–95% by 2040 (ACT Planning Strategy, 2024). Major emission sources include transport and natural gas in buildings, prompting strategic investment in solar, wind, and battery storage technologies.

To reduce transportation emissions, Canberra is expanding its light rail network, enhancing public transit, and improving pedestrian and cycling infrastructure. Climate resilience efforts target the urban heat-island effect through "living infrastructure" such as trees, parks, and waterways, which also support biodiversity and community wellbeing.

Water management is another priority. The Canberra Water Strategy 2014–44 guides efforts to improve stormwater retention, manage flood risks, and implement water-sensitive urban design standards in new developments. These initiatives aim to optimize the water cycle and reinforce Canberra's identity as a sustainable, livable city.

Geneva

Effective waste management in Geneva encompasses transport, treatment, and storage, with significant cost and environmental implications due to the volume and weight of materials involved. Coordinated recycling between construction sites presents a promising approach to reducing landfill dependency, conserving natural resources, and limiting transportation demands. The report *Le recyclage des matériaux de construction à Genève* (GESDEC, 2006) promotes this model by advocating for the reuse of excavated materials and the use of recycled aggregates in concrete.

To facilitate this process, digital material exchange platforms have emerged as essential tools for improving waste management in construction. These platforms—such as those used in the cantons of Vaud and Jura and Salza.ch—enable direct transactions between suppliers and buyers, fostering resource circularity without requiring intermediary storage (Varesano et al., 2018). Geneva's Materrium initiative, classified as a "ressourcerie," combines online and physical systems to display, manage, and resell second-hand construction materials, further expanding access to sustainable resources.

The Smart Geneva Initiative, led by the State of Geneva and aligned with the UN's U4SSC framework, adopts a holistic and human-centered smart city approach. It integrates sustainability, inclusivity, and ethical innovation while emphasizing public health, environmental stewardship, and resilience. Drawing inspiration from global leaders such as Amsterdam and Vienna, Smart Geneva encourages civic participation through an online portal where residents can propose urban projects. Examples include *Légumes perches* and *Potager des Nations* (urban agriculture), *Géovélo* and *OVO* (sustainable mobility), and *HUG@home* (telemedicine) (Geneve.ch., 2024).

The initiative also promotes "Green IT" to reduce the environmental impact of digital infrastructure. Its governance model involves collaboration across cantonal agencies, municipalities, academia, and utilities, coordinated by the Department of Territory's Information Division.

Geneva's broader sustainability strategy aims for 100% renewable energy by 2050, supported by solar, geothermal, and building retrofitting projects. From 2010 to 2017, these efforts reduced energy use in municipal buildings by 10% and CO_2 emissions by 25% (Geneve.ch., 2024). The city's Master Plan emphasizes balanced residential and employment development, while responsible mobility is advanced through public transit promotion and reduced car dependency, despite challenges in implementing speed reduction zones.

Geneva also invests heavily in waste reduction, recycling, and composting, with rising recycling rates since 2009. Water conservation, pollution control, and biodiversity protection efforts are similarly prioritized through urban greening and environmental inventories. Collectively, these actions position Geneva as a model for integrated, sustainable urban development.

Singapore

Singapore is internationally recognized as a leading example of smart city innovation, consistently securing top positions in global smart city rankings the IMD Smart City Index (2024). In the 2018 Smart City Ranking by ABI Research, Singapore was praised for its forward-thinking applications in freight and mobility services, notably Freight as a Service (FaaS) and Mobility as a Service (MaaS), as well as for leveraging advanced technology to solve complex urban challenges (Shamsuzzoha et al., 2021). This success continued in 2019, with Singapore being ranked first by the International Institute for Management Development (IMD) in their inaugural Smart City Index, underscoring its prominence in smart city initiatives globally (International Institute for Management Development, 2019).

The cornerstone of Singapore's smart city development is the Smart Nation initiative, launched in 2014 under the management of the Infocomm Media Development Authority (IMDA). The initiative focuses on establishing comprehensive infrastructure that integrates both physical (hard) and digital (soft) systems, including nationwide IoT sensor networks and advanced data analytics capabilities. The Smart Nation Platform created under this initiative serves as a foundation for continuous innovation, enabling both government agencies and private entities to develop and deliver advanced public services more effectively.

Transport remains a critical area within Singapore's Smart Nation strategy, driven by the need to optimize limited urban space through safer, more efficient, and sustainable transportation systems. Autonomous vehicle trials feature prominently, ranging from sedan-sized self-driving cars to autonomous shuttle buses and electric minibuses serving specific urban locations such as university campuses and public gardens. Additionally, trials of driverless trucks and platooning technology for heavy vehicles have demonstrated Singapore's proactive stance in transportation innovation (Shamsuzzoha et al., 2021).

Singapore's smart mobility achievements are substantial, particularly highlighted by its efficient Mass Rapid Transit (MRT) system, which significantly reduces reliance on private cars and improves commuting convenience (Sukawan & Rachmawati, 2021). Public bus services are further enhanced by dedicated bus lanes and intelligent traffic signal prioritization, increasing their appeal and reliability. Additionally, Singapore's pioneering Electronic Road Pricing (ERP) system effectively manages traffic congestion by dynamically charging road users during peak periods, incentivizing the use of public transportation (Wolniak & Grebski, 2023).

The adoption of Intelligent Transport Systems (ITS) has further optimized traffic management in Singapore. Real-time traffic monitoring, adaptive signal controls, and user-friendly travel applications collectively contribute to smoother traffic flow, reduced congestion, and an enhanced commuting experience (Zhong et al., 2016). Shared mobility services, including bike-sharing and e-scooter programs, complement these initiatives by providing accessible, eco-friendly transportation options. These services not only reduce traffic congestion but also encourage active lifestyles and support Singapore's vision of becoming a "car-lite" society (Ferro-Escobar et al., 2022; Joo, 2023).

Energy efficiency represents another significant aspect of Singapore's smart city objectives, especially within the residential sector. According to Singapore Energy Statistics (Bhati et al., 2017), residential energy usage constitutes approximately 15% of total national electricity consumption. Public housing, representing 80% of housing units, accounts for around 60% of residential energy use. Notably, air conditioners, water heaters, and refrigerators collectively account for approximately 76% of household energy consumption.

Research into residential energy consumption highlights several influential factors. Household growth and shifts toward larger residential units are primary contributors to increased energy use, particularly in environment control systems and household appliances. Moreover, external factors like noise and pollution significantly impact energy consumption behaviors, as households often respond by increasing their use of air-conditioning to mitigate these discomforts. Agarwal et al. (2016) highlighted a measurable increase in electricity usage attributable to construction-related noise pollution in public housing, demonstrating that such impacts persist even after the cessation of the disturbances.

Contrary perspectives suggest that assumptions about uniform consumption behaviors may overlook important psychological and social dimensions. Ho (2015) argues for a more nuanced understanding of energy consumption, emphasizing that households are influenced by social practices, cultural values, and behavioral norms. Similarly, He and Kua (2013) identify demographic factors, psychological attitudes, and environmental values as key drivers in adopting energy-saving practices.

Singapore's commitment to sustainable energy management is encapsulated in its National Energy Policy Framework, formulated by the Ministry of Trade and Industry (MTI) in 2007 (Bhati et al., 2017). This framework addresses energy security, environmental sustainability, and economic competitiveness through liberalizing the electricity retail market, fostering competition, and providing consumers with diverse energy choices. The gradual liberalization of the market, especially the lowering of contestability thresholds, enables more consumers to select competitive electricity packages, enhancing efficiency and consumer choice.

Moreover, Singapore is actively diversifying its energy sources to reduce dependence on imported fossil fuels and to mitigate associated risks. Renewable energy options are increasingly considered viable alternatives, fostering healthy market competition beneficial to residential consumers.

Legislation and regulatory standards further support Singapore's energy efficiency objectives. Initiatives like the Mandatory Energy Labelling Scheme, introduced in 2008 and expanded in 2015, ensure consumers can make informed decisions about appliance efficiency. Additionally, the Minimum Energy Performance Standards, established in 2011, systematically eliminate low-efficiency appliances from the market, driving a significant increase in the adoption of high-efficiency models (Bhati et al., 2017).

Table 1 presents a comparative overview of five leading smart cities—Zurich, Oslo, Canberra, Geneva, and Singapore—highlighting their strategic priorities, key sustainability initiatives, technological innovations, and governance approaches. By examining each city's strengths and focus areas, the table provides insight into diverse models of smart urban development and offers a foundation for understanding how different urban contexts adopt technology-driven solutions to enhance quality of life, environmental performance, and civic engagement.

Table 1. A comparison between Zurich, Oslo, Canberra, Geneva, and Singapore.

City	Smart City Focus Areas	Environmental Initiatives	Technological Innovations	Citizen Engagement	Key Challenges
Zurich	Sustainable urban growth, green infrastructure, digital twins, 2000-watt Society initiative	Urban densification, air quality improvements, green spaces, sustainable mobility	Digital twins, smart waste management, ICT for planning	Multi-stakeholder planning, community involvement in sustainability projects	High costs, need for broader participation, limited quantitative data
Oslo	Cross-sectoral collaboration, EV promotion, biogas transit, data-driven planning	EV incentives, car-free zones, emission reduction targets, renewable energy integration	Big data analytics for planning, climate-smart environments	Public participation in planning, data transparency	High cost of zero-emission practices, limited practical implementation
Canberra	People, mobility, livability, governance, education, net-zero emission goals	100% renewable electricity target, green infrastructure, water and climate resilience	Climate Adaptation Strategy, solar and wind projects, battery storage	High civic participation, cultural and educational involvement	Water and solar project performance, environmental indicators
Geneva	Human-centric smart city, circular economy, digital sustainability, collaborative innovation	100% renewable energy goal by 2050, biodiversity protection, green urban planning	Material exchange platforms, digital governance tools, open innovation	Public proposal platforms, collaborative governance committees	Slow implementation of speed reduction zones, data protection concerns
Singapore	Smart Nation infrastructure, autonomous transport, ITS, ERP, energy efficiency	Energy-efficient housing, smart transport, energy policy framework, renewable diversification	IoT networks, real-time traffic monitoring, smart appliances, energy labeling	Consumer energy choice, social behavior insights for energy saving	Noise-related energy behavior, assumption-based consumption modeling

SMART CITIES AND SUSTAINABLE DEVELOPMENT GOALS (SDGS)

This section explores the interconnections between smart city solutions and the United Nations SDGs. It highlights how the implementation of innovative urban technologies can directly contribute to achieving global sustainability targets by addressing challenges related to energy, mobility, health, infrastructure, and social inclusion. Through practical examples, the section illustrates how smart city initiatives can serve as powerful tools for advancing environmental, economic, and social dimensions of sustainable development. As noted in the literature (Bhati et al., 2017), SDGs 11, 7, 8, 12, 3, 13, 10, and 6 have garnered considerable focus, reflecting their strong alignment with smart city development. Consequently, due to space limitations, the discussion in this study is concentrated on these particular goals.

SDG 3: Good Health and Wellbeing

Smart cities play an increasingly vital role in supporting Sustainable Development Goal 3 (Good Health and Well-being), especially as urban populations face risks from pollution, sedentary lifestyles, overcrowding, and inadequate water and food quality (Wu et al., 2020). Although few studies explicitly link smart city efforts to SDG 3, embedded technologies offer substantial potential to enhance public health outcomes. For instance, the use of electric vehicles and renewable energy sources helps reduce urban emissions, while machine learning and Building Information Modeling (BIM) contribute to improved indoor air and water quality (Liu et al., 2021; De Las Heras et al., 2020). Mobility solutions like bike-sharing encourage physical activity and lower traffic congestion (Moreno et al., 2021), and smart lighting enhances safety and mental well-being—though it must be designed to avoid ecological disruption (Tavares et al., 2021).

In healthcare, smart platforms facilitate real-time data exchange and remote monitoring for vulnerable groups, improving access and responsiveness (Hussain et al., 2015; Ismagilova et al., 2019; Liaqat et al., 2021). Blockchain applications further safeguard medical data privacy and improve pharmaceutical supply chain reliability (Mora et al., 2021). Smart waste management and food-sharing technologies reduce health risks from poor sanitation and urban hunger (Hannan et al., 2020). During the COVID-19 pandemic, tools like contact tracing apps and blockchain-based tracking enhanced disease monitoring while protecting user anonymity (Balfaqih et al., 2023a; Alzhrani et al., 2024; Khayyat et al., 2024).

Despite these advances, scholars caution against excessive reliance on technology without adequate ethical and legal safeguards. Dominance by tech giants in urban governance may erode public accountability and autonomy (Mann et al., 2022).

Therefore, integrating transparency, ethical oversight, and participatory governance is essential for smart city health technologies to fully support SDG 3.

SDG 6: Clean Water and Sanitation

Sustainable Development Goal 6 (SDG 6), which focuses on ensuring clean water and sanitation for all, is frequently referenced in academic literature; however, direct analysis of how smart technologies address this goal remains limited (Ismagilova et al., 2019). Nonetheless, technologies such as Artificial Intelligence (AI) and blockchain have demonstrated considerable potential in indirectly advancing SDG 6 by improving water efficiency, supporting wastewater management, raising environmental awareness, and strengthening water security and access (Blasi et al., 2022; Mora et al., 2021; Vinuesa et al., 2020). For instance, blockchain can enable peer-to-peer water rights trading and streamline water-related contracts, enhancing transparency and reducing delays (Parmentola et al., 2022; Mora et al., 2021). These technologies also support autonomous water distribution systems that optimize delivery in real time, improving equity and resource conservation. Platforms like Thewaterproject.org and Charitywater.org have adopted blockchain for cryptocurrency-based donations, promoting financial transparency.

Moreover, smart city platforms contribute to water quality monitoring by detecting contamination sources, tracking leaks, and assessing usage patterns—functions essential for timely interventions and the sustainability of urban water systems (Kutty et al., 2020; Blasi et al., 2022). However, concerns remain regarding the commercial motives of smart infrastructure developers. Kutty et al. (2020) warn that marginalized communities, particularly those in informal settlements, may be excluded from such services, as seen during the COVID-19 pandemic. A failure to prioritize equitable water access within smart city agendas can undermine not only SDG 6 but also interconnected goals such as SDGs 2, 3, 7, 11, 12, and 13. Therefore, inclusive policy frameworks and governance mechanisms are essential to ensure that smart water technologies benefit all populations fairly.

SDG 7: Affordable and Clean Energy

Smart cities play a key role in advancing Sustainable Development Goal 7 (SDG 7), which seeks to ensure access to affordable, reliable, sustainable, and modern energy. Aligned with global agendas like the Paris Agreement, smart city initiatives increasingly integrate renewable energy systems (Zaheer, 2020). Technologies such as AI, IoT, and blockchain contribute to energy sustainability by supporting demand forecasting, minimizing peak loads, improving storage, enabling decentralized generation, and facilitating real-time energy monitoring (Lazaroiu & Roscia, 2018; Liaqat et al., 2021; Mauladdawilah et al., 2025).

Projects like SmartCity Malaga in Spain, which cut electricity use by 25%, demonstrate the tangible benefits of smart grids and integrated energy systems (De Las Heras et al., 2020). Yet, scholars caution that optimization-focused models can overlook ecological complexities and underestimate rebound effects (Giampietro & Mayumi, 2018; Loh et al., 2020).

Advanced tools like neural networks and peer-to-peer energy systems, especially when integrated with PV technologies—as seen in studies from Iran and India—enhance resilience and access (Ghadami et al., 2021; Bardhan et al., 2020). Blockchain-based platforms like SolarCoin and NRGcoin facilitate decentralized, transparent energy trading among prosumers, reducing grid dependency (Mora et al., 2021).

Big data analytics and machine learning further support real-time consumption forecasting and efficiency strategies such as Appliance Level Energy Characterization (ALEC), aiding decarbonization (De Las Heras et al., 2020; Liaqat et al., 2021). Declining costs of solar panels and batteries also broaden access to clean energy (Allam & Allam, 2020; Dhunny et al., 2019).

Smart grids promote data sharing for better energy governance, and AI can synchronize usage with renewable availability, enhancing sustainability (Vinuesa et al., 2020). Paired with ALEC, smart meters raise user awareness and support responsible consumption, positioning prosumers as central to energy decentralization (Hua et al., 2022). However, inclusive governance is essential. Without equitable policy support, renewable adoption may widen social inequalities, especially among disadvantaged groups (Liaqat et al., 2021). Moreover, rebound effects could offset energy savings. Hence, regulatory frameworks must ensure that smart energy solutions are both sustainable and socially inclusive.

SDG 8: Decent Work and Economic Growth

The COVID-19 pandemic significantly disrupted global economies, hindering progress toward Sustainable Development Goal 8 (SDG 8), which promotes inclusive economic growth, productive employment, and decent work. Widespread job losses, supply chain delays, and financial instability deepened vulnerabilities, especially among marginalized groups, and affected progress on related goals such as SDG 5 (gender equality) and SDG 12 (responsible consumption and production) (Dong et al., 2021). Ongoing geopolitical tensions further exacerbate barriers to sustainable economic development.

To address these challenges, many nations have embraced Circular Economy (CE) and Green Economy (GE) frameworks, aligning with smart city principles to promote sustainable economic growth through resource efficiency and digital innovation (Savchenko & Borodina, 2020). Smart cities foster knowledge-based industries, enhance productivity, and support equitable service delivery while minimizing environmental impact (Ismagilova et al., 2019; Shepelev et al., 2023).

However, the pandemic exposed digital divides. While digital economies enabled e-commerce and accelerated technological adoption, they also entrenched socio-economic inequalities. Platforms increasingly mediate economic and cultural interactions, raising concerns over exclusion (Poell et al., 2019; Barns, 2019).

Smart technologies like blockchain offer opportunities for financial inclusion, especially for SMEs in developing regions, by enabling decentralized transactions and microcredit outside traditional banking systems (Lazaroiu & Roscia, 2018; Truby, 2020). These tools also enhance supply chain transparency and ethical labor practices (Mora et al., 2021). Yet, equitable deployment demands robust governance. Initiatives such as the EU's Markets in Crypto-Assets Regulation (MiCA) illustrate the importance of consumer protection and financial oversight. Similar frameworks are essential globally to ensure inclusive participation in digital economies.

Operationally, smart technologies can reduce costs—for example, through digital monitoring of municipal services like waste collection—freeing resources for broader economic initiatives (Hannan et al., 2020). Nonetheless, inclusive planning and digital literacy programs remain critical to preventing further socio-economic disparities (Truby, 2020).

SDG 10: Reduced Inequalities

Reducing inequality is central to social sustainability and a core objective of Sustainable Development Goal 10 (SDG 10). However, global challenges such as the COVID-19 pandemic have significantly disrupted progress, exacerbating social and economic disparities (UN, 2022). Emerging technologies offer promising pathways to mitigate these effects by enhancing inclusive governance, equitable service access, and institutional transparency (Ismagilova et al., 2019). For instance, ICT tools and artificial intelligence (AI) can identify disparities in access to services and track inequality over time (Batty et al., 2012; Vinuesa et al., 2020).

Blockchain technology has shown particular promise in reducing inequality. It increases transparency in charitable giving, enabling donors to trace the use of their funds and strengthening trust in social programs (Mora et al., 2021). Blockchain also facilitates low-cost, cross-border remittances, empowering underserved communities, as seen with Bitpesa in Kenya. In Estonia, blockchain-based digital IDs have enabled migrants and refugees to access healthcare, financial services, and employment—addressing critical inclusion gaps (Mora et al., 2021).

Despite these benefits, concerns remain. Scholars warn against over-reliance on technology, which may reinforce systemic biases, especially in automated decision-making systems (Luusua et al., 2023; Balfaqih et al., 2023b). Algorithmic opacity and lack of accountability can lead to unjust outcomes, particularly in recruitment and public services. Moreover, rapid urban digitization risks deepening the digital divide, excluding those without digital literacy or access to resources (Blasi et al., 2022; Kolotouchkina et al., 2022).

Toli and Murtagh (2020) question whether smart city initiatives truly serve marginalized groups or primarily benefit digitally privileged populations. Increased automation may also threaten jobs in sectors vulnerable to technological displacement (Vinuesa et al., 2020). To address these risks, scholars advocate for inclusive, context-specific smart city strategies that ensure equitable technological integration, especially in developing regions with limited infrastructure (Acilar & Sæbø, 2023; Balfaqih & Alharbi, 2022).

SDG 11: Sustainable Cities and Communities

Cities play a pivotal role in advancing sustainable development, especially as urbanization accelerates and smart technologies become integral to city management. Sustainable Development Goal 11 (SDG 11) highlights the need for inclusive, safe, resilient, and sustainable cities, and smart city solutions have become central to achieving this vision (Jain et al., 2023). Technologies such as IoT and advanced ICT facilitate environmental protection, resource optimization, and infrastructure modernization while promoting socio-economic development (Almohammedi et al., 2023; Balfaqih, 2023; Alagirisamy & Ramesh, 2022).

Smart cities rely on data-driven technologies like AI, blockchain, and deep learning to support urban planning, resilience, and governance. Cities such as Casablanca, New Delhi, and New Castle have successfully deployed these tools to manage energy, water, and environmental quality (Parmentola et al., 2022; Thangavelu, 2020). Moreover, 3D modeling has improved urban insights by visualizing pollution patterns and socio-economic disparities (Santos et al., 2021; Balfaqih et al., 2021). However, concerns remain about big data usage—particularly related to algorithmic bias, opacity, and limited accountability—necessitating transparent and equitable digital governance (Rittenbruch et al., 2022; Al-Khatib et al., 2020).

In the realm of urban mobility, smart platforms like FiarBike support shared and low-emission transport through cryptocurrency integration and mobile applications (Garber et al., 2019; Mora et al., 2021). These initiatives reduce emissions and enhance safety through sensor technologies (El Hamdani et al., 2020; Huang et al., 2022). When effectively managed, automated transport can improve accessibility for remote and underserved communities (Grindsted et al., 2022). However, poor implementation may lead to rebound effects, congestion, and environmental setbacks (Blasi et al., 2022; Grindsted et al., 2022).

Smart governance frameworks are also evolving, with blockchain improving transparency, reducing bureaucracy, and enabling secure service delivery and revenue collection (Mora et al., 2021; Makani et al., 2022). Still, critics caution that an overemphasis on data efficiency may marginalize vulnerable populations, promote surveillance, and weaken democratic oversight—especially amid expanding digital infrastructure post-COVID-19 (Sheikh et al., 2023; Mora et al., 2021).

Ultimately, the pandemic has highlighted both the risks and opportunities of digital urban transformation. For smart cities to fulfill their potential, they must prioritize inclusivity, ecological sustainability, and democratic governance, ensuring technology enhances—not undermines—urban equity and resilience.

SDG 12: Responsible Consumption and Production

Sustainable Development Goal 12 (SDG 12) emphasizes responsible consumption and production, playing a central role in the sustainable management of natural resources and environmental preservation. Strongly interlinked with goals such as SDG 11 (sustainable cities), SDG 13 (climate action), SDG 7 (affordable and clean energy), and SDG 9 (industry and innovation), SDG 12 fosters innovation under the broader agenda of sustainable development. Smart cities, by leveraging ICT and Artificial Intelligence of Things (AIoT), facilitate efficient monitoring, planning, and optimization of resource usage across urban systems (Balfaqih, 2024; Irajifar et al., 2023). These technologies not only contribute to more effective environmental management but also support climate goals aligned with SDG 13 (Ismagilova et al., 2019).

AIoT applications in smart cities have accelerated circular economy (CE) practices. Examples include the automation of rainwater harvesting, smart grids that conserve energy, product-sharing platforms to reduce material use, and virtual platforms that promote plastic recycling (Siddiqui & Pandit, 2021). Solid waste management is another area benefiting from smart solutions. Hannan et al. (2020) introduced a mixed-integer linear programming model that integrates fixed and variable routing optimization with smart bin monitoring. This system improved waste collection efficiency, supporting SDGs 7, 11, 12, and 13.

Moreover, Bibri and Krogstie (2020) emphasize the synergistic value of smart technologies—such as smart grids, meters, buildings, appliances, and environmental monitoring—in promoting energy efficiency and reducing pollution. These tools provide real-time feedback on energy and material flows, encouraging more responsible consumption (Liaqat et al., 2021).

The concept of urban metabolism (UM), which captures the flow of energy and materials within cities, is vital to circular economy models. When enhanced with advanced technologies, it evolves into smart urban metabolism (SUM), offering dynamic insights for sustainable urban management (Bibri & Krogstie, 2020). Within this context, BIM emerges as a powerful tool. BIM improves construction efficiency, reduces waste, and fosters collaboration and data exchange among stakeholders (Liu et al., 2021).

Blockchain technology also contributes significantly to SDG 12. It supports sustainable production by enabling traceability in supply chains, enhancing transparency, and reducing inefficiencies and resource losses (Mora et al., 2021; Parmentola et al., 2022). It also plays a role in environmental monitoring, from groundwater tracking to EV charging oversight, and can help inform consumers about inefficient appliances (Liaqat et al., 2021).

Despite these innovations, challenges remain. The high energy demands of AI, particularly when powered by non-renewable sources, could hinder progress on climate-related goals, underscoring the need for carbon-neutral energy solutions in smart city initiatives (Vinuesa et al., 2020). Therefore, while smart cities offer powerful tools for advancing SDG 12, careful consideration of energy sources and system design is essential for truly sustainable development.

SDG 13: Climate Action

Sustainable Development Goal 13 (SDG 13) calls for urgent climate action, a critical priority for cities due to their major share in global greenhouse gas (GHG) emissions (Bai et al., 2022). The United Nations Framework Convention on Climate Change (UNFCCC) underscores the role of improved data management by local governments in driving urban decarbonization. Although the intersection of ICT and climate change has long been acknowledged, its integration into smart city frameworks has only recently gained prominence, particularly in sectors such as energy, transport, industry, construction, and waste management.

Bibri and Krogstie (2020) propose a data-driven strategy that emphasizes energy efficiency and pollution reduction to support smart, climate-responsible cities. Their roadmap targets urban emissions of less than one ton per capita by 2030 and envisions fossil fuel-free, climate-positive cities by 2050. Complementing this, urban models like the 15-Minute City enhance accessibility, minimize commuting, and reduce emissions (Moreno et al., 2021).

AI plays a significant role in climate action by optimizing waste collection (Hannan et al., 2020), improving climate modeling, supporting resource management, and facilitating renewable energy integration (Vinuesa et al., 2020). Real-time data sharing enables dynamic energy management and aligns urban operations with sustainability objectives (Liaqat et al., 2021).

Blockchain technology is also advancing SDG 13 by enabling transparent emissions trading, accurate GHG accounting, and civic engagement through tokenized carbon credits (Parmentola et al., 2022; Mora et al., 2021). When coupled with IoT, blockchain enhances emissions tracking accuracy and reduces fraud risks (Mora et al., 2021).

However, the environmental cost of digitalization must be addressed. AI-driven systems demand high energy consumption, with data centers contributing significantly to GHG emissions, potentially counteracting climate goals (Vinuesa et al., 2020). Moreover, technocratic smart city approaches have been critiqued for neglecting underlying socio-environmental issues (Yigitcanlar et al., 2019). To ensure holistic climate resilience, Fieuw, Foth, and Caldwell (2022) advocate for governance models that move beyond technological fixes. Addressing systemic challenges such as inequality, overconsumption, and environmental degradation is essential for building inclusive and sustainable urban futures.

LESSONS LEARNED FROM PIONEERING SMART CITIES

The shift toward smart cities represents a strategic response to modern urban challenges such as rapid population growth, resource constraints, and climate change. By integrating advanced technologies and fostering inclusive stakeholder collaboration, smart cities aim to improve governance, service delivery, and overall quality of life. Drawing on global case studies, this section outlines key lessons learned from leading smart city initiatives to inform future urban transformation.

Technology Adoption and Emerging Challenges

Smart cities use technologies such as AI, IoT, and big data to enhance urban operations and improve living standards. However, these innovations bring challenges related to cybersecurity, data privacy, and energy consumption. To ensure sustainability, cities must adopt energy-efficient computing models and power infrastructure with renewable sources. Solutions such as low-power AI and optimized digital systems can reduce the ecological footprint while maintaining efficiency.

Governance and Public Engagement

Effective governance and participatory models are essential for successful smart city implementation. Some cities adopt top-down planning, while others promote grassroots engagement through digital tools, forums, and consultations. Inclusive participation enhances policy relevance, equity, and transparency. International cooperation also plays a key role, with cities forming alliances to share innovations and overcome regional disparities.

Strategic and Adaptive Management

Resilient smart city models benefit from flexible, locally responsive planning. Best practices from American and European cities emphasize the value of education, continuous innovation, citizen feedback, and strong digital infrastructure. These elements are crucial for transitioning from traditional smart cities to "smart sustainable cities" aligned with long-term development goals.

Improved Data Collection and Monitoring

Accurate, real-time data is critical for evaluating the success of smart city initiatives. Investments in sensor networks, analytics platforms, and monitoring tools support evidence-based policymaking and adaptive management. These technologies enable cities to measure sustainability performance, identify gaps, and optimize resources.

Policy and Financial Support

Robust policy frameworks and targeted funding are essential to scale sustainability efforts. Governments should allocate resources for clean transport, green infrastructure, and digital innovation. Financial incentives and public-private partnerships can accelerate technology deployment and ensure equitable access to benefits.

Cross-City Collaboration and Knowledge Sharing

Global collaboration accelerates smart city progress. Sharing experiences, tools, and strategies fosters innovation and supports the replication of successful models. Joint initiatives addressing energy efficiency, circular economies, and climate resilience build collective capacity for sustainable urban development.

Standardized Evaluation Frameworks

Establishing clear metrics for sustainability performance enhances accountability and transparency. Cities should adopt standardized indicators to track emissions, energy use, and environmental quality. Regular reporting and evaluation promote continuous improvement and informed decision-making.

Together, these insights highlight the importance of integrated planning, inclusive governance, and strategic investment in realizing the full potential of smart cities as engines of sustainable and resilient urban futures.

CONCLUSION

This chapter has offered a comprehensive exploration of smart cities as dynamic frameworks for addressing the complex challenges of urbanization, climate change, and socio-economic inequality through the lens of technological innovation and sustainable development. By analyzing foundational smart city components, evaluating comparative strategies across leading global cities, and mapping their contributions to selected SDGs, the chapter underscores the transformative potential of integrated digital infrastructures to enhance urban functionality, resilience, and inclusivity. The insights gained from pioneering examples demonstrate that while smart technologies play a critical role in shaping future cities, their success depends equally on inclusive governance, cross-sector collaboration, and evidence-based policy frameworks. Ultimately, the path toward intelligent, equitable, and sustainable urban environments is not solely about adopting new technologies but about embedding innovation within a broader vision of human-centered, environmentally responsible urban development.

REFERENCES

Acilar, A., & Sæbø, Ø. (2023). Towards understanding the gender digital divide: A systematic literature review. *Global Knowledge, Memory and Communication, 72*(3), 233–249.

Agarwal, S., Satyanarain, R., Sing, T. F., & Vollmer, D. (2016). Effects of construction activities on residential electricity consumption: Evidence from Singapore's public housing estates. *Energy Economics, 55*, 101–111.

Ahad, M. A., Paiva, S., Tripathi, G., & Feroz, N. (2020). Enabling technologies and sustainable smart cities. *Sustainable Cities and Society, 61*, 102301.

Al-khatib, A. A., Mohammed, B., & Abdelmajid, K. (2020). *A survey on outlier detection in Internet of Things big data. Big Data-Enabled Internet of Things*. IET.

Alagirisamy, B., & Ramesh, P. (2022). Smart sustainable cities: Principles and future trends. In *Sustainable cities and resilience: Select proceedings of VCDRR 2021* (pp. 301–316). Springer Singapore.

Allam, Z., & Allam, Z. (2020). Revisiting energy policy and planning in future living cities. *Biotechnology and Future Cities: Towards Sustainability, Resilience and Living Urban Organisms*, 35-52.

Allwinkle, S., & Cruickshank, P. (2011). Creating Smart-er Cities: An Overview. *Journal of Urban Technology, 18*(2), 1–16. DOI: 10.1080/10630732.2011.601103

Almohammedi, A. A., Balfaqih, M., Nahas, S., Bokhari, A., & Alqudsi, A. (2023). Design and Implementation of IoT-Enabled Intelligent Fire Detection System Using Neural Networks. In International Conference on AI and Mobile Services (pp. 63-70). Cham: Springer Nature Switzerland.

Alzhrani, A. A., Balfaqih, M., Alsenani, F., Alharthi, M., Alshehri, A., & Balfagih, Z. (2024). Design and Implementation of an IoT-Integrated Smart Locker System utilizing Facial Recognition Technology. *Engineering, Technology &. Applied Scientific Research, 14*(4), 16000–16010.

Andersen, B., & Skrede, J. (2017). Planning for a sustainable Oslo: The challenge of turning urban theory into practice. *Local Environment, 22*(5), 581–594.

Angelidou, M. (2014). Smart city policies: A spatial approach. *Cities (London, England), 41*(1), S3–S11. DOI: 10.1016/j.cities.2014.06.007

Bai, X., Blanco, H., Gurney, K. R., Kılkış, Ş., Lucon, O., Murakami, J., & White, L. (2022). Urban systems and other settlements. In *Climate Change 2022-Mitigation of Climate Change* (pp. 861–952). Cambridge University Press.

Balfaqih, M. (2023). Design and Implementation of IoT-enabled Intelligent Irrigation System Using Machine Learning. In *2023 28th Asia Pacific Conference on Communications (APCC)* (pp. 436-441). IEEE.

Balfaqih, M. (2024). Enhancing security and flexibility in smart locker systems: A multi-authentication approach with IoT integration. In *2024 21st Learning and Technology Conference (L&T)* (pp. 325-329). IEEE.

Balfaqih, M., & Alharbi, S. A. (2022). Associated information and communication technologies challenges of smart city development. *Sustainability*, *14*(23), 16240.

Balfaqih, M., Alharbi, S. A., Alfarsi, F., Alharbi, O., & Alsalmi, F. (2021). An Efficient Method for Rapid Fabrication Using Low Cost 3D Printer. In *Research and Innovation Forum 2020: Disruptive Technologies in Times of Change* (pp. 537-544). Springer International Publishing.

Balfaqih, M., Altwaim, A., Almohammedi, A. A., & Yusof, M. H. M. (2023b). An intelligent movies recommendation system based facial attributes using machine learning. In *2023 3rd International Conference on Emerging Smart Technologies and Applications (eSmarTA)* (pp. 1-6). IEEE.

Balfaqih, M., Balfagih, Z., Almohammedi, A. A., & Alfawaz, K. M. (2023a). A Smart and Privacy-Preserving Logistics System Based on IoT and Blockchain Technologies. In *2023 1st International Conference on Advanced Innovations in Smart Cities (ICAISC)* (pp. 1-5). IEEE.

Bardhan, R., Debnath, R., Gama, J., & Vijay, U. (2020). REST framework: A modelling approach towards cooling energy stress mitigation plans for future cities in warming Global South. *Sustainable Cities and Society*, *61*, 102315. PMID: 33014694

Barns, S. (2019). *Platform urbanism: Negotiating platform ecosystems in connected cities*. Springer Nature.

Batty, M., Axhausen, K. W., Giannotti, F., Pozdnoukhov, A., Bazzani, A., Wachowicz, M., & Portugali, Y. (2012). Smart cities of the future. *The European Physical Journal. Special Topics*, *214*, 481–518.

Bhati, A., Hansen, M., & Chan, C. M. (2017). Energy conservation through smart homes in a smart city: A lesson for Singapore households. *Energy Policy*, *104*, 230–239.

Bibri, S. E., & Krogstie, J. (2020). Environmentally data-driven smart sustainable cities: Applied innovative solutions for energy efficiency, pollution reduction, and urban metabolism. *Energy Informatics*, *3*(1), 29.

Blasi, S., Ganzaroli, A., & De Noni, I. (2022). Smartening sustainable development in cities: Strengthening the theoretical linkage between smart cities and SDGs. *Sustainable Cities and Society*, *80*, 103793.

De Las Heras, A., Luque-Sendra, A., & Zamora-Polo, F. (2020). Machine learning technologies for sustainability in smart cities in the post-covid era. *Sustainability*, *12*(22), 9320.

Dhunny, A. Z., Allam, Z., Lobine, D., & Lollchund, M. R. (2019). Sustainable renewable energy planning and wind farming optimization from a biodiversity perspective. *Energy*, *185*, 1282–1297.

Dong, L., Liu, Z., & Bian, Y. (2021). Match circular economy and urban sustainability: Re-investigating circular economy under sustainable development goals (SDGs). *Circular Economy and Sustainability*, *1*, 243–256.

Duygan, M., Fischer, M., Pärli, R., & Ingold, K. (2022). Where do Smart Cities grow? The spatial and socio-economic configurations of smart city development. *Sustainable Cities and Society*, *77*, 103578.

El Hamdani, S., Benamar, N., & Younis, M. (2020). Pedestrian support in intelligent transportation systems: Challenges, solutions and open issues. *Transportation Research Part C, Emerging Technologies*, *121*, 102856.

Ferro-Escobar, R., Vacca-González, H., & Gómez-Castillo, H. (2022). Smart and sustainable cities in collaboration with IoT: The Singapore success case. In *Machine Learning for Smart Environments/Cities: An IoT Approach* (pp. 213–243). Springer International Publishing.

Fieuw, W., Foth, M., & Caldwell, G. A. (2022). Towards a more-than-human approach to smart and sustainable urban development: Designing for multispecies justice. *Sustainability*, *14*(2), 948.

Garber, M. D., Watkins, K. E., & Kramer, M. R. (2019). Comparing bicyclists who use smartphone apps to record rides with those who do not: Implications for representativeness and selection bias. *Journal of Transport & Health*, *15*, 100661. PMID: 31938687

Geneve.ch. (2024). GenevaSustainable city: Environmental commitments. Available online: https://www.geneve.ch/geneve-ville-durable-engagements-lien-environnement [08 April 2025]

Ghadami, N., Gheibi, M., Kian, Z., Faramarz, M. G., Naghedi, R., Eftekhari, M., & Tian, G. (2021). Implementation of solar energy in smart cities using an integration of artificial neural network, photovoltaic system and classical Delphi methods. *Sustainable Cities and Society*, *74*, 103149.

Giampietro, M., & Mayumi, K. (2018). Unraveling the complexity of the Jevons Paradox: The link between innovation, efficiency, and sustainability. *Frontiers in Energy Research*, *6*, 349753.

Gillanders, R., & Whelan, K. (2014). Open For Business? Institutions, Business Environment and Economic Development. *Kyklos*, *67*(4), 535–558. DOI: 10.1111/kykl.12067

Grindsted, T. S., Christensen, T. H., Freudendal-Pedersen, M., Friis, F., & Hartmann-Petersen, K. (2022). The urban governance of autonomous vehicles–In love with AVs or critical sustainability risks to future mobility transitions. *Cities (London, England)*, *120*, 103504.

Guelzim, T., Obaidat, M. S., & Sadoun, B. (2016). Introduction and overview of key enabling technologies for smart cities and homes. In *Smart cities and homes* (pp. 1–16). Morgan Kaufmann.

Hannan, M. A., Begum, R. A., Al-Shetwi, A. Q., Ker, P. J., Al Mamun, M. A., Hussain, A., & Mahlia, T. M. I. (2020). Waste collection route optimisation model for linking cost saving and emission reduction to achieve sustainable development goals. *Sustainable Cities and Society*, *62*, 102393.

He, H. Z., & Kua, H. W. (2013). Lessons for integrated household energy conservation policy from Singapore's southwest Eco-living Program. *Energy Policy*, *55*, 105–116.

Ho, E. (2015). Bound by ethical complexities and socio-material histories: An exploration of household energy consumption in Singapore. *Energy Research & Social Science*, *10*, 150–164.

Hofstad, H., & Torfing, J. (2017). Towards a climate-resilient city: Collaborative innovation for a 'green shift'in Oslo. In *Carbon Footprint and the Industrial Life Cycle: From Urban Planning to Recycling* (pp. 221–242). Springer International Publishing.

Hua, W., Chen, Y., Qadrdan, M., Jiang, J., Sun, H., & Wu, J. (2022). Applications of blockchain and artificial intelligence technologies for enabling prosumers in smart grids: A review. *Renewable & Sustainable Energy Reviews*, *161*, 112308.

Huang, B., Thomas, T., Groenewolt, B., & van Berkum, E. C. (2022). An investigation of the motivators and barriers of smartphone app incentives for encouraging cycling. *Decision Analytics Journal*, *5*, 100127.

Hussain, A., Wenbi, R., Da Silva, A. L., Nadher, M., & Mudhish, M. (2015). Health and emergency-care platform for the elderly and disabled people in the Smart City. *Journal of Systems and Software*, *110*, 253–263.

International Institute for Management Development. (2019). Singapore tops new 'citizen-centric' global smart city index. International Institute for Management Development. Available online: https://www.imd.org/news/competitiveness/updates-singapore-tops-new-citizen-centric-global-smart-city-index/ [08 April 2025]

Irajifar, L., Chen, H., Lak, A., Sharifi, A., & Cheshmehzangi, A. (2023). The nexus between digitalization and sustainability: A scientometrics analysis. *Heliyon*, *9*(5). PMID: 37153424

Ismagilova, E., Hughes, L., Dwivedi, Y. K., & Raman, K. R. (2019). Smart cities: Advances in research—An information systems perspective. *International Journal of Information Management*, *47*, 88–100.

Jain, A., Gue, I. H., & Jain, P. (2023). Research trends, themes, and insights on artificial neural networks for smart cities towards SDG-11. *Journal of Cleaner Production*, *412*, 137300.

Joo, Y. M. (2023). Developmentalist smart cities? The cases of Singapore and Seoul. *International Journal of Urban Sciences*, *27*(sup1), 164-182.

Julsrud, T. E., & Krogstad, J. R. (2020). Is there enough trust for the smart city? exploring acceptance for use of mobile phone data in oslo and tallinn. *Technological Forecasting and Social Change*, *161*, 120314. PMID: 32981976

Khayyat, M., Balfaqih, M., Balfaqih, H., & Ismail, M. (2024). Challenges and Factors Influencing the Implementation of Green Logistics: A Case Study of Saudi Arabia. *Sustainability*, *16*(13), 5617.

Kolotouchkina, O., Barroso, C. L., & Sánchez, J. L. M. (2022). Smart cities, the digital divide, and people with disabilities. *Cities (London, England)*, *123*, 103613.

Komninos, N. (2008). *Intelligent Cities and Globalisation of Innovation Networks*. Routledge.

Komninos, N. (2011). Intelligent Cities: Towards Interactive and Global Innovation Environments. *International Journal of Innovative Research and Development*, *1*, 337–355. DOI: 10.1504/IJIRD.2009.022726

Kutty, A. A., Abdella, G. M., Kucukvar, M., Onat, N. C., & Bulu, M. (2020). A system thinking approach for harmonizing smart and sustainable city initiatives with United Nations sustainable development goals. *Sustainable Development, 28*(5), 1347–1365.

Lazaroiu, C., & Roscia, M. (2018, October). Smart resilient city and IoT towards sustainability of Africa. In *2018 7th International Conference on Renewable Energy Research and Applications (ICRERA)* (pp. 1292-1298). IEEE.

Liaqat, R., Sajjad, I. A., Waseem, M., & Alhelou, H. H. (2021). Appliance level energy characterization of residential electricity demand: Prospects, challenges and recommendations. *IEEE Access : Practical Innovations, Open Solutions, 9*, 148676–148697.

Liu, Z., Yang, Z., & Osmani, M. (2021). The relationship between sustainable built environment, art therapy and therapeutic design in promoting health and well-being. *International Journal of Environmental Research and Public Health, 18*(20), 10906. PMID: 34682646

Loh, S., Foth, M., Caldwell, G. A., Garcia-Hansen, V., & Thomson, M. (2020). A more-than-human perspective on understanding the performance of the built environment. *Architectural Science Review, 63*(3-4), 372–383.

Luusua, A., Ylipulli, J., Foth, M., & Aurigi, A. (2023). Urban AI: Understanding the emerging role of artificial intelligence in smart cities. *AI & Society, 38*(3), 1039–1044. PMID: 36196366

Mahmood, M. (2018). *The three regularities in development: growth, jobs and macro policy in developing countries*. Palgrave Macmillan.

Makani, S., Pittala, R., Alsayed, E., Aloqaily, M., & Jararweh, Y. (2022). A survey of blockchain applications in sustainable and smart cities. *Cluster Computing, 25*(6), 3915–3936.

Mann, M., Mitchell, P., & Foth, M. (2022). Between surveillance and technological solutionism: A critique of privacy-preserving apps for COVID-19 contact-tracing. *New Media & Society, 26*(7), 4099–4117.

Mauladdawilah, H., Balfaqih, M., Balfagih, Z., Gago, E. J., & Pegalajar, M. C. (2025, January). Optimization of Photovoltaic Power Forecasting: A Comparative Study of Deep Learning Architectures, Optimization Techniques, and Evaluation Metrics. In *2025 22nd International Learning and Technology Conference (L&T)* (Vol. 22, pp. 109-114). IEEE.

Menendez, M., & Ambühl, L. (2022). Implementing design and operational measures for sustainable mobility: Lessons from Zurich. *Sustainability, 14*(2), 625.

Mora, H., Mendoza-Tello, J. C., Varela-Guzmán, E. G., & Szymanski, J. (2021). Blockchain technologies to address smart city and society challenges. *Computers in Human Behavior, 122*, 106854.

Moreno, C., Allam, Z., Chabaud, D., Gall, C., & Pratlong, F. (2021). Introducing the "15-Minute City": Sustainability, resilience and place identity in future post-pandemic cities. *Smart cities, 4*(1), 93-111.

Nam, T., & Pardo, T. A. (2020). Conceptualizing smart city with dimensions of technology, people, and institutions. *Proceedings of the 12th Annual International Digital Government Research Conference on Digital Government Innovation in Challenging Times - Dg.o '11*. https://doi.org/DOI: 10.1145/2037556.2037602

Nikolov, N. (2024, June). Smart cities as a tool for environmental sustainability: opportunities and challenges. In *ENVIRONMENT. TECHNOLOGIES. RESOURCES. Proceedings of the International Scientific and Practical Conference* (Vol. 1, pp. 261-266).

Parmentola, A., Petrillo, A., Tutore, I., & De Felice, F. (2022). Is blockchain able to enhance environmental sustainability? A systematic review and research agenda from the perspective of Sustainable Development Goals (SDGs). *Business Strategy and the Environment, 31*(1), 194–217.

Perić, A., Jiang, Y., Menz, S., & Ricci, L. (2023). Green Cities: Utopia or Reality? Evidence from Zurich, Switzerland. *Sustainability, 15*(15), 12079.

Peric Momcilovic, A., De Blust, S., Jiang, Y., Guariento, N., & Wälty, S. (2022). Urban Strategies for Dense and Green Zurich: From Healthy Neighbourhoods towards Healthy Communities? In *Proceedings of the 58th ISOCARP World Planning Congress* (pp. 1676-1691). ISOCARP.

Planning Strategy, A. C. T. (2024). Sustainable and resilient territory. Available online: https://www.planning.act.gov.au/act-planning-strategy/strategic-directions/sustainable-and-resilient-territory [08 April 2025]

Poell, T., Nieborg, D., & Van Dijck, J. (2019). Platformisation. *Internet Policy Review, 8*(4), 1–13.

Rittenbruch, M., Foth, M., Mitchell, P., Chitrakar, R., Christensen, B., & Pettit, C. (2022). Co-designing planning support systems in urban science: The questions they answer and the questions they raise. *Journal of Urban Technology, 29*(2), 7–32.

Rodrigues, M., & Franco, M. (2019). Measuring the urban sustainable development in cities through a Composite Index: The case of Portugal. *Sustainable Development, 28*(4), 507–520.

Ruggieri, R., Ruggeri, M., Vinci, G., & Poponi, S. (2021). Electric mobility in a smart city: European overview. *Energies, 14*(2), 315.

Santos, T., Deus, R., Rocha, J., & Tenedório, J. A. (2021). Assessing sustainable urban development trends in a dynamic tourist coastal area using 3D spatial indicators. *Energies, 14*(16), 5044.

Savchenko, A. B., & Borodina, T. L. (2020). Green and digital economy for sustainable development of urban areas. *Regional Research of Russia, 10*, 583–592.

Savithramma, R. M., Sumathi, R., & Sudhira, S. H. (2022). Variability Analysis of Isolated Intersections Through Case Study. *International Journal of Engineering Trends and Technology, 70*(8), 359–374. DOI: 10.14445/22315381/ijett-v70i8p237

Schaffers, H., Komninos, N., Pallot, M., Trousse, B., Nilsson, M., & Oliveira, A. (2011). Smart Cities and the Future Internet: Towards Cooperation Frameworks for Open Innovation. *Future Internet, 6656*, 431–446. DOI: 10.1007/978-3-642-20898-0_31

Schrotter, G., & Hürzeler, C. (2020). The digital twin of the city of Zurich for urban planning. *PFG–Journal of Photogrammetry. Remote Sensing and Geoinformation Science, 88*(1), 99–112.

Senior, C., Jowkar, M., Temeljotov-Salaj, A., & Johansen, A. (2021, March). Empowering citizens in a smart city project one step at a time: A Norwegian case study. In *2021 IEEE European Technology and Engineering Management Summit (E-TEMS)* (pp. 10-15). IEEE.

SERVICE CANTONAL DE GESTION DES DECHETS – GESDEC. (2006). *Le recyclage des matériaux de construction à Genève*. Etat de Genève.

Shamsuzzoha, A., Nieminen, J., Piya, S., & Rutledge, K. (2021). Smart city for sustainable environment: A comparison of participatory strategies from Helsinki, Singapore and London. *Cities (London, England), 114*, 103194.

Sheikh, H., Mitchell, P., & Foth, M. (2023). More-than-human smart urban governance: A research agenda. *Digital Geography and Society, 4*, 100045.

Shepelev, V., Glushkov, A., Slobodin, I., & Balfaqih, M. (2023). Studying the relationship between the traffic flow structure, the traffic capacity of intersections, and vehicle-related emissions. *Mathematics, 11*(16), 3591.

Siddiqui, A., & Pandit, R. K. (2021). Smart Cities in India: Linkages with circular economy. In *Sustainability in the built environment in the 21st century: Lessons learned from India and the region* (pp. 185-200). Springer International Publishing.

Sukawan, H. A. R., & Rachmawati, R. (2021, March). MyTransport. SG as a new communication platform in implementing smart mobility in Singapore. []. IOP Publishing.]. *Journal of Physics: Conference Series, 1834*(1), 012006.

Tariq, M. A. U. R., Faumatu, A., Hussein, M., Shahid, M. L. U. R., & Muttil, N. (2020). Smart city-ranking of major Australian cities to achieve a smarter future. *Sustainability, 12*(7), 2797.

Tavares, P., Ingi, D., Araújo, L., Pinho, P., & Bhusal, P. (2021). Reviewing the role of outdoor lighting in achieving sustainable development goals. *Sustainability, 13*(22), 12657.

Thangavelu, M. (2020). Outer space activities and city evolution. In *ASCEND 2020* (p. 4059).

Toh, C. K. (2022). Smart city indexes, criteria, indicators and rankings: An in-depth investigation and analysis. *IET Smart Cities, 4*(3), 211–228.

Toli, A. M., & Murtagh, N. (2020). The concept of sustainability in smart city definitions. *Frontiers in Built Environment, 6*, 77.

Tranos, E., & Gertner, D. (2012). Smart networked cities? *Innovation (Abingdon), 25*(2), 175–190. DOI: 10.1080/13511610.2012.660327

Truby, J. (2020). Governing artificial intelligence to benefit the UN sustainable development goals. *Sustainable Development, 28*(4), 946–959.

Türetken, O., Grefen, P., Gilsing, R., Adali, E., & Ozkan, B. (2019). Business model innovation in the smart mobility domain. In Lu, M. (Ed.), *Cooperative Intelligent Transport Systems* (Vol. 25, pp. 63–86). IET Transportation Series., DOI: 10.1049/PBTR025E_ch4

UN. (2022). The Sustainable Development Goals Report, United Nations. Available online: https://unstats.un.org/sdgs/report/2022/ [09/04/2025].

Varesano, D., Amos, E., Domer, B., Laferrière, F., Redaelli, D., & Viviani, M. (2018). Towards a" comprehensive" and smart platform for coordinated waste management in construction in Geneva?: a case study. In *Proceedings of 20. Status-Seminar" Forschen für den Bau im Kontext von Energie und Umwelt", Zurich, Switzerland, 6-7 September 2018*. 6-7 September 2018.

Vinuesa, R., Azizpour, H., Leite, I., Balaam, M., Dignum, V., Domisch, S., & Fuso Nerini, F. (2020). The role of artificial intelligence in achieving the Sustainable Development Goals. *Nature Communications*, *11*(1), 233. PMID: 31932590

Wolniak, R., & Grebski, W. (2023). *Smart mobility in smart city–Singapore and Tokyo comparison.* Zeszyty Naukowe. Organizacja i Zarządzanie/Politechnika Śląska.

Wu, C., Zheng, P., Xu, X., Chen, S., Wang, N., & Hu, S. (2020). Discovery of the environmental factors affecting urban Dwellers' mental health: A data-driven approach. *International Journal of Environmental Research and Public Health*, *17*(21), 8167. PMID: 33167348

Yigitcanlar, T., Foth, M., & Kamruzzaman, M. (2019). Towards post-anthropocentric cities: Reconceptualizing smart cities to evade urban ecocide. *Journal of Urban Technology*, *26*(2), 147–152.

Zaheer, A. (2020). Sustainability and resilience in megacities through energy diversification, land fragmentation and fiscal mechanisms. *Sustainable Cities and Society*, *53*, 101841.

Zavratnik, V., Podjed, D., Trilar, J., Hlebec, N., Kos, A., & Stojmenova Duh, E. (2020). Sustainable and Community-Centred Development of Smart Cities and Villages. *Sustainability*, *12*(10), 3961. DOI: 10.3390/su12103961

Zhong, C., Batty, M., Manley, E., Wang, J., Wang, Z., Chen, F., & Schmitt, G. (2016). Variability in regularity: Mining temporal mobility patterns in London, Singapore and Beijing using smart-card data. *PLoS One*, *11*(2), e0149222. PMID: 26872333

Chapter 13
Data Privacy and Security in Smart Cities:
Avoiding Problems and Resolution

Sandip Dey
https://orcid.org/0000-0002-1005-3304
Sukanta Mahavidyalaya, Dhupguri, India

ABSTRACT

This paper aims to describe the relationships between data privacy and security issues and smart cities, discussing data as the driver of the city development. The author explains what smart cities are, why they are important and what the research is going to focus on. This work explores various aspects of data privacy including the type of gathered data, privacy issues, case studies where breaches occurred and the legal and ethical analysis. It continues with a discourse of security concerns, which gives information about typical risks, hostile elements of structures, actual events, and legislations. Proposed interventions are data protection/encryption, distributed ledger technologies such as blockchain, privacy by design, security roles of Artificial Intelligence (AI) etc. The future direction for this paper discusses future advancement in technology and policy suggestions to comprehensively establish a smart city solutions framework. A call to action encourages stakeholders to focus on privacy and security to make smart cities the secure, efficient, and trustworthy part of the future city.

INTRODUCTION

In its simplified form, Smart cities are urban centers adopting Internet of Things (IoT), AI and big data technologies for improving residents' standard of living. Its

DOI: 10.4018/979-8-3373-0154-9.ch013

goals are to increase productivity throughout the transport, energy, medical, and municipal industries by processing enormous amounts of information. These data enable city officials to have adequate information and make right decisions in managing resources and solving unexpected problems like traffic jams, pollution, and insecurity among others. Therefore, it is crucial to understand that the value from investing into smart cities is found in the possibility of making cities future proof. For instance, smart energy management will prevent wastage of energy and smart transportation will also avoid congested traffic and emissions (Syed et al., 2021). In addition, these cities can provide citizens with service updates for better functioning of public services and, in case of an emergency, make the community safer and more informed. However, as more and more information is collected, privacy and security issues emerge increasing the need for solutions for it to avoid having the positives of smart cities beaten by the negatives. The security of individual information and the ability to control free access and use of such data and information are critical issues in smart cities due to fear of impersonality or data hacking and loss of confidence among the residents. By harnessing these dimensions, smart cities can work towards fulfilling its primary objectives of efficiency, profitability as well as environment-sustainability with adequate strain on people's right to privacy and security (Braun et al., 2018).

Data is core business unto itself in smart cities where it is useful as the supporting framework of several services and innovations that shape life in urban centers. These cities use many data gained from sensors, devices, and people to coordinate resources, enhance public utilities, and promote sustainability (Kitchin, 2014). This data is used to manage the traffic flow to minimize energy use, increase safety, and facilitate environmental surveillance (Batty et al., 2012). For instance, the flow of traffic in a city can be managed through signals, based on up-to-date information, thus avoiding jam, and cutting the time to be spent on the road. Likewise, smart meters supply information on energy consumption that play a role in its regulation and result in savings for municipalities and occupants (Andreani et al., 2019). But relying on data is not done without its drawbacks, it raises highly sensitive issue of privacy and security. To avoid conciliation of this data and concession on the citizens' rights to protection from misuse of their data, measures should be put in place for enhancing the trust of users in the benefits associated with smart cities (Rathore et al., 2018).

The research interest of the study concerns data privacy and security issues in smart cities. With modern urban services and amenities drawing on information and communications technologies to become smarter, protective of the environment, and productive for the people, information security and privacy concerns arise. The goals of this research are to determine the most critical threat factors of privacy and security related to data collection and usage in smart cities, examine current controls, and recommend effective palliative measures for implementation. The aim of this study is to develop a case that would define how smart city systems can be deployed and operated in a manner that would protect the information of citizens without negating progress. It also examines how data privacy and security challenges can be solved legally, ethically, and socially, based on policy and thoughts for policy makers, technology producers and even city designers. This chapter aims to use case studies and evaluate global approaches to enrich knowledge about best practices and identify key lessons towards building safe and sensitive smart cities (Kitchin, 2014; Gassmann et al., 2019). Several AI and IoT based applications in contest of security of smart city are presented in Figure 1.

Figure 1. AI and IoT Applications in contest of security of smart city

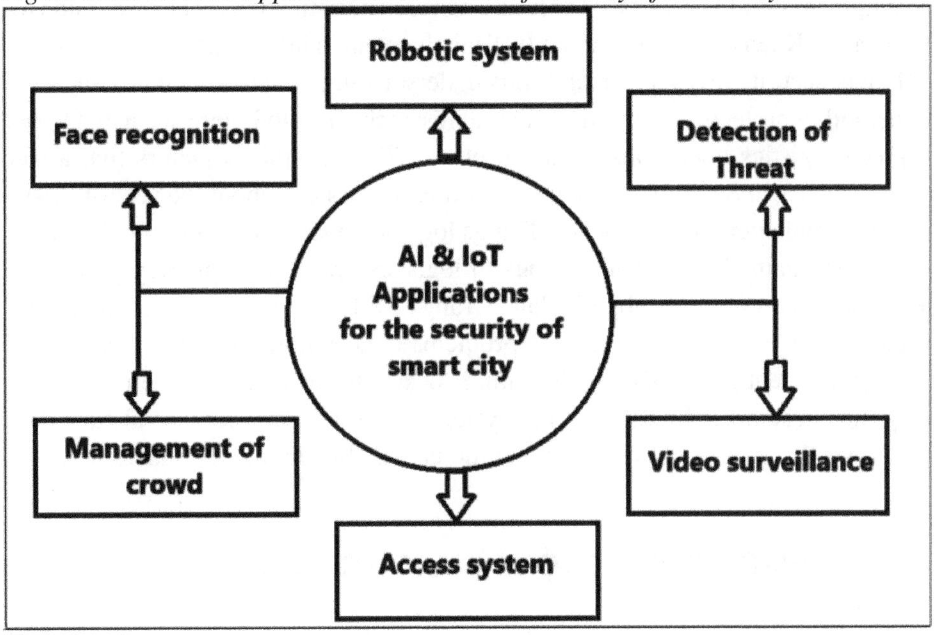

RELATED WORKS

Academic scholars have conducted thorough investigations into data privacy and security aspects in smart cities. One needs to assess the current scholarly works carefully to find research gaps that will guide future studies. Various research investigations try to resolve smart city security issues, but these studies face important restrictions with their scope range and study methods and real-world application capacity. The discussion requires examination of additional studies which concentrate on Information and Communication Technologies (ICT) challenges in smart city development together with green logistics and blockchain-based logistics systems.

Numerous studies about smart city security face challenges due to their fragmented solution approaches. Single security aspects like encryption and blockchain and IoT security are addressed in current literature without a unified framework which integrates different security methodologies. The research conducted by Rathore et al. (Rathore et al., 2024) examines big data analytics in smart cities without establishing any privacy protections for safe data transactions. The AI-based intrusion detection system does not solve the real-time adaptability issues along with scalability needs of contemporary smart city infrastructure (Sowmya & Anita, 2024).

New information and telecommunications technologies that arise during smart city development should be the basis of future research investigations. The discussion by Khan et al. (Khan et al. 2014) about ICT involvement in urban transformation lacks sufficient security analysis which limits understanding of network communication vulnerability protection methods. Existing research tends to ignore security factors in green logistics operations. Khayyat et al. (2024) examines factors that affect green logistics in Saudi Arabia while omitting discussions about the security risks linked to implementing Internet-of-Things logistics systems (Khayyat et al., 2024).

A blockchain and IoT applications in logistics systems for privacy protection demonstration is presented by Li, Han & Tang (2021). The reported security model of blockchain encounters scalability problems in decentralized authentication according to recent analysis by their team. Blockchains encounter difficulties when integrated in smart cities due to these systems processing enormous real-time data operations that require effective computational handling for operational excellence.

DATA PRIVACY CHALLENGES IN SMART CITIES

The risks associated with data collection, surveillance, and cybersecurity create immense challenges in smart cities. These may include illegal data access without taking user consent, third-party data sharing and so forth. GDPR may have a crucial role in ensuring citizen trust and preserve sensitive personal information. These issues are briefly illustrated in the following subsections.

Types of Data Collected

Within smart cities, a large amount of diverse data is gathered to advance the quality of life in such cities and optimize the processes. This information can be sub-categorized into personal information, geographical information, and structural information. Personal data encompasses identifiers including names and addresses, from different residents, as well as health records gotten from healthcare digital services, social networks etc. (Kitchin, 2014). Environmental data is physical information related to the ambient conditions in and around an environment including air quality, weather information and pollution level and often collected through sensors and IoT objects (Batty et al., 2012). Infrastructural data includes information about transportation networks, electricity and water consumption and the traffic pattern within cities which are essential to management of the cities (Allwinkle & Cruickshank, 2011). The collection of this data argues with privacy as personal information problems can be exploited if not well protected. There is a need to adopt strict policies to regulate the use of personal data collected from residents to prevent misuse of individual information and endorse acceptable use of data and information (Gellman, 2009). The issue of how much data can be used to improve a city in question while also preserving its inhabitants' privacy remains a focal point in smart city creation.

Privacy Concerns in Data Collection

Privacy issues are always a concern when it comes to data acquisition within smart cities because of many people's and personal data collected from different sources throughout the cities including sensors, cameras, and connected devices. This data may contain place information, personal characteristics, and even behavior patterns of people, which raises serious concerns about such things as hacking, abuse and possibly surveillance (Kitchin, 2014). The amalgamation of data from various sources raises the danger of re-identification of anonymized data; that is, data sets can be linked to profile related personal data (Ohm, 2009). These issues are worsened by the lack of sufficient laws and very poor and erratic enforcement of privacy policies that Permit residents to know how their information is being used (Cavoukian, 2021). As technology advances further in this direction of smart cities, these privacy concerns need to be managed to ensure that citizens gain confidence in adopting the smart technology that powers the smart city, without in effect, sacrificing their fundamental right to privacy.

Case Studies of Privacy Breaches

Failures in privacy issues for smart cities are demonstrated by analyzing case studies; the scale of data collection and management is large. A good example is Sidewalk Labs, owned by Alphabet Inc which was working on the smart city project referred to as Quayside in Toronto; the project received much backlash over data Collection and Use that can trigger misuse of Personal Information (Dadayan, 2019).

The absence of clear principles and guidelines to protect participants' data breaches and failing to provide clear descriptions of data collection practices the project was later shutdown by the critics. Another case features the outflow of personal data in the smart city of Songdo in South Korea due to ill-protected information and vulnerability of entrance to residents' personal and financial data (Robinson & Ji, 2022). This breach brought out the importance of strong measures and showed that lax security measures increase the danger. Further, the data breaches of Aadhaar system in India in 2018 and which compromises the data of more than one billion people highlight the difficulties associated with protecting large database system. That is why these episodes emphasize how important it is to introduce strong DPS in smart cities to avoid the leakage of residents' data and gain the population's confidence.

Legal and Ethical Implications

Legal and ethical issues in relation to smart cities raise concerns over data privacy since the use of technology involves large amounts of personal data in simultaneous collection and its processing, which is unlawful and unethical. From a legal perspective, smart cities face challenges that range from compliance with the European Union's GDPR on data protection policy and regulation which limits and controls the collection, processing, and use of data, among citizens (Regulation, 2016). Failure to observe such measures would result in fines and of course legal consequences. Also, privacy laws remain different in different locations making it challenging to focus on equal data protection measures (Kitchin, 2014). The ethical issue that smart cities have is to determine how such data gathering and analysis will not infringe the rights of citizens or individuals in society. Aspects like informed consent, the client's right to know and the right to be forgotten emerging as very important in the preservation of trust (Solove, 2005). Another issue that I must consider ethics for, is the issue of data misuse and probable discrimination, in case the information received is mishandled (Cavoukian, 2021). For example, in predictive policing, algorithms arising from working with biased datasets may mean prejudicial treatment of some sections of the population and thus increase social injustice (Eubanks, 2018). Finding the balance between innovative and private infringement is possible but needs strong ethical principles, and legal regulations to support smart city development without infringing people's freedom. Few common privacy concerns in Smart Cities are presented in Table 1.

Table 1. Common privacy concerns in smart cities

Privacy Concern	Description	Examples
Data Collection	Conduction of broad data collection from various sources	CCTV, sensors, smart meters
Data Sharing	Information exchange between the different organizational departments of a city and other utilizing parties	Public transport data shared with apps
Data Storage	Keeping of sensitive data over a long time	Cloud storage of personal data
Surveillance	Prolonged observation of areas of interest in terms of social activity	Surveillance cameras, drones
Consent and Transparency	Insufficient evidence of user's permission and awareness of data consumption	Terms of service agreements

Significance of Selected Challenges, Solutions, and Case Studies

The three components of IoT and AI and big data analytics combine to improve urban life in smart cities. Enhanced data protection alongside user privacy becomes difficult to achieve in smart city applications due to modern technological advancements. The research has centered its analysis on privacy and security problems together with solutions and real-life examples which exert major effects on smart city frameworks. The importance of unauthorized data access and cyber threats along with ethical questions regarding surveillance technologies are highlighted because people increasingly use connected systems (Almeida, 2023). The protection of citizen privacy together with data security represents an essential requirement for both earning public confidence and satisfying GDPR requirements (Rao et al., 2023).

A method of challenge selection bases its decision on practical effects combined with typical occurrence rates and safety risks. The combination of smart grid data breaches with surveillance system breaches represents privacy threats while security concerns mainly involve hacking attempts and denial-of-service (DoS) attacks. The chapter studies mitigating measures which include blockchain security solutions and federated learning approaches to protect privacy. This analysis presents Toronto's Sidewalk Labs project as an assessment case study because it demonstrates privacy security problems (Arcila, 2024) in addition to successful and unsuccessful case experiments that show fundamental concepts.

This chapter examines data security challenges and solutions in smart cities through a comprehensive analysis to develop reasonable improvement strategies. Current data from selected research facilitates the application of theoretical principles through practical solutions. The development of smart city frameworks needs to resolve these matters because it ensures sustainable infrastructure development together with citizen rights protection.

DATA SECURITY ISSUES IN SMART CITIES

Nowadays, several data security issues like cyberattacks, unlawful access, and data breaches are very much common. Risks may include surveillance susceptibilities, poor encryption, and exploitation of several IoT devices. To ensure cybersecurity, it may necessitate sturdy encryption, protected networks, and monitoring. Few issues with regards to data security in smart cities are presented below.

Common Security Threats

The general security challenges of smart cities entail different risks attributable to the numerous connections in this setup and the use of digital technologies. Terrorism in cities might be cyber, and it includes threats such as hacking, phishing, and malware since their consequences could sever consequences for urban functioning (Zhang et al., 2017). A common threat is constantly conducting Distributed Denial of Service (DDoS) attacks on cities systems which reduces the traffic flow control and public safety services (Sookhak et al., 2018). Further, smart cities experience security threats that can arise from IoT devices, which are usually exposed to several threats due to their poor standard security features (Sicari et al., 2015). An equally well-known risk is data breaches in which confidential information is extracted and stolen, putting the citizens' identity and their financial credit scores in jeopardy (Chourabi et al., 2012). Further, main types of ransom attacks like, city system is locked and requires a ransom, have become more common and threaten the basic societal structures and the very citizens (Alcaraz & Zeadally, 2015). Smart cities are integrated where a security breach of one component will affect the others and therefore paralyses several services (Wang, Lin & Zhang, 2010). To curb these security threats there is needed development of various cybersecurity measures, consistent system update and the proper legislation governing urban digital systems.

Vulnerabilities in Smart City Infrastructure

Smart city infrastructure is therefore exposed to number of risks and threats because of the connected environment. These infrastructures use many devices and sensors for data monitoring and transferring and may become exposed to cyber threats (Lee & Lee, 2018). The first one can be attributed to the susceptibility of the system to use of insecure or NOS-compatible devices, which can be used by the hackers to intervene in the system unlawfully (Roman, Zhou, & Lopez, 2018). Moreover, the use of many technologies from various manufacturers in aviation causes compatibility problems and insecure interfaces (Ma, 2021). One of the key features of smart cities which also creates potential vulnerabilities for the system is the accumulation of a huge amount of data, collecting which results in numerous threats, ranging from identity theft to major financial losses and social issues (Zhang et al., 2017). Moreover, these kinds of security flaws are magnified due to the absence of uniform protocols implemented in various aspects of the smart city framework, because such differences can be manipulated by attackers (Ahmad et al., 2022). In this regard, there is a need to adopt good security practices for smart cities such as using updated security lapses, conducting security testing and exercises, and adopting advanced encryption methods that are recommended by Hossain, Fotouhi, and Hasan (2015).

Real-World Incidents of Security Failures

The cases of security violations and attacks on smart cities showcase severe lacking within their frameworks – thus proving just how important it is to have strong security guarantees. A classic example is the black out of Ukraine's power grid in December 2015 where different areas in the country experienced partial or full power blackouts due to cyber-attack (Case, 2016). Another event that took place in working in Atlanta in the year 2018 where a ransomware attack affected different services in the city inclusive of payments to law enforcement (Atlanta Ransomware Attack, 2018). Likewise, in 2020, in Oldsmar Florida, a water supply system was attacked, the hacker tried to change sodium hydroxide level in a water reservoir to a toxic level that would harm many people (Greenberg, 2021). To such extremes such cases prove that in highly connected systems proper security measures cannot be overlooked and that proper cybersecurity solutions must always be employed. They also show the risks of exploitation which entail breakdown of safety, monitory loss and loss of credibility (Hamid et al., 2019). To address these issues faithful cooperation of governments, providers of smart technologies, and security specialists is needed on the further reinforcement of safeguards and countermeasures against new and modified threats to smart city systems (Wenge et al., 2014).

Regulatory Frameworks and Security Standards

Smart cities highly depend on a huge quantity of data gathered by sensors, appliances, and networks for managing city and improving the living standard of the citizens. However, the large number of data use creates a big concern of security thus approaching regulatory frameworks and meeting set industry securities are major in protecting data.

Regulatory Frameworks

General Data Protection Regulation (GDPR) which was set by the European Union (EU) can be well regarded as an essential part of the regulations. The GDPR outlines very stringent provisions how data is collected, processed, and stored with an information member having complete control over their own data. Any smart city in EU area should comply with GDPR: personal data should be processed transparently, collected for specific, legitimate purpose and with the consent given (Protection, 2018).

Similarly, the United States has such other various sector-specific regulation like California Consumer Privacy Act (CCPA) provides California residents right to know, delete and opt out (Shatz & Chylik,, 2019). Nonetheless, for smart city ecosystems, compliance with the CCPA guarantees conspicuous data collection from public and private networks respects user data privacy.

In Asia to date Singapore has developed the Personal Data Protection Act (PDPA), that calls for organization to observe on compliance of proper governance of personal data (Chik, 2013).

Security Standards

International standards are particularly important in that they help set a norm or a base reference for stakes in smart cities as pertains to security. Other international standards include the International Organization for Standardization (ISO) standards, of which ISO/IEC 27001 is on information security, which delivers a measurement to control information security risks by adopting an ISMS (Culot et al., 2021).

This is also true for the National Institute of Standards and Technology (NIST) in the United States that gives detailed guidance by its Cybersecurity Framework starting with the process of identifying and protecting, then detecting, computing, responding, and recovering from a cyber threat (Cybersecurity, 2018).

Challenges and Recommendations

However, compliance of smart cities is an issue because operational overviews of the data processes are intricate and the applicability of robust institutional frameworks and standards is limited. The adoption of these frameworks across different sectors and making sure that the organisation is operating in real-time with the framework is a massive investment that needs the right kinds of skills and abilities.

To overcome the above pitfalls, 'smart city' deployment must include a multi-layered security approach coupled with periodic audits, widespread collaboration with key stakeholders and deployment of innovative technologies such as blockchain for protected data exchange and AI for security threat detection. However, there is a continuous need to train all the stakeholders on the changes taking place in the regulatory practices and key security matters. Table 2 presents a few important examples of security risks in smart city infrastructure.

Table 2. Key security risks in smart city infrastructure

Security Risk	Description	Potential Impact
Cyber Attacks	This occurrence or any other unauthorized access or attack on the city's systems	Service disturbance, data theft
Malware and Ransomware	Virus and malware attacks on smart city equipment	System lockdown, ransom demands
Insider Threats	Occupational threats resulting from employees or contractual workers	Data leaks, sabotage
IoT Vulnerabilities	The security issues commonly referred to as the vulnerability of IoT devices	Unauthorized device control
Network Interception	Interception of data over city networks	Data breaches, information leakage

SOLUTIONS FOR DATA PRIVACY AND SECURITY

Data privacy and security can be ensured by implementing several criteria. These are described in the following subsections.

Data Encryption and Anonymization Techniques

As the world goes digital, it is important that persons, organizations, and companies protect information they deem confidential. One is data encryption, and the other is anonymization, and both have been considered crucial in this course. Such techniques make it possible for data to be secured from unauthorized individuals, in addition, to user's privacy when data sharing is compulsory as it is in the health, financial, and research departments.

Data Encryption

Encryption is a technique in which information in its normal form is transformed to another form which is not intelligible and can be understood or processed only by an individual with a decryption key. This technique makes it that even if data therapies intercepted, they will not be able to access it in the first instance without the right Code (Stamp, 2011). Encryption can be categorized into two key types: of both, the symmetric and the asymmetric encryption. Unlike asynchronous, which has two subkeys- an encryption and decryption, symmetric encryption is faster because the same subkey is used both in encryption and decryption, but the drawback is the need to distribute the subkey securely. Asymmetric encryption is done using a single entity of keys which include public key for encryption and private key for decryption making the security strong but having a problem with its efficiency of computation (Katz & Lindell, 2020).

AES is one of the most extensively used symmetric encryption algorithms because these criteria are optimal for such a method's use: improved security and performance. The widely known method in secure data transfer is RSA, an acronym for Rivest-Shamir-Adleman, and conjointly SSL/TLS protocols (Menezes et al., 2018).

Data Anonymization

Data anonymization is the process of minimising or eradicating the possibility of a third party on identifying specific individuals in each population from data sets. Some methods include k-anonymity, which refer to the data minimization technique that is used to generalize or eliminate information in a database so that each record is no more distinguishable than are at least k-1 other records (Sweeney, 2002). Another approach is differential privacy where a little amount of statistical noise is added to data to prevent one's data from being identified while the analysts still get data about all of them (Dwork, 2006).

Nevertheless, anonymization is not entirely possible. One major attack on de-anonymization is when raw data stored in the anonymized databases are potentially linked with data in other publicly available databases. Hence, the usefulness of the anonymization techniques is an ongoing process being threatened by ever advancing threats (Narayanan & Shmatikov, 2008).

Since either encryption and anonymization are significant steps towards the protection of data privacy and security, it is possible that the two steps are still relevant despite the fairly young age of big data. While data encryption protects data both as it travels across networks and when stored locally, anonymization protects data while it is being shared and analysed. The use of these techniques involves assessing the context of application and the associated risks with reference to ongoing monitoring and enhancement in response to emergent threats.

Implementation of Blockchain for Secure Data Management

This is true because; blockchain technology provides a strong platform for ensuring the integrity of data besides ensuring proper anonymization of data through strong cryptographic methods. This section focuses the main approaches of how blockchain strengthen the data protection with special references to the encryption and anonymity.

Data Encryption Techniques

Encryption can be said to be one of the foundations of blockchain's security approaches. Data in blockchain related systems is secured by means of cryptographic algorithms preventing unauthorized access to data. The Bitcoin takes shape with SHA-256 (Secure Hash Algorithm 256- bit) for securitizing the transaction data. This cryptographic hash function alters its input data to a fixed-sized hash; it remains crystal clear that even the slightest change in the input data produces a vastly unique hash.

The other major element to the security of blockchain technology is the public key cryptography, also known as asymmetric encryption. This form of the technique comprises of a single pair of keys that is the public key that is used in strengthening and the private key that is used in signing. Public key cryptography is designed to allow one, and only the other, the ability to decode the given information. To provide more security, the variants like the Ethereum use the Elliptic Curve Digital Signature Algorithm (ECDSA) to confirm the transactions (Wood, 2014).

Anonymization Techniques

Privacy in blockchain means use of methods that help in the concealment of users' identification details while at the same time affirming the validity of a transaction. For example, Zcash uses zk-SNARKs (zero-knowledge succinct non-interactive arguments of knowledge) where one party can demonstrate knowledge of certain data without having to divulge that data to the second party. Being an off-blockchain technique, this approach affords transaction data complete privacy and anonymity of the users.

An is there another important approach, which we can call as mixers or tumblers wherein the link between the sender address and the receiver address is disrupted, and the received amount of bitcoin is also rearranged in a random manner. Services like CoinJoin offer the anonymity of transaction detail without sacrificing the transparency of the records (Wahrstätter et al., 2024).

Privacy-by-Design in Smart City Planning

The Privacy-by-Design (PbD) is an effective strategy that addresses and implements privacy and data protection in smart city initiatives without having to retrofit protection to the concepts. Cartooned by Ann Cavoukian, Ravictim recognizes this concept focuses on integrating confidentiality into the design and structure of technologies and systems (Cavoukian, 2021).

Implementation in Smart Cities

The process of implementing IT into the urban cities is called smart cities.

PbD, if applied in smart cities, requires coordination between different actors at the state and organizational levels as well as individual citizens. Legal requirement of PbD is evident; legal frameworks like the GDPR have pedalled for the adoption of PbD, especially in data intensive projects (Tzanou, 2020).

For instance, the DECODE project has adopted PbD in Barcelona where the citizens have an empowered control on their data passing through decentralised technologies as the privacy of the users is preserved as smart services are delivered (Calzada, 2021).

Role of Artificial Intelligence in Enhancing Security

There is another specific problem when elaborating the AI systems for smart cities that rely upon numerous interconnected systems and enormous data flows: data security needs to be improved. In smart cities, AI is used to secure personal and organizational data apart from guaranteeing the functionality of urban structures. For purpose of this note, relative scholarly works explain AI's role in managing data protection risks as elaborated below.

AI provides a way of enhancing threat intelligence to correlatively forecast security threats through processing Big Data in real-time. In a smart city environment, there is a collective ability of the AI systems to detect suspicious activities within networks and respond as a group before the threats build up into a large problem. These systems require learning from experience so that the future detection of abnormalities is improved with time (Batra et al., 2021).

Furthermore, AI improves the safety of all the IoT devices on which smart city functionalities rely on. AI algorithms can also carry out the security of IoT devices against people by checking in the authorization process. AI can be used to detect weaknesses within IoT systems and consequently protect data in addition to device reliability.

Through AI, smart city also provides safeguards of data encryption and data transmission to other networks. Calculating machines or AI can apply algebraic equations to multiply or add to improve encryption standards for cyber defense (Ogala et al., 2023). Moreover, with the help of Al powered predictive analytics this can predict possible security threats to the infrastructure, which will give city administrators the possibility to organize preventive measures and actions.

Another important application of AI to the process of improving security is automated incident handling. AI systems can automatically attack detected threats and as a result reduce response time and the effect of threats on the system. AI response mechanisms help smart cities to continue the functioning of the organization while responding to security threats effectively (Chintala, 2024).

However, as with any system integration, there are achievements and barriers to using AI in smart city security systems. Challenges like machine learning and its problems like the issues of bias in algorithms; vulnerabilities in the AI algorithms leading to adversarial attacks need to be put into consideration to warrant the dependability and efficacy of using AI-based security mechanisms. In addition, the ethical question about AI and surveillance and particularly the protection of individual data becomes a very important area of explaining the rights and freedoms of citizens against the background of enhancing security.

AI prepares smart cities for better protection of their data by analyzing the recent threats, improving protection of IoT devices, increasing efficiency of encryption and aiding in automation of actions in case of security threats. Thus, it is important to point out that the great opportunities to use AI in enhancing the security of smart city's infrastructures relate to several potential inherent problems and ethical issues.

Community and Stakeholder Engagement

Participation of the community and key stakeholders is very important when it comes to managing data security challenges of smart cities. They therefore rely profoundly of data collection, analysis, and dissemination to facilitate optimum delivery of urban services and enhancement of infrastructure. On the same note, reliance on data brings on board the data privacy and security issues that require the engrossment of all the stakeholders, especially the government and private players and the public.

The inclusion of the community makes its data collection, use together with its protection transparent hence fostering community trust. Due to engagement from the residents, they can sound concerns and expectations about privacy, thus influencing the data governance policies put in place in their best interest (Braun et al.,

2018). Stakeholder engagement also facilitates cohesiveness that see various sectors in harmonious sharing of the most desirable practices and innovation of even more efficient security solutions (Clement et al., 2022).

Central and local government agencies are responsible for the coordination of pliant engagement through holding meetings, seminars, and foci. They allow stakeholder conferences on identifying and managing possible risks and strategies for the same. For example, the city of Barcelona has used participatory processes to ensure that smart city plans respect the image that citizens have of privacy.

Furthermore, while private companies are profit driven and are responsible for managing some of the utmost important facets of smart city operations, they must protect people's data and declare their data policies. Such an approach can assist these companies in bringing the operational mode into compliance with public considerations, improve their basis for admissible referral upward, and thus, diminish the technological disruptions' resistances (Jacobs et al., 2020).

Therefore, one can firmly highlight the role of the community and stakeholders in overcoming the problems of data security in smart cities. It confirms the key principles necessary to make smart city data governance frameworks less exclusive and more transparent to achieve the public's expectations and thus enable safe and effective SMART CITY solutions. Few popular technologies for enhancing security in Smart Cities is listed in Table 3.

Table 3. Technologies for enhancing security in smart cities

Technology	Function	Use Case
Blockchain	Distributed ledger to ensure security to every transaction.	Secure data sharing, identity management
AI and Machine Learning	Security Threat Awareness	Intrusion detection systems
Biometric Authentication	Employment of biological data in the grant of secure access	Access control in smart buildings
Cybersecurity Frameworks	Security recommendations	NIST Cybersecurity Framework
IoT Security Platforms	Stand-alone IoT security platforms for IoT equipment	Device monitoring, firmware updates

CASE STUDIES AND BEST PRACTICES

Few case studies along with best practices are described through the following subsections.

Successful Implementation of Privacy Solutions

Privacy solutions can be a tricky and unavoidable addition to smart cities as they are aimed at protecting the data of the inhabitants of the city while improving its functioning. Promising examples identify a few general lessons that can be learned and followed by new smart cities to avoid or minimize privacy threats.

A good example is Barcelona, which has incorporated privacy by design as a culture within the smart city projects it endorses. The city has adopted a distributed data architecture that puts data ownership back in the hands of its inhabitants (Hassebo & Tealab, 2023). In this case, the application of blockchain system makes Barcelona provide a platform through which residents can engage in transparent and secure data transaction, hence, earning their confidence. This approach has been adopted in compliance with GDPR international data protection standards.

Similarly, the city of Singapore currently has the Smart Nation Sensor Platform (SNSP) with data anonymization and secure multiparty computation. These measures cannot allow strangers to access this data and guarantee safety of data in cases where data analysis is called for. It has thus played a very important role of ensuring that high levels of public confidence are developed, which is very significant for the implementation of smart city projects.

Privacy by Design, realized in the city of Toronto's Sidewalk Labs project, is another example of this intervention strategy, as is the enforcement of privacy-enhancing technologies (PETs). Application of differential privacy and robust encryption means that data obtained from urban sensors could not be used to profile the individuals that were being monitored (Dadayan, 2019). Despite the public scrutiny that the project received, the use of open data management policies to enhance control of the privacy factors also went a long way in resolving the challenge.

Stakeholders' engagement appeared to be crystallized and confirmed in the best practices drawn from these case studies across the planning and actual implementation. Public participation together with comprehensible communication regarding usage and protection of data is imperative to build public confidence. In addition, constant surveillance and improvement of privacy technologies are required to ensure that the threat perpetuates can be addressed (Goldsmith & Crawford, 2014).

Lessons Learned from Security Incidents

Regarding the topic of smart cities, security breaches are informative given data protection and security augmenting models. It shows trends and shared weaknesses allowing for improved protective steps and safeguard techniques after studying previous offenses.

Incident Analysis and Common Vulnerabilities

This is the reason why it is crucial to apply software updates and patches as was highlighted in the 2018 Atlanta ransomware attack (Armstrong et al., 2020). Thus, many breaches target outdated systems that are likely to have unpatched flaws on record. Equally, the Baltimore ransomware attack in 2019 expanded the poor data backup measures that led to a long-lasting disruption (Gawazah et al., 2024).

Response and Mitigation Strategies

Another good lesson is effective incident response plans. The Singapore health data breach of 2018 was an excellent example of the need for organizations to quickly detect and respond to a Cybersecurity event. The incidence involved 1.5 million patients' personal data and was further compounded by slow reaction and low threat detection (Wong & Jie, 2019). Recommendations coming from post-incident analysis was to draw up additional controls and staff training to reduce response time as much as possible while improving correctness.

Public-Private Collaboration

Accidents have also pointed to the benefits of partnership between the government and agencies and PSCs. Thus, the probability of cyber events during the Tokyo 2020 Olympics was minimized by the synergy of government departments and cybersecurity companies (Ishikawa & Kryvoi, 2023). This cooperation helped the residents in the city to switch threat information and enhance protection.

Policy and Governance Improvements

Cases of security breaches have, therefore, stimulated policy changes to enhance data management. Due to some violations of personal data, the GDPR of the European Union was adopted to tighten up the data protection regulation (Voigt & Von dem Bussche, 2017). This has formed the benchmark that other regions aspiring to enhance data protection have used.

Best Practices Adoption

Such incidents, hence, have raised outstanding practices like security audit, multi-factor authentication and the adoption of a zero-trust model as some of the ways of creating better security. Other smart cities such as Helsinki and Barcelona have embraced such practices hence lowering their risks of cyber threats (Sookhak, 2018).

Studying these lessons help smart cities to design better resilient and secure infrastructure to protect the people living in such cities.

Comparative Analysis of Global Smart City Initiatives

Global smart city comparison identifies distinct ways in which different signatures tend to adopt or consider data privacy and security features as they attend to urban issues. Specifically, strategies implemented in Singapore, Catalonia, and the Netherlands can help to understand the key lessons from the smart city initiative.

Singapore plans to become a smart nation via the Smart Nation Program which focuses on strong data management and cybersecurity practices. Today's Singapore has a good data privacy law thanks to the PDPA that can help maintain strict rules (Bokolo, 2023). The government's Smart Nation Sensor Platform (SNSP) shows one of many how secure IoT structures may be integrated in management of cities.

The Smart City Model applied by the city of Barcelona is characterized for being open and focusing on citizens. Barcelona uses the DECODE project to give people technology tools to manage their data and enhance confidence in smart city applications (Roche et al., 2012). That is why the use of open-source platforms by the city guarantees that data protection is integrated into its IT environment.

Smart City Framework is implemented in Amsterdam to ensure security where the private sectors and government work closely. Amsterdam Data Exchange (AMDEX) underlines ethical data sharing with an approach to data protection based on privacy by design principles that have been identified by Hollands in 2020. It also ensures that data minimization is employed and that secure access to data is achieved (Hollands, 2020).

These cases illustrate the applicability of solutions based on specific urban needs but also the cruciality of data protection and security. These lessons highlight inadequacies in existing legal systems, a lack of focus on citizens and their privacy and the need for collective smart city governance.

FUTURE DIRECTIONS AND RECOMMENDATIONS

The future directions with regards to data privacy and security in smart cities along with few recommendations are described in the following subsections.

Emerging Technologies for Data Privacy and Security

With integration of smart technologies and software forming the whole framework of smart cities, protection of data shall be of paramount importance. These have called for merged expertise and creativity where emerging technologies have been the solution providers with inclined awareness to both value-added functionality and privacy.

There is the concept of blockchain technology which is the decentralized and unalterable system where information is recorded and transmitted. Blockchain technology can then again be applied in smart cities to ensure that collections of data held by various parties involved can only be shared in a secure and transparent way, improving both trust and security at the same time (Zheng et al., 2017). Further, blockchain helps to develop enhanced management of identities by excluding opportunities for personal data misuse and hacker attacks.

Another promising assurance technology is homomorphic encryption which enables processing of data in the encrypted form. This means that one can analyze and use sensitive information and never decrypted it thus retaining proper security even when processing data (Gentry, 2009). This technology is most suitable to be applied in smart cities where people's big data is constantly being processed.

Security threats are also detected and dealt with by AI and ML. They permit the identification of abnormal activity and potential cyber threats, have predictive security actions (Nguyen et al., 2020). Other use cases are scenario-based data obfuscation, where data to be masked must be masked depending on the context and the level of access required.

Other levels of security are possible, for example, differential privacy in data analytics. In the system level, differential privacy adds noise into data queries and guarantees that precise individual entry cannot be extracted from group data, thus protecting individual privacy while allowing useful analysis to be performed on the data (Dwork, 2006).

Policy Recommendations for Governments and Organizations

Within smart cities, therefore, it is more important to guarantee reliable data security and privacy to prevent emerging problems and address current concerns. Policymakers and organizations should consider enforcement of an elaborate strategy that would adequately protect citizens' data and at the same time create trust and innovation.

Data Governance Frameworks

Extend thorough regulations on data governance that pinpoint ways we need to acquire, store, process, and share data. Such frameworks should also respect international regulation standards like – GDPR, thus achieving compliance and compatibility (Voigt & von dem Bussche, 2017).

Privacy-by-Design

Call for the adoption of privacy-innovative mechanisms in the designing and implementation of smart city solutions. This approach entails assimilating privacy attributes into the system structure, thus reducing exposure to data compromise (Cavoukian, 2021).

Enhanced Cybersecurity Measures

Incorporate IT security principles such as the use of passwords, encryption of data, using two or more factors in logging in to a system, and carrying out security checks on the system. They remain important in mimicking secure ways by which unauthorized people and cybercriminals cannot access delicate information (Kshetri, 2017).

Public Awareness and Education

Policy makers ought to introduce awareness creation platforms which aim at creating awareness of citizens concerning their data privacy and the need to embrace cyber security. This initiative can help rebuild public trust and promote responsible dissemination of data (Bertot et al., 2012).

Collaborative Efforts

Promote cooperation of various stakeholders in the private and public domains in terms of sharing of effective evidence-informed ways of increasing data privacy and security. Collaborations between the private and public sectors can bring solutions on the scale necessary for smart city development (Zheng et al., 2014).

The present policy recommendations can help governments and organizations resolve data privacy and security issues to support the development of smart sustainable cities.

Building a Resilient Smart City Ecosystem

Enhance Data Security Frameworks: Smart cities need to enhance the privacy of data as cities collect considerable amounts of information. This comprises; the use of measures such as complex encryption mechanisms and the use of multiple factor identification (Jan et al., 2019).

Promote Public Awareness

It is, therefore, indispensable for teaching the citizens on data privacy and security. At its purely informative level, awareness campaigns enable people to know their rights and the significance 1of personal data protection (Marune & Hartanto, 2022).

Implement Stronger Regulations

Much still needs to be done in this regard, governments should develop and implement tough legal regulations pertaining to the protection of Smart City data to meet the needed legal standards (Vojković & Katulić, 2021).

Foster Collaboration

Developing enablement and solutions for smart cities require government, private enterprises, and academic unions to seek an appropriate answer to data protection issues (Siddiqui et al., 2023).

Regular Security Audits

Moreover, in the current investigation, the authors stress that evaluating the safety hazards of smart city systems depending on annual security inspections proves valuable for improving safety governance (Khan, Pervez and Abbasi, 2017).

CONCLUSION

Smart urban development's use data-based innovation to deliver operational efficiency along with various privacy and protection issues regarding data collection and security. The research analyzed various complex issues concerning personal data and infrastructure data collection with their following utilization and protection within smart city environments. The examination revealed essential privacy protections and security requirements because it both analyzed vulnerabilities together with real-life data break-ins. The presented analysis included several useful approaches consisting of encryption methods combined with anonymization and blockchain solutions and privacy-by-design principles and AI technology analysis along with stakeholder participation. Various case analyses demonstrated both protective measures and protocols which various nations implemented to prevent cyber risks. Urban areas of the future will achieve resilience through technological implementation and rigid standards of data management between public-private sectors which result in secure urban construction. Security development and the creation of trustful stakeholder relationships and privacy protection serve as the cornerstones for maintaining smart city progress. Permanent development of secure private smart cities with public sector alongside private entities requires continuous joint work and inclusiveness.

REFERENCES

Atlanta Ransomware Attack https://coverlink.com/case-study/city-of-atlanta-ransomware/ [Accessed on 26th December, 2025].

Ahmad, K., Maabreh, M., Ghaly, M., Khan, K., Qadir, J., & Al-Fuqaha, A. (2022). Developing future human-centered smart cities: Critical analysis of smart city security, Data management, and Ethical challenges. *Computer Science Review*, *43*, 100452.

Alcaraz, C., & Zeadally, S. (2015). Critical infrastructure protection: Requirements and challenges for the 21st century. *International Journal of Critical Infrastructure Protection*, *8*, 53–66.

Allwinkle, S., & Cruickshank, P. (2011). Creating smart-er cities: An overview. *Journal of Urban Technology*, *18*(2), 1–16.

Almeida, F. (2023). Prospects of cybersecurity in smart cities. *Future Internet*, *15*(9), 285.

Andreani, S., Kalchschmidt, M., Pinto, R., & Sayegh, A. (2019). Reframing technologically enhanced urban scenarios: A design research model towards human centered smart cities. *Technological Forecasting and Social Change*, *142*, 15–25.

Arcila, B. B. (2024). How the Law Makes Smart Cities Unaccountable, and How to Start Making it Better: Lessons from Sidewalk Toronto. [-pages.]. *Vanderbilt Journal of Entertainment and Technology Law*, 58.

Armstrong, M. E., Jones, K. S., Namin, A. S., & Newton, D. C. (2020). Knowledge, skills, and abilities for specialized curricula in cyber defense: Results from interviews with cyber professionals. [TOCE]. *ACM Transactions on Computing Education*, *20*(4), 1–25.

Batra, R., Shrivastava, V. K., & Goel, A. K. (2021). Anomaly Detection over SDN Using Machine Learning and Deep Learning for Securing Smart City. In *Green Internet of Things for Smart Cities* (pp. 191–204). CRC Press.

Batty, M., Axhausen, K. W., Giannotti, F., Pozdnoukhov, A., Bazzani, A., Wachowicz, M., Ouzounis, G., & Portugali, Y. (2012). Smart cities of the future. *The European Physical Journal. Special Topics*, *214*, 481–518.

Bertot, J. C., Jaeger, P. T., & Hansen, D. (2012). The impact of polices on government social media usage: Issues, challenges, and recommendations. *Government Information Quarterly*, *29*(1), 30–40.

Bokolo, A. J. (2023). Data driven approaches for smart city planning and design: A case scenario on urban data management. *Digital Policy. Regulation & Governance*, *25*(4), 351–367.

Braun, T., Fung, B. C., Iqbal, F., & Shah, B. (2018). Security and privacy challenges in smart cities. *Sustainable Cities and Society*, *39*, 499–507.

Calzada, I. (2021). Data co-operatives through data sovereignty. *Smart Cities*, *4*(3), 1158–1172.

Case, D. U. (2016). Analysis of the cyber attack on the Ukrainian power grid. *Electricity information sharing and analysis center (E-ISAC), 388*(1-29), 3.

Cavoukian, A. (2021). *Privacy by design: The seven foundational principles*. IAPP Resource Center.

Chik, W. B. (2013). The Singapore Personal Data Protection Act and an assessment of future trends in data privacy reform. *Computer Law & Security Review*, *29*(5), 554–575.

Chintala, S. (2024). Harnessing AI and BI for Smart Cities: Transforming Urban Life with Data Driven Solutions. [IJSR]. *International Journal of Scientific Research*, *13*(9), 337–342.

Chourabi, H., Nam, T., Walker, S., Gil-Garcia, J. R., Mellouli, S., Nahon, K., Pardo, T. A., & Scholl, H. J. (2012, January). Understanding smart cities: An integrative framework. In *2012 45th Hawaii international conference on system sciences* (pp. 2289-2297). IEEE.

Clement, J., Manjon, M., & Crutzen, N. (2022). Factors for collaboration amongst smart city stakeholders: A local government perspective. *Government Information Quarterly*, *39*(4), 101746.

Culot, G., Nassimbeni, G., Podrecca, M., & Sartor, M. (2021). The ISO/IEC 27001 information security management standard: Literature review and theory-based research agenda. *The TQM Journal*, *33*(7), 76–105.

Cybersecurity, C. I. (2018). Framework for improving critical infrastructure cybersecurity. *URL:* https://nvlpubs. nist. gov/nistpubs/CSWP/NIST. *CSWP, 4162018*, 7 [Accessed on 24th December, 2025].

Dadayan, P. (2019). *City of Surveillance? The Implications of Sidewalk Labs' Resolution to Build a Smart City in Toronto* (Doctoral dissertation, Toronto Metropolitan University).

Dwork, C. (2006, July). Differential privacy. In *International colloquium on automata, languages, and programming* (pp. 1-12). Berlin, Heidelberg: Springer Berlin Heidelberg.

Eubanks, V. (2018). *Automating inequality: How high-tech tools profile, police, and punish the poor*. St. Martin's Press.

Gassmann, O., Böhm, J., & Palmié, M. (2019). *Smart cities: Introducing digital innovation to cities*. Emerald Publishing Limited.

Gawazah, L., Rondla, A., & Balhareth, M. S. A. (2024). To pay or not to pay: the us colonial pipeline ransomware attack.

Gellman, R. (2009, February). Privacy in the clouds: Risks to privacy and confidentiality from cloud computing. In *World privacy forum* (Vol. 23).

Gentry, C. (2009). *A fully homomorphic encryption scheme*. Stanford university.

Goldsmith, S., & Crawford, S. (2014). *The responsive city: Engaging communities through data-smart governance*. John Wiley & Sons.

Greenberg, A. (2021). A hacker tried to poison a Florida city's water supply, officials say. *Wired. com, 2*.

Hamid, B., Jhanjhi, N. Z., Humayun, M., Khan, A., & Alsayat, A. (2019, December). Cyber security issues and challenges for smart cities: A survey. In *2019 13th International Conference on Mathematics, Actuarial Science, Computer Science and Statistics (MACS)* (pp. 1-7). IEEE.

Hassebo, A., & Tealab, M. (2023). Global models of smart cities and potential IoT applications: A review. *IoT*, *4*(3), 366–411.

Hollands, R. G. (2020). Will the real smart city please stand up?: Intelligent, progressive or entrepreneurial? In *The Routledge companion to smart cities* (pp. 179–199). Routledge.

Hossain, M. M., Fotouhi, M., & Hasan, R. (2015, June). *Towards an analysis of security issues, challenges, and open problems in the internet of things. In 2015 ieee world congress on services*. IEEE.

Ishikawa, T., & Kryvoi, Y. (Eds.). (2023). *Public and Private Governance of Cybersecurity: Challenges and Potential*. Cambridge University Press.

Jacobs, N., Edwards, P., Markovic, M., Cottrill, C. D., & Salt, K. (2020). Who trusts in the smart city? Transparency, governance, and the internet of things. *Data & Policy*, *2*, e11.

Jan, M. A., Zhang, W., Usman, M., Tan, Z., Khan, F., & Luo, E. (2019). SmartEdge: An end-to-end encryption framework for an edge-enabled smart city application. *Journal of Network and Computer Applications*, *137*, 1–10.

Katz, J., & Lindell, Y. (2020). Introduction to modern cryptography crc press.

Khan, Z., Ludlow, D., Loibl, W., & Soomro, K. (2014). ICT enabled participatory urban planning and policy development: The UrbanAPI project. *Transforming Government: People. Process and Policy*, *8*(2), 205–229.

Khan, Z., Pervez, Z., & Abbasi, A. G. (2017). Towards a secure service provisioning framework in a smart city environment. *Future Generation Computer Systems*, *77*, 112–135.

Khayyat, M., Balfaqih, M., Balfaqih, H., & Ismail, M. (2024). Challenges and Factors Influencing the Implementation of Green Logistics: A Case Study of Saudi Arabia. *Sustainability*, *16*(13), 5617.

Kitchin, R. (2014). The data revolution: Big data, open data, data infrastructures and their consequences. *Sage (Atlanta, Ga.)*.

Kshetri, N. (2017). The economics of the Internet of Things in the Global South. *Third World Quarterly*, *38*(2), 311–339.

Lee, I., & Lee, K. (2018). The Internet of Things (IoT): Applications, investments, and challenges for enterprises. *Business Horizons*, *58*(4), 431–440.

Li, H., Han, D., & Tang, M. (2021). A privacy-preserving storage scheme for logistics data with assistance of blockchain. *IEEE Internet of Things Journal*, *9*(6), 4704–4720.

Ma, C. (2021). Smart city and cyber-security; technologies used, leading challenges and future recommendations. *Energy Reports*, *7*, 7999–8012.

Tzanou, M. (2020). *Health Data Privacy under theGDPR - Big Data Challenges and RegulatoryResponses*. Routledge.

Marune, A. E. M. S., & Hartanto, B. (2022). Strengthening personal data protection, cyber security, and improving public awareness in Indonesia: Progressive legal perspective. *International Journal of Business, Economics, and Social Development*, *2*(4), 143–152.

Menezes, A. J., Van Oorschot, P. C., & Vanstone, S. A. (2018). *Handbook of applied cryptography*. CRC press.

Narayanan, A., & Shmatikov, V. (2008, May). Robust de-anonymization of large sparse datasets. In *2008 IEEE Symposium on Security and Privacy (sp 2008)* (pp. 111-125). IEEE.

Nguyen, G., Dlugolinsky, S., Tran, V., & Garcia, A. L. (2020). Deep learning for proactive network monitoring and security protection. *ieee. Access : the Newsmagazine of the American Dental Hygienists' Association, 8,* 19696–19716.

Ogala, J. O., Ahmad, S., Shakeel, I., Ahmad, J., & Mehfuz, S. (2023). Strengthening KMS Security with Advanced Cryptography, Machine Learning, Deep Learning, and IoT Technologies. *SN Computer Science, 4*(5), 530.

Ohm, P. (2009). Broken promises of privacy: Responding to the surprising failure of anonymization. *UCLA l. Rev., 57,* 1701.

Protection, F. D. (2018). General data protection regulation (GDPR). *Intersoft Consulting, Accessed in October, 24*(1).

Rao, P. M., & Deebak, B. D. (2023). Security and privacy issues in smart cities/industries: Technologies, applications, and challenges. *Journal of Ambient Intelligence and Humanized Computing, 14*(8), 10517–10553.

Rathore, M. M., Paul, A., Hong, W. H., Seo, H., Awan, I., & Saeed, S. (2018). Exploiting IoT and big data analytics: Defining smart digital city using real-time urban data. *Sustainable Cities and Society, 40,* 600–610.

Rathore, S. P. S., Dalabhai, C. V., Patel, C. K. B., Sharma, R., Mathur, A., & Yadav, A. (2024, December). Big Data Analytics for Smart Cities. In *2024 1st International Conference on Advances in Computing, Communication and Networking (ICAC2N)* (pp. 1289-1294). IEEE.

Regulation, P. (2016). Regulation (EU) 2016/679 of the European Parliament and of the Council. [eu]. *Regulation, 679,* 2016.

Robinson, T., & Ji, M. (2022). Smart City Seoul: Solving the Urban Puzzle. In *Sustainable, Smart and Solidary Seoul: Transforming an Asian Megacity* (pp. 99–134). Springer International Publishing.

Roche, S., Nabian, N., Kloeckl, K., & Ratti, C. (2012, May). Are 'smart cities' smart enough. In *Global geospatial conference* (pp. 215-235).

Roman, R., Zhou, J., & Lopez, J. (2018). On the features and challenges of security and privacy in distributed internet of things. *Computer Networks, 57*(10), 2266–2279.

Shatz, S., & Chylik, S. E. (2019). The California consumer privacy act of 2018: A sea change in the protection of California consumers' personal information. *Bus. LAw., 75*, 1917.

Sicari, S., Rizzardi, A., Grieco, L. A., & Coen-Porisini, A. (2015). Security, privacy and trust in Internet of Things: The road ahead. *Computer Networks, 76*, 146–164.

Siddiqui, S., Hameed, S., Shah, S. A., Khan, A. K., & Aneiba, A. (2023). Smart contract-based security architecture for collaborative services in municipal smart cities. *Journal of Systems Architecture, 135*, 102802.

Solove, D. J. (2005). A taxonomy of privacy. *U. Pa. l. Rev., 154*, 477.

Sookhak, M., Tang, H., He, Y., & Yu, F. R. (2018). Security and privacy of smart cities: A survey, research issues and challenges. *IEEE Communications Surveys and Tutorials, 21*(2), 1718–1743.

Sowmya, T., & Anita, E. M. (2024). A comprehensive review of AI based intrusion detection system. *Measurement. Sensors, 28*, 100827.

Stamp, M. (2011). *Information security: principles and practice*. John Wiley & Sons.

Sweeney, L. (2002). k-anonymity: A model for protecting privacy. *International Journal of Uncertainty, Fuzziness and Knowledge-based Systems, 10*(05), 557–570.

Syed, A. S., Sierra-Sosa, D., Kumar, A., & Elmaghraby, A. (2021). IoT in smart cities: A survey of technologies, practices and challenges. *Smart Cities, 4*(2), 429–475.

Voigt, P., & Von dem Bussche, A. (2017). The eu general data protection regulation (GDPR). *A Practical Guide, 1st Ed., Cham: Springer International Publishing, 10*(3152676), 10-5555.

Vojković, G., & Katulić, T. (2021). Data protection and smart cities. In *Handbook of smart cities* (pp. 903–928). Springer International Publishing.

Wahrstätter, A., Taudes, A., & Svetinovic, D. (2024). Reducing Privacy of Coin-Join Transactions: Quantitative Bitcoin Network Analysis. *IEEE Transactions on Dependable and Secure Computing*.

Wang, Y., Lin, W., & Zhang, T. (2010, October). Study on security of wireless sensor networks in smart grid. In *2010 International Conference on Power System Technology* (pp. 1-7). IEEE.

Wenge, R., Zhang, X., Dave, C., Chao, L., & Hao, S. (2014). Smart city architecture: A technology guide for implementation and design challenges. *China Communications, 11*(3), 56–69.

Wong, G., & Jie, W. J. (2019). Singapore in 2018: Between uncharted waters and old ghosts. *Southeast Asian Affairs, 2019*(1), 296–323.

Wood, G. (2014). Ethereum: A secure decentralised generalised transaction ledger. *Ethereum project yellow paper, 151*(2014), 1-32.

Zhang, K., Ni, J., Yang, K., Liang, X., Ren, J., & Shen, X. S. (2017). Security and privacy in smart city applications: Challenges and solutions. *IEEE Communications Magazine*, *55*(1), 122–129.

Zheng, Y., Capra, L., Wolfson, O., & Yang, H. (2014). Urban computing: Concepts, methodologies, and applications. [TIST]. *ACM Transactions on Intelligent Systems and Technology*, *5*(3), 1–55.

Zheng, Z., Xie, S., Dai, H., Chen, X., & Wang, H. (2017, June). *An overview of blockchain technology: Architecture, consensus, and future trends. In 2017 IEEE international congress on big data (BigData congress).* IEEEE.

Chapter 14
From Smart to Cognitive Cities:
The Next Evolution in Urban Intelligence

Zain Balfagih
https://orcid.org/0000-0002-0847-6447
Effat University, Saudi Arabia

ABSTRACT

As urban environments become increasingly digitized, the evolution from smart cities to cognitive cities marks a transformative shift in how cities sense, learn, and adapt. While smart cities rely on data collection and automation to optimize urban systems, cognitive cities integrate artificial intelligence, machine learning, and cognitive computing to create self-aware, context-sensitive, and continuously learning ecosystems. This chapter explores the theoretical underpinnings and enabling technologies of cognitive cities, distinguishing them from their smart city predecessors. It examines key applications in mobility, public services, and citizen engagement, and highlights real-world case studies from leading global cities. Ethical, social, and governance implications are discussed, including issues of data privacy, algorithmic transparency, and participatory design. The chapter concludes by identifying current challenges and outlining future research directions.

INTRODUCTION

Cognitive cities represent the latest paradigm in urban development, evolving beyond the smart city concept by integrating advanced artificial intelligence, real-time data analytics, and human-centric design to create urban environments that are

DOI: 10.4018/979-8-3373-0154-9.ch014

not only intelligent, but also adaptive, proactive, and committed to the well-being of their citizens. While smart cities focus on utilizing technology such as IoT networks, digital infrastructure, and automation for more efficient city management, cognitive cities aim for deeper integration of technology with human behavior, social processes, and participatory governance to foster continuous learning, adaptation, and meaningful engagement.

A cognitive city can be defined as an urban ecosystem that leverages cognitive computing and artificial intelligence to organize, analyze, and act upon a continuous stream of data from physical infrastructure, digital services, and human interactions. In this model, cities "think" by learning from historical and real-time data, optimizing operations, predicting needs, and evolving to new challenges. Cognitive cities are marked by their ability to process information not just passively, but also proactively—anticipating problems before they arise, and personalizing solutions for citizens, businesses, and institutions (Pacheco Rocha et al., 2022)

Distinct from their smart city predecessors, cognitive cities emphasize the "human factor," involving cognition, creativity, learning, and active citizen participation in shaping their evolution. Rather than focusing narrowly on technology deployment, cognitive cities recognize citizens as critical components: people are both sensors and actors, feeding back experiences and data while also helping set priorities and inform governance processes. This results in cities that are not only efficient but also adaptable, equitable, and resilient to change.

The transition from smart to cognitive cities is driven by the recognition that urban environments are increasingly complex and dynamic, requiring more than static, technology-driven solutions. Smart cities, while leveraging connected sensors and automated systems, often remain reactive, responding to issues as they occur (Zhang et al., 2018). Cognitive cities, by contrast, use technologies such as digital twins, AI, and big data analytics to anticipate urban needs, simulate outcomes, and deploy adaptive responses in advance.

Academic literature highlights several key differences between these paradigms. Cognitive cities integrate real-time situational awareness, collaborative intelligence (between humans and machines), and co-creation with stakeholders to solve challenges ranging from mobility and sustainability to governance and citizen well-being. This shift necessitates new frameworks for inclusive decision-making, adaptive policy, and human-centric service design—a move from technocratic management towards participatory, self-evolving systems that empower residents.

Contemporary cities face mounting pressures from climate change, rapid urbanization, and social inequalities, all of which require solutions that can adapt to shifting circumstances, learn from feedback, and anticipate emerging needs. Cognitive cities respond to these challenges by embedding flexibility and foresight into urban infrastructure and governance, creating environments that continuously learn, adapt, and reconfigure to serve diverse populations.

Moreover, a human-centric approach ensures that technological advancements are not ends in themselves, but means to foster well-being, inclusivity, and resilience. This orientation is reflected in recent guidelines and literature advocating for people-centered cities, where co-designed solutions and the lived experiences of residents drive the evolution of urban spaces.

In general, cognitive cities build on the intelligence of smart cities by embracing adaptability, proactive problem-solving, and a fundamentally human focus, setting the stage for urban environments that can thrive amid complexity and change.

CORE PRINCIPLES AND TENETS OF COGNITIVE CITIES

The concept of Cognitive Cities builds upon and extends the principles of Smart Cities by introducing a deeper level of perception, understanding, learning, and adaptation in urban systems. While smart cities optimize services through automation and data analytics, cognitive cities are self-aware, context-sensitive, and capable of evolving over time through continuous interaction with their environment and inhabitants. In this section, we will present the core principles of the cognitive cities as shown in Figure 1.

Figure 1. Cognitive city principles

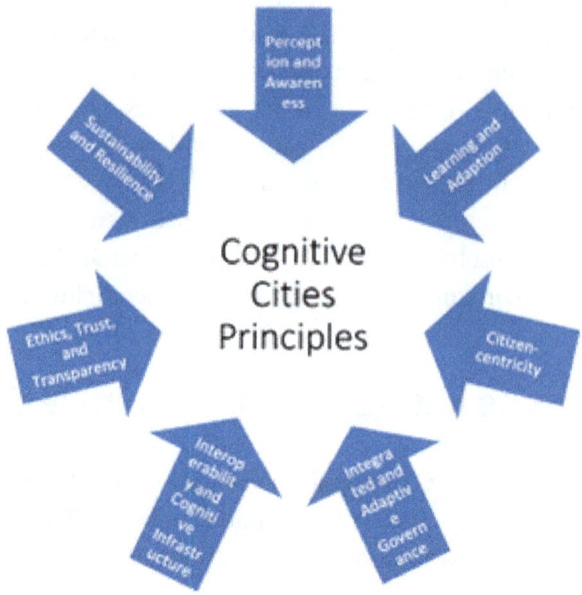

Perception and Awareness

One of the foundational pillars of cognitive cities is perception and awareness which is the ability of the urban system to sense, interpret, and maintain situational awareness of the physical, social, and contextual dynamics of the city. While smart cities often focus on data collection and sensing infrastructure, cognitive cities demand a richer, layered, contextually sensitive perception that supports real-time interpretation, feedback, and situational understanding. t its simplest, perception in a cognitive city begins with sensors and IoT devices that monitor variables such as traffic flow, air quality, energy usage, public transport status, pedestrian density, and health infrastructure. However, raw sensing alone is insufficient; the city must interpret the sensed data within its context. The notion of context-awareness (i.e. recognizing not just what is happening but under what circumstances) is central to this transition. A recent systematic review of context-aware applications in smart city infrastructures underscores that many urban applications are built to respond dynamically based on contextual information (e.g., mobility, health, alert systems) rather than static rules (Pacheco Rocha et al., 2022). A cognitive city's perceptual capability must integrate multiple modes:

- Physical sensing: IoT sensors for environment (e.g. temperature, humidity, pollution), infrastructure (e.g. structural health), mobility (traffic, public transport), utilities (water, power).
- Visual perception: Cameras, computer vision, remote sensing (e.g. satellite, aerial), street-view imagery. For instance, Zhang et al. (2018) developed a machine learning approach to quantify human perceptions of urban places across large-scale urban settings, correlating visual features with public sentiment (e.g., beauty, safety) to infer perceptual qualities (Zhang et al., 2018)
- Crowdsourced and social-data sensing: Citizens' smartphones, geotagged social media, participatory sensing platforms, text commentaries. Jang & Kim (2019) propose crowd-sourced cognitive mapping by aggregating georeferenced textual data and mapping collective place identity; this method enhances perceptual awareness via citizens' subjective views (Jang & Kim, 2019).
- Cognitive/perceptual networks: Pasolini et al. (2020) introduce the concept of a Cognitive Perceptual Internet (CPI) that supports crowd-based perception of the physical world, enabling systems to incorporate human perceptual judgments into urban awareness loops (Pasolini et al., 2020).

By combining these modalities, cognitive cities can derive a richer representation of the urban state, not merely numeric metrics, but interpreted meaning, sentiment, anomalies, and trends.

Perception must lead into awareness which is -in cognitive cities context- an understanding of how elements in the environment relate, evolve, and may influence future states. In human factors literature, the concept of situation awareness (SA) is often phrased in three levels: (1) perception of elements, (2) comprehension of their meaning, and (3) projection of their status into the near future (i.e. how the situation might evolve). While this model is rooted in human operators, the analogy is useful for cognitive city systems: a city must perceive, understand, and project to act proactively.

In practice, the system should detect, for example, that pedestrian density plus weather conditions plus events calendar may indicate a hazard or congestion risk. It should then project where crowding or bottlenecks might occur and preemptively adapt routes, services, or communications.

To raise perception beyond raw data, cognitive cities adopt semantic layering which is linking sensed data to higher-level concepts, ontologies, and knowledge graphs. Through semantic models, the city interprets low-level signals into meaningful categories (e.g., "rush hour congestion," "hazardous air alert," "unusual crowd formation"). This link enables reasoning: e.g. that a cluster of anomalous temperature, humidity, and pollution values might trigger an asthma alert in a district.

Knowledge graphs, linked data, and urban ontologies help the system maintain consistency, manage heterogeneity, and facilitate interoperability across subsystems. While direct peer-reviewed work on full-scale cognitive cities is emergent, research on context-aware systems has long emphasized semantic modeling to bridge perception and action (Vallejo-Correa et al., 2021).

Perception and awareness sit at the base of the data → knowledge → action → feedback lifecycle in cognitive cities. Without robust perceptual models, higher layers (e.g. decision-making, adaptation, citizen interaction) cannot function reliably. The perceptual layer enables the city to sense anomalies, understand evolving patterns, and form situational hypotheses. Figure 2 shows the perceptual pipeline in cognitive cities.

Figure 2. Perceptual pipeline in cognitive cities

Learning and Adaption

The principle of Learning and Adaptation is central to making a smart city truly cognitive: not merely reacting to predefined rules, but evolving over time, refining models, and reshaping behavior based on experience and feedback. In a cognitive city, learning bridges the gap between perception and action. Through continuous exposure to new data, the system refines its internal models, discovers novel correlations, and updates its decision policies to better align with shifting dynamics. Unlike static smart-city systems, which tend to rely on offline analytics or fixed heuristics, cognitive cities strive for incremental, continual, and context-aware learning. As noted in reviews of the cognitive city concept, the city "learns and adapts its behavior based on past experiences" via hybrid flows of human and machine intelligence (Machin & Solanas, 2018).

One enabling paradigm is continual learning (also called lifelong learning), where models are updated as fresh data arrives, without forgetting previously acquired knowledge. In the domain of urban intelligence, this is especially relevant because data distributions (e.g., traffic patterns, mobility, energy usage) shift over time. The continual learning methods can be applied in transportation, environmental monitoring, and public health in urban settings to adapt models in real time (Yang et al., 2024). Another useful approach is transferring learning or meta-learning, where insights from one city or region can inform modeling in another, accelerating adaptation under limited data. For example, Yao et al. propose a meta-learning framework for spatiotemporal prediction, enabling a model trained on multiple cities to adapt quickly to a new city with limited data (Yao et al., 2019) Adaptation in a cognitive city can occur at different levels and timescales:

a. Short-term adaptation: Real-time adjustments in system behavior based on immediate context (e.g. dynamic traffic signal timing, congestion rerouting). Reinforcement learning and online learning methods are suitable here.
b. Mid-term adaptation: Daily or weekly pattern adjustment, e.g. refining demand forecasts, updating energy dispatch policies, or changing waste collection schedules. These use batch or incremental retraining.
c. Long-term structural adaptation: Strategic reconfiguration of infrastructure, algorithmic upgrades, and policy shifts in response to persistent trends (e.g. climate shifts, demographic change). Simulation-based learning, scenario planning, and adaptive pathways come into play.

A cognitive city's learning system is not unidirectional: it embeds feedback loops. Actions taken by the system (e.g., changing a bus route) influence outcomes; those new observations feed back into the perception and learning layers. Over time, the system refines its policies, uncovers latent causality, and adjusts both models and behaviors. This intelligence amplification loop—alternating between action and learning—underpins adaptability (Machin et al., 2021). These loops also require mechanisms for uncertainty quantification, model validation, and risk-aware learning so that adaptation does not lead to instability or oscillation.

Citizen-Centricity

In contrast to technology-driven smart city visions that treat citizens as passive recipients of services, citizen-centricity places users at the center of design, decision, and feedback loops. In a cognitive city paradigm, this principle entails that urban systems not only optimize efficiency and resilience, but also respond to citizens' experiential, behavioral, and participatory needs. The goal is to co-create, co-govern, and continuously adapt services to human expectations, preferences, and values.

A cognitive city must transcend pure technical ideals to maintain legitimacy, social acceptance, and sustainable uptake of its systems. When citizens perceive that systems respect their agency, voice, and lived experience, they are more likely to engage, trust, and contribute data and feedback. Empirical studies on smart city adoption underscore that perceived usefulness, ease of use, trust, and social influence are significant predictors of citizens' intention to participate in smart city services (Nusir et al., 2023). Moreover, citizen discontent can arise when technological visions neglect actual user needs. Twist et al. highlights that top-down implementations often lead to dissatisfaction, lack of participation, or withdrawal, especially when systems feel opaque or disconnected from daily realities (Van Twist et al., 2023). Citizen-centricity can happen across several engagement modalities:

Participatory Design and Co-Creation

Citizens are invited to help design services, interfaces, and policies, rather than merely being consulted after design (Zakzak, 2019).

Participatory Sensing and Opportunistic Sensing

Citizens contribute data actively (reporting, feedback apps) or passively (device sensors), enriching the perceptual layer. Delmastro et al. propose "people-centric computing" in smart cities, integrating participatory sensing, mobile social networks, and opportunistic sensing as user-driven data streams (Delmastro et al., 2016)

Digital and Hybrid Feedback Loops

Platforms that show citizens how their inputs result in action build transparency and trust. For example, the Traffy Fondue platform in Bangkok lets citizens report urban issues (e.g., potholes, drainage) via a chatbot, and tracks resolution; the study finds that citizen participation via the platform contributes significantly to the success of urban projects (Hansen & Dahiya, 2025).

Civic Digital Twins and Social Modeling

In more advanced forms, citizen behavior, preferences, and feedback are modeled within urban digital twins, enabling co-simulations and collaborative planning. The Civic Digital Twin concept is being tested in Bologna, where citizens, organizations, and civic actors are invited to engage via simulations and scenario co-design (Luca et al., 2024).

In a cognitive city, citizen-centricity is not an add-on; it forms a critical feedback and learning loop. Citizen data, preferences, corrections, and normative values become part of the system's learning layer and adaptation logic. In effect, citizens are co-sensors, co-evaluators, and co-shapers of city cognition. The system must constantly re-balance efficiency goals with human satisfaction, equity, legitimacy, and behavioral adoption.

Integrated and Adaptive Governance

Integrated and adaptive governance is a key enabler of a truly cognitive city. It denotes governance systems that are holistic, multi-level, dynamic, and capable of learning and evolving in response to shifting urban conditions. Rather than static top-down hierarchies or siloed bureaucracies, cognitive cities require governance architectures that blend integration across domains (transport, energy, water, health, etc.) with adaptability to uncertainty, feedback, and change. In the literature of smart cities, governance is often highlighted as a challenge due to fragmentation, weak institutional coordination, and limited adaptability (Ruhlandt, 2018). Cities are complex systems where infrastructure, social systems, ecology, and economy interlock. Governance that treats domains in isolation or uses rigid rules is likely to fail when confronted with emergent interactions or unexpected perturbations. In cognitive cities, governance must integrate across domains (e.g. transport, energy, mobility, environment), coordinate across scales (local, municipal, regional, national), adapt continuously, and learn from feedback. So, the major components or dimensions that cognitive city governance should have:

Polycentric and Multilevel Structure

Adaptive governance theory suggests that polycentric governance, where multiple overlapping decision centers operate at different scales, enhances robustness and responsiveness. In urban contexts, this means empowering neighborhood-level bodies, city-wide agencies, and regional authorities to interact and self-organize. Birkmann et al. explore the challenges of adaptive urban governance in highly dynamic environments, pointing to the need to rethink planning systems and tools to support multi-level coordination (Birkmann et al., 2014).

Institutional Innovation and Flexibility

Rigid institutional structures struggle with novelty or crises. Institutional innovation—creating new rules, flexible mandates, hybrid public–private entities, and adaptive norms—becomes essential. Patterson & Huitema (2019) discuss institutional innovation in urban climate governance, explaining how new formal and informal structures can help cities respond to evolving climate risks (Patterson & Huitema, 2019). Adaptive policy design is another manifestation: Sha et al. advocate for adaptive policy packages that integrate instruments capable of adjusting over time (e.g. dynamic subsidies, regulatory sandboxes), rather than fixed one-size policies (Sha, K., & Taeihagh, 2024).

Collaborative and Co-Creative Governance

Integration is not only technocratic but relational. Shared governance, co-creation, participatory planning, and collaboration among government, citizens, civil society, and private sectors are central. Bradley et al. examined how integrated collaborative governance structures evolve in cities, particularly in projects for nature-based urban transformation, and how shared governance contributes to adaptive capacities over time (Bradley et al., 2022).

Data-Driven Monitoring, Feedback, and Reflexivity

For governance to adapt, it needs feedback loops: measurement, evaluation, reflection, and adjustment. Cognitive governance mandates robust metrics, real-time sensor data, dashboards, and evaluation systems that feed into decision-making. Over time, policies are refined based on successes or failures, rather than being immutable (Balfaqih & Alharbi, 2022).

Interoperability, Policy Alignment, and Normative Coherence

Integration requires that policies, regulations, IT systems, data architecture, and institutional procedures are aligned. Smart city governance research emphasizes the importance of interoperability (technical, legal, organizational) in achieving cohesive governance. Hardi et al. discuss how interoperability attributes in smart city governance influence outcomes and resilience (Hardi et al., 2025).

Integrated and adaptive governance operates as the control and meta-learning layer in the cognitive city stack. It binds together perception, learning, citizen engagement, and infrastructure decisions. Governance must mediate trade-offs, arbitrate conflicting objectives, and govern the evolution of adaptation itself (i.e. governance of governance). Adaptive governance in cognitive cities ensures that decisions are not reactive patchworks but coherent, evolving strategies aligned with long-term urban goals.

Interoperability and Cognitive Infrastructure

One of the foundational principles underpinning a cognitive city is interoperability, tightly coupled with the notion of cognitive infrastructure. For a city to reason, learn, and act across multiple systems, the underlying infrastructure must enable seamless connectivity, semantic alignment, modularity, and scalable intelligence across heterogeneous domains. Without interoperability, silos emerge, and the promise of a unified cognitive urban brain remains unattainable. Interoperability ensures that diverse components (sensors, legacy systems, data platforms, analytics engines, decision modules, and governance interfaces) can communicate, share, and interpret data without friction. In a cognitive city, where perception, inference, adaptation, and citizen engagement must be tightly orchestrated, interoperability is the glue that binds these layers into a coherent system (Frascella et al., 2018; Pliatsios et al., 2023).

Furthermore, governance-focused research shows that interoperability is not just a technical issue but a governance concern. Interoperable systems have been shown to strengthen smart city governance and human security by enabling cross-sector collaboration and aligning policy, systems, and user engagement (Hardi et al., 2025). To operate interoperability within cognitive cities, there are several interlocking dimensions to consider:

a. Technical Interoperability which covers the protocol, interface, and data exchange compatibility (e.g. REST APIs, MQTT, standardized messaging).
b. Syntactic Interoperability ensures that shared data adheres to compatible data formats (XML, JSON, etc.), field naming conventions, and schema compatibility.
c. Semantic Interoperability requires ontologies, vocabulary, and canonical data models so that "vehicle_count" in one system aligns with a "traffic_flow" concept in another (Pliatsios et al., 2023).
d. Organizational / Institutional Interoperability especially that systems reside in different agencies, departments, or private sectors.
e. Process Interoperability which icludes workflows, decision logic, and business processes must align or interface across domains. For example, a flood warning generated by the environment department may trigger alerts by transportation, emergencies, and public services; the processes must interlock smoothly (Prabowo et al., 2023).

Interoperability and cognitive infrastructure underpin the "plumbing" of a cognitive city: they enable perception modules, learning modules, decision modules, and citizen engagement modules to interoperate. Without this backbone, cognitive functions remain siloed, and the city cannot realize the synergies of integrated intelligence.

Ethics, Trust, and Transparency

In a cognitive city, autonomous systems, data-driven decision-making, and adaptive feedback loops manage critical urban functions. This deep embedding of algorithmic logic into urban life raises profound ethical questions. The principle of Ethics, Trust, and Transparency thus demands that the design, deployment, and governance of cognitive city systems be both morally grounded and socially legitimate. Without accountable, transparent, and trustworthy systems, the social license for cognition in the city may erode, breeding skepticism, resistance, or misuse (Kitchin, 2016).

Deploying AI systems in urban contexts can exacerbate bias, discrimination, or uneven impact if models are trained on biased data. Sanchez et al. (2025) survey AI ethics in urban planning, highlighting issues such as discriminatory allocation of resources, opaque algorithms in infrastructure decisions, and uneven burdens of surveillance across social groups (Sanchez et al., 2025).

Trust is the bridge between citizens and cognitive systems. Without it, data inputs might shrink, participation may wane, and systems may be resisted. In smart city contexts, transparency in IoT deployments and clarity on data management are critical to building trust in urban governance (Jacobs et al., 2020). Trust is not only a function of correctness or performance, but of legitimacy, perceived fairness, accountability, and responsiveness. Cognitive city systems must be designed in ways that are trustworthy not by default, but by design (Lai et al., 2025).

Transparency is the mechanism through which trust is earned. That entails making visible the workings of the system (its data flows, decision logic, feedback loops, error models, and governance oversight). But transparency is multi-layered: not all users need full internal detail; rather, different stakeholders should have access to appropriate levels of explanation and audit. One technical tool to support transparency is decision provenance—that is, tracking and exposing the chain of inputs, transformations, and outputs within algorithmic pipelines. Singh et al. propose that provenance techniques can expose data lineage and support auditability of decision systems (Singh et al., 2018). In the other side, full transparency may expose sensitive personal data or system vulnerabilities. The design challenge is enabling auditability without compromising privacy.

Generally, ethics and trust must be designed from the outset, not bolted on afterward. In effect, the cognitive city becomes a "trustworthy urban brain" whose recommendations and operations are grounded in accountable systems and reflexive governance.

Sustainability and Resilience

The principle of Sustainability and Resilience ensures that cognitive cities not only optimize present-day performance but also endure and adapt in the face of disruptions (environmental, social, or technological) while maintaining ecological balance and intergenerational equity. Sustainability and Resilience are complementary goals in cognitive cities. Sustainability focuses on the long-term balance between economic development, social equity, and environmental protection, ensuring that resource use does not exceed ecological limits. In the other side, resilience emphasizes the capacity of the urban system to absorb, adapt, recover, or transform in reaction to stressors or shocks (e.g. climate extremes, pandemics, infrastructure failure). In cognitive cities, these two goals must be integrated: sustainability constrains the baseline (e.g. emissions, resource use) while resilience ensures that thresholds and disturbances do not push the system into failure. Studies of smart city contributions to the UN Sustainable Development Goals (SDGs) show that while many smart-city projects aim for sustainability, resilience is often the weaker partner—suggesting the need for stronger integration (Sharifi et al., 2024; Chandran et al. 2025).

For cognitive cities, embedding sustainability and resilience requires that planning, governance, and systems architecture be designed holistically rather than in silos. It is common that smart city projects focus on technology deployment rather than long-term resilience or sustainable practice (Almulhim, 2025). Moreover, smart city planning must adopt an adaptive cycle mindset (prepare, absorb, recover, transform) in order to adjust trajectories as conditions change (Buzási & Csizovszky, 2023).

ENABLING TECHNOLOGIES AND INFRASTRUCTURE

The evolution from smart to cognitive cities requires a paradigm shift in how urban infrastructure is conceptualized—not merely as static, physical systems but as dynamic, intelligent ecosystems capable of sensing, learning, adapting, and responding. This transformation is powered by a network of enabling technologies, including Internet of Things (IoT) devices, cloud and edge computing, artificial intelligence (AI), digital twins, and secure data infrastructures. These technologies collectively form the nervous system and brain of a cognitive city, enabling real-time perception, reasoning, and action.

In contrast to traditional smart cities that emphasize automation and efficiency, cognitive cities demand infrastructure that supports context-awareness, continual learning, citizen engagement, and resilience. This requires building platforms that are modular, interoperable, secure, and ethically aligned with human-centered values. This section explores the foundational and next-generation infrastructures that enable the cognitive capabilities of modern cities.

Core Infrastructure Layers:

The infrastructure of the cognitive city can be classified to three layers as the following:

Sensing and IoT Infrastructure

A cognitive city's ability to perceive its surroundings hinges critically on a robust sensing and IoT infrastructure—a dense network of devices, sensors, actuators, and connectivity links that continuously capture, relay, and sometimes preprocess urban data. These sensors monitor environmental phenomena (e.g. air quality, temperature, noise), infrastructure health (e.g. structural strain, vibration), mobility flows (traffic, pedestrian counts), utilities (water, energy), and social signals (via crowd sensed contributions) (Zaman et al., 2024). In a cognitive city, this infrastructure is not passive: edge nodes and gateways often perform initial processing, filtering, or anomaly detection locally to reduce latency and bandwidth demands (Balfaqih, 2023). Wireless sensor networks (WSNs) are prominent in urban deployments, though urban environments pose challenges such as multi-path interference, energy constraints, and node failures especially in terms of power, coverage, reliability, and data integrity (Aburukba & El Fakih, 2025).

Scalability, heterogeneity, and interoperability are ongoing challenges. The sheer variety of sensing hardware, communication protocols, and data modalities demands modular, interoperable frameworks. Moreover, the concept of "sensor cities" illustrates how best practices in large-scale IoT deployments emphasize not only technical robustness but also citizen participation, cross-domain integration, and sustainability (Omrany et al., 2024).

Data and Cloud Infrastructure

Data and cloud infrastructure is the backbone of cognition in urban systems, enabling storage, processing, analytics, and interoperable access to urban-scale datasets. In cognitive cities, raw sensor data must become structured, stored, indexed, and made queryable across domains and time. Romualdo-Suzuki and Finkelstein propose that data should be treated as infrastructure that is, a shared, durable public good rather than a byproduct of operations. Their framework emphasizes aligning technical, governance, and organizational dimensions around data platforms so that multiple stakeholders can reliably reuse data across services (Romualdo-Suzuki, & Finkelstein, 2020).

Data integration across silos—transport, energy, health, environment—is key. Traditional smart city architectures highlight the need for data management layers that bridge heterogenous sources. In the other side, cloud platforms provide scalable computing, elastic storage, and high-availability services. In cognitive cities, cloud infrastructure supports batch analytics, long-span historical modeling, cross-domain correlation, and heavy-machine-learning workloads. Cloud computing enables integration of city systems by making servers, storage, and databases accessible over networks, thus reducing coupling and friction among domain systems (Dener, 2019). However, latency, bandwidth limitations, and privacy concerns motivate hybrid cloud-edge models. Some core capabilities of robust data-cloud infrastructure include:

- Data pipelines and streaming ingestion
- Data lakes and warehouses
- Metadata, catalogs, and governance
- APIs and platform services
- Resilience and fault tolerance

Challenges remain especially in data governance, ownership, and privacy—especially in public sector contexts. Also, data silos, heterogeneity, and semantic mismatch pose integration barriers. Interoperability across schemas and context definitions is nontrivial. Finally, cost, energy, and sustainability: large cloud infrastructures consume significant energy; cognitive city designs must optimize for green data architectures (Lnenicka et al., 2024).

Cybersecurity and Privacy Infrastructure

In cognitive cities, where sensing, data flows, and automated decision-making are pervasive, cybersecurity and privacy infrastructure become critical pillars to maintain system integrity, citizen trust, and legal compliance. Without robust protective layers, cognitive cities are vulnerable to attacks, data breaches, and surveillance abuses that can undermine their legitimacy. Urban cyber-physical systems combine digital and physical assets, making attacks potentially harmful in both cyberspace and real space. Common threats include unauthorized access, denial-of-service attacks, data tampering, spoofing, and cascading failures in interdependent systems (Xia et al., 2023).

A proper cybersecurity and privacy infrastructure in a cognitive city spans multiple layers:

- Device and endpoint security: Ensuring IoT sensors, gateways, and actuators are hardened, authenticated, and updatable. Weak firmware or default credentials often become entry points.
- Network security and segmentation: Using secure protocols (TLS, IPsec), virtual LANs, software-defined networking (SDN) segmentation, zero-trust architectures, and intrusion detection/prevention systems.
- Access control and identity management: Strong authentication, role-based access control (RBAC), attribute-based access control (ABAC), and federation across city domains.

- Data encryption, anonymization, and masking: Encrypting data in transit and at rest, employing anonymization or pseudonymization for privacy protection, and differential privacy when data is published.
- Privacy-by-design and legal compliance: Embedding privacy in system architecture from the start—minimizing data collection, limiting retention, respecting citizen consent, and providing transparency.
- Audit, logging, and forensic capabilities: Maintaining immutable logs, provenance metadata, and forensic audit trails to detect anomalies, trace malicious actions, and support accountability.

Cybersecurity and privacy infrastructure in cognitive cities must be baked in, not bolted on. They are essential to preserving system safety, citizen dignity, and the long-term viability of urban intelligence.

AI and Cognitive Computing

Artificial Intelligence (AI) and cognitive computing form the decision and reasoning core of a cognitive city, transforming raw data into insights, predictions, and autonomous or semi-autonomous actions. While smart cities use AI to optimize discrete services (e.g., traffic routing, energy dispatch), cognitive cities embed AI more deeply—enabling cross-domain inference, adaptive policy, and human-machine co-reasoning. Cognitive computing refers to systems that mimic human cognitive functions such as reasoning, learning, and perception in an integrated way, often combining symbolic reasoning, probabilistic inference, and neural learning (Elnagar et al., 2023).

In the urban domain, cognitive computing systems (CCSs) can bridge policy, planning, and real-time operations. CCSs enable a city to sense, reason, and act in feedback loops enhancing resilience by aligning technological operations with evolving policy goals (Degirmenci et al., 2024). AI underpins many core cognitive city functions:

- Predictive Analytics & Forecasting: Using historical and real-time data to forecast traffic jams, energy demand, air quality, or public health trends. AI models, especially deep learning and spatio-temporal networks, are increasingly used.
- Autonomous & Adaptive Systems: AI agents that dynamically adjust infrastructure settings—traffic lights, HVAC systems, waste collection routes—based on context.

- Contextual Reasoning & Decision Support: Integrating symbolic models, knowledge graphs, or rules with learned models to reason about "why" and "what if."
- Human–Machine Interaction: Natural language interfaces, chatbots, and AI assistants support citizen engagement, participatory sensing, and explanation of decisions. This enables citizens to query city systems, co-interpret predictions, and influence adaptation.

Next-Generation Digital Infrastructure

Several digital technologies are playing pivotal role in cognitive cities such as edge computing and digital twins. Edge computing brings computation, storage, and analytics closer to where data are generated (i.e. at sensor nodes, gateways, base stations), thereby reducing latency, bandwidth use, and reliance on centralized cloud services. This is particularly critical in cognitive cities, where real-time responsiveness and low-lag decision loops matter (e.g. traffic control, emergency response) (Trigka & Dritsas, 2025). Edge computing also enhances privacy (by local anonymization), resilience (local autonomy when cloud links fail), and scalability. Jararweh et al. emphasize that trustworthy and sustainable smart city services often depend on edge deployment to localize processing and reduce security exposure at central nodes (Jararweh et al., 2020). However, deploying edge infrastructure introduces challenges: resource constraints on edge nodes, heterogeneity of hardware, orchestration of workloads between edge and cloud, and dynamic load balancing. In a cognitive city, edge nodes may host lightweight AI models (e.g. for anomaly detection, initial inference), pre-filter data streams, run real-time control loops, and feed summarized information upward to cloud or higher cognitive layers.

Another promising technology in cognitive cities is digital twins. Digital twins are virtual replicas of physical assets, systems, or entire urban environments, continuously updated through real-time data streams. In the cognitive city frame, digital twins enable simulation, "what-if" experiments, predictive foresight, scenario testing, and closed-loop augmentation. Digital twins' technology supports monitoring, control, simulation, and decision support functions. It also has the capability to integrate multi-domain data (infrastructure, mobility, environment) and support sustainable, well-planned urban development (Peldon et al., 2024). Further, digital twins contribute to resilience by, for example, simulating disaster scenarios or stress tests and enabling proactive planning (Therias & Rafiee, 2023). When combined with edge architecture, digital twins can be updated in close to real time, making them not static models but continuously evolving cognitive assets. Some frameworks (e.g. Snap4City) integrate real-time multidata flows into digital twin architectures accessible to planners and citizens (Adreani et al., 2024).

In a cognitive city, edge computing and digital twins are not separate silos but synergistic components. Edge nodes collect and preprocess data fed into the digital twin; the twin runs simulations whose results guide adaptation and control loops back to the edge (or physical infrastructure) (Yao et al., 2025). This layered interplay allows cognitive systems to perceive, simulate, predict, and act with minimal latency and maximal contextual awareness. However, realizing this synergy requires solving issues of interoperability, versioning, consistency between real and virtual states, scalability of twin models, and resource orchestration across edge, fog, and cloud tiers.

Challenges and Considerations

Despite their promise, cognitive city technologies face several complex challenges that must be carefully managed. One major hurdle is interoperability—the integration of diverse systems, platforms, and data standards across city domains remains technically and institutionally difficult. Many cities still operate in silos, and achieving seamless, cross-domain coordination requires not just technical tools but shared governance frameworks.

Scalability is another pressing issue. As more sensors and devices come online, the volume and velocity of data can overwhelm systems not designed for large-scale, real-time processing. Ensuring consistent performance across city-wide deployments requires robust infrastructure planning and dynamic resource allocation.

Data governance and ethics present further concerns. Questions around data ownership, consent, surveillance, and digital rights become increasingly relevant as urban intelligence deepens. Without strong ethical frameworks, cognitive systems risk undermining public trust.

There are also resource constraints—both in terms of funding and human capacity. Implementing AI, edge computing, and digital twins requires significant investment and skilled personnel, which not all municipalities can sustain.

Finally, cities must grapple with legacy systems, organizational inertia, and regulatory bottlenecks. Transitioning from traditional to cognitive infrastructure requires a phased, adaptive approach that balances innovation with risk management and inclusivity.

APPLICATIONS AND USE CASES

While cognitive cities are built on advanced infrastructures and intelligent systems, their true value is realized through tangible applications that improve quality of life, governance, and sustainability. These applications bring together data, AI, human feedback, and adaptive decision-making to create responsive urban experiences that evolve with citizen needs and environmental pressures. Unlike traditional smart city solutions, which often focus on siloed optimizations (e.g., traffic flow, lighting systems), cognitive city applications are characterized by cross-domain intelligence, continuous learning, and human-centered adaptation. This section highlights key application domains where cognitive capabilities are being realized and evaluated, grounded in some use cases.

AI-Driven Emergency Response and Public Safety

In cognitive cities, AI augments emergency response and public safety by enabling real-time detection, predictive anticipation, and autonomous coordination. Two illustrative peer-reviewed deployments reveal how cognitive systems are already transforming urban safety.

Yao et al. deployed an AI-enabled Smart Video Solution (SVS) across 16 existing cameras in a campus environment. Their system processed pose-based data and anomaly detection to issue alerts to stakeholders (Aboualola et al., 2023). Over a 21-hour continuous run, it maintained ~16.5 frames per second and an average end-to-end latency of 26.76 seconds from anomaly detection to notification, showcasing real-world viability of cognitive safety systems.

A review of edge-AI for disaster and hazard detection illustrates multiple urban use cases: flood forecasting from sensor networks, real-time earthquake detection using distributed models, and fire detection via thermal imaging at edge nodes (Aboualola et al., 2023). These edge-enabled initiatives reduce reliance on remote cloud processing, enabling faster local reactions even amid network disruptions.

Personalized Citizen Services

In cognitive cities, personalized citizen services leverage AI, behavioral modeling, and context awareness to tailor public services—healthcare, social welfare, education, and utility provisioning—to individual needs and constraints (Hansen et al., 2017). These systems move beyond one-size-fits-all offerings to dynamic, anticipatory, and adaptive service delivery.

One illustrative case is home care prediction in Copenhagen, where an ensemble-learning model analyzed demographic and service usage data to forecast which citizens would demand significantly more home care hours. The system enabled proactive allocation of resources and personalized interventions (Hansen et al., 2017).

Another domain is urban issue reporting platforms. The Traffy Fondue platform implemented in Bangkok allows citizens to report problems like flooding, waste, and infrastructure damage via a conversational chatbot. The system then matches resolution channels to the reported issue, routing it to appropriate agencies and giving feedback. It achieved an 84 % citizen satisfaction rate and substantial cost savings (Hansen & Dahiya, 2025).

More broadly, Bastos et al. reviewed numerous smart city applications enabling citizens to report issues, monitor environmental parameters, and share opinions—forming a personalized feedback loop that makes services more responsive and individually relevant (Bastos et al., 2022).

Cognitive Mobility

Cognitive mobility systems fuse real-time sensing, predictive models, and control loops to dynamically optimize traffic flows and coordinate autonomous transport, thereby improving efficiency, safety, and sustainability. Below are two peer-reviewed case studies that illuminate how cognitive mobility is being realized in urban settings.

Cognitive mobility extends beyond static optimization in smart cities. For example, an adaptive traffic signals in China's 100 most congested cities showed that integrating real-time data analytics into signal control led to an 11 % reduction in peak-hour trip times and an 8 % reduction in off-peak times—translating into substantial time savings and lower emissions (Wu et al., 2025). This exemplifies how cognitive systems can refine existing infrastructure operations.

In the domain of Autonomous Mobility-on-Demand (AMoD), Alotaibi et al. present ten case studies across different cities deploying fleets of driverless vehicles orchestrated by AI. These systems adapt routes, rebalancing, and availability in response to real-time demand patterns. The results demonstrated reductions in required vehicle fleet size, decreased parking demand, and increased service responsiveness.

Additionally, research on integrating cognitive intelligence with vehicular ad hoc networks (VANETs) proposes systems in which vehicles and infrastructure communicate continuously to detect congestion, share situational awareness, and adjust signal timings or route suggestions in real time. Their approach shows improved traffic flow and adaptability to changing congestion conditions (Shepelev et al., 2023).

Cognitive Economy

In a cognitive city, economic growth is catalyzed not merely by infrastructure or investment, but by intelligent use of urban data, predictive analytics, and strategic interventions that foster innovation, competitiveness, and inclusive prosperity.

The cognitive economy application seeks to convert city data assets into economic insight, spur new business models, and align public policy with market signals.

One compelling example is Málaga City case. Málaga leverages cognitive and analytical tools to promote economic diversification, urban innovation, and poverty reduction. The city uses data-driven intelligence to identify growth clusters, guide startup incubation, and dynamically allocate resources to catalyze local economic development (Pérez & Toledo, 2017).

Another case is embedded in the "Data-Driven Smart Eco-City framework" by Bibri, which includes empirical exemplars of London and Barcelona (Bibri, 2023). These cities harness big data analytics for predictive modeling of consumption, service demands, and urban trends, thereby informing economic policy, investment planning, and innovation zones. Through data-driven decision support, local economies are better tuned to emerging opportunities.

Cognitive Environment

In cognitive cities, the cognitive environment application focuses on leveraging data, models, and adaptive decisions to optimize resource use, monitor environmental dynamics, and support sustainability goals. By integrating sensing, predictive analytics, and feedback loops, cities can balance ecological limits with human needs in domains such as energy, water, air quality, waste, and climate resilience.

One example is the use of City Digital Twins (CDTs) to bolster resilience and environmental management. Therias & Rafiee (2023) examined how a digital twin of Paris is used to simulate flooding scenarios, infrastructure stress, and recovery strategies, enabling planners to anticipate hazards and adaptively allocate resources across systems (Therias & Rafiee, 2023). Another case is the deployment of urban digital twins in Singapore and Dubai: modeling environmental behavior, optimizing energy and water use, and informing urban planning to reduce emissions and resource waste (Mazzetto, 2024).

Additionally, recent work on data-driven critical infrastructure planning shows how digital twin frameworks help manage climate-sensitive urban systems. Zhu et al. present cases where UDTs have been applied to simulate infrastructure under climate stress, guide adaptive resource allocation, and coordinate cross-domain responses (e.g. energy, water, drainage) in cities facing environmental change (Zhu & Jin, 2025).

FUTURE DIRECTIONS AND OPPORTUNITIES

As cognitive cities continue to evolve, their future lies in moving beyond siloed innovations to develop holistic, ethically grounded, and citizen-centric ecosystems. While current deployments focus on infrastructure, data integration, and localized AI capabilities, the next wave of advancements is expected to revolve around adaptive intelligence, human-AI collaboration, and urban co-evolution.

One promising direction is the convergence of cognitive cities with neuromorphic and affective computing. Future urban environments may not only sense physical and behavioral data but also interpret emotional cues and cognitive states. This will enable emotionally responsive public services, empathetic interfaces, and social robots integrated into urban spaces. As such technologies mature, cognitive cities may shift toward becoming emotionally intelligent environments—capable of understanding and responding to the needs of diverse populations in contextually aware ways.

Another key opportunity is in urban AI ethics frameworks. As cities become increasingly autonomous in decision-making, there is a growing need to establish clear, participatory frameworks that embed fairness, accountability, and transparency. Future research must explore mechanisms for algorithmic auditing, citizen oversight, and co-governance of urban AI systems. In parallel, legislative and institutional frameworks must evolve to define rights and responsibilities in cognitive environments.

The fusion of cognitive cities with the metaverse and XR (extended reality) offers another frontier. Immersive technologies can provide real-time overlays of city data, enable virtual planning consultations, and simulate long-term development scenarios. These tools could empower both policymakers and citizens to co-design urban futures, improving participation and foresight.

From an operational perspective, the deployment of self-configuring, self-healing systems; enabled by federated learning and swarm intelligence; can enhance the resilience of cognitive infrastructure in the face of disruptions. Cities could learn and adapt collectively, sharing data and models across jurisdictions to improve coordination in disaster management, pandemics, or large-scale mobility systems.

Finally, international collaboration is essential. The cognitive city vision demands global frameworks for data interoperability, urban AI ethics, and sustainability benchmarking. Collaborative innovation hubs, open data ecosystems, and global testbeds will be crucial for scaling cognitive systems across diverse contexts.

CONCLUSION

Cognitive cities represent a transformative evolution beyond the traditional smart city paradigm by integrating perception, learning, reasoning, and adaptation to create environments that are truly aware, responsive, and citizen-centric. This chapter explored the foundational principles of cognitive cities, highlighting how AI, digital infrastructure, and human-centric governance can converge to build intelligent urban ecosystems. Through concrete applications in safety, mobility, economic development, and sustainability, cognitive cities demonstrate their potential to reshape urban life in equitable, efficient, and resilient ways. Yet, as these systems grow in complexity, critical attention must be paid to ethical governance, transparency, and inclusion to ensure trust and long-term viability. Looking ahead, the future of cognitive cities lies in fostering continuous learning, not only by machines and systems, but by societies themselves. By embracing cognitive principles, cities can become dynamic partners in co-creating sustainable futures with their inhabitants at the center.

REFERENCES

Aboualola, M., Abualsaud, K., Khattab, T., Zorba, N., & Hassanein, H. S. (2023). Edge technologies for disaster management: A survey of social media and artificial intelligence integration. *IEEE Access : Practical Innovations, Open Solutions*, *11*, 73782–73802.

Aburukba, R., & El Fakih, K. (2025). Wireless Sensor Networks for Urban Development: A Study of Applications, Challenges, and Performance Metrics. *Smart Cities*, *8*(3), 89.

Adreani, L., Bellini, P., Fanfani, M., Nesi, P., & Pantaleo, G. (2024). Smart city digital twin framework for real-time multi-data integration and wide public distribution. *IEEE Access : Practical Innovations, Open Solutions*, *12*, 76277–76303.

Almulhim, A. I. (2025). Building Urban Resilience Through Smart City Planning: A Systematic Literature Review. *Smart Cities (2624-6511), 8*(1).

Balfaqih, M. (2023). A Hybrid Movies Recommendation System Based on Demographics and Facial Expression Analysis using Machine Learning. *International Journal of Advanced Computer Science and Applications*, *14*(11).

Balfaqih, M., & Alharbi, S. A. (2022). Associated information and communication technologies challenges of smart city development. *Sustainability*, *14*(23), 16240.

Bastos, D., Fernández-Caballero, A., Pereira, A., & Rocha, N. P. (2022, October). Smart city applications to promote citizen participation in city management and governance: A systematic review. [). MDPI.]. *Informatics (MDPI)*, *9*(4), 89.

Bibri, S. E. (2023). Data-driven smart eco-cities of the future: An empirically informed integrated model for strategic sustainable urban development. *World Futures*, *79*(7-8), 703–746.

Birkmann, J., Garschagen, M., & Setiadi, N. (2014). New challenges for adaptive urban governance in highly dynamic environments: Revisiting planning systems and tools for adaptive and strategic planning. *Urban Climate*, *7*, 115–133.

Bradley, S., Mahmoud, I. H., & Arlati, A. (2022). Integrated collaborative governance approaches towards urban transformation: Experiences from the CLEVER cities project. *Sustainability*, *14*(23), 15566.

Buzási, A., & Csizovszky, A. (2023). Urban sustainability and resilience: What the literature tells us about “lock-ins”? *Ambio*, *52*(3), 616–630. PMID: 36520412

Chandran, M. S., Chandran, R., Das, D., & Vinodkumar, K. (2025). Examining the role of smart cities and sustainability research in achieving SDGs through bibliometric lens. *Discover Sustainability*, *6*(1), 679.

Degirmenci, K., Desouza, K. C., & Watson, R. T. (2024). Urban Resilience Through Cognitive Computing Systems. *Journal of Urban Technology*, *31*(2), 117–124.

Delmastro, F., Arnaboldi, V., & Conti, M. (2016). People-centric computing and communications in smart cities. *IEEE Communications Magazine*, *54*(7), 122–128.

Dener, M. (2019). The role of cloud computing in smart cities. *The Eurasia Proceedings of Science Technology Engineering and Mathematics*, *7*, 39–43.

Elnagar, S., Thomas, M. A., & Osei-Bryson, K. M. (2023). What is Cognitive Computing? An Architecture and State of The Art. *arXiv preprint arXiv:2301.00882*.

Frascella, A., Brutti, A., Gessa, N., De Sabbata, P., Novelli, C., Burns, M., & He, L. (2018). A minimum set of common principles for enabling Smart City Interoperability. *Techne : Journal of Technology for Architecture and Environment*, 56–61. PMID: 39474009

Hansen, C., Hansen, C., Alstrup, S., & Lioma, C. (2017, November). Smart city analytics: Ensemble-learned prediction of citizen home care. In *Proceedings of the 2017 ACM on Conference on Information and Knowledge Management* (pp. 2095-2098).

Hansen, M. M., & Dahiya, B. (2025). Traffy Fondue: A smart city citizen engagement. *Frontiers in Sustainable Cities*, *7*, 1491621.

Hardi, R., Nurmandi, A., Purwaningsih, T., & Manaf, H. A. (2025). Smart city governance and interoperability: Enhancing human security in Yogyakarta and Makassar, Indonesia. *Frontiers in Political Science*, *7*, 1553177.

Jacobs, N., Edwards, P., Markovic, M., Cottrill, C. D., & Salt, K. (2020). Who trusts in the smart city? Transparency, governance, and the internet of things. *Data & Policy*, *2*, e11.

Jang, K. M., & Kim, Y. (2019). Crowd-sourced cognitive mapping: A new way of displaying people's cognitive perception of urban space. *PLoS One*, *14*(6), e0218590. PMID: 31220148

Jararweh, Y., Otoum, S., & Al Ridhawi, I. (2020). Trustworthy and sustainable smart city services at the edge. *Sustainable Cities and Society*, *62*, 102394.

Kitchin, R. (2016). The ethics of smart cities and urban science. *Philosophical Transactions. Series A, Mathematical, Physical, and Engineering Sciences*, *374*(2083), 20160115. PMID: 28336794

Lai, C. M. T., Cole, A., & Stivas, D. (2025). Understanding Trust in the Smart City Through Focus Groups in Hong Kong. *Urban Science (Basel, Switzerland)*, *9*(4), 123.

Lnenicka, M., Hervert, P., & Horak, O. (2024). Understanding big data and data protection measures in smart city strategies: An analysis of 28 cities. *Urban Governance*, *4*(4), 255–273.

Luca, M., Lepri, B., Gallotti, R., Paolazzi, S., Bigi, M., & Pistore, M. (2024). Towards civic digital twins: Co-design the citizen-centric future of bologna. *arXiv preprint arXiv:2412.06328*.

Machin, J., Batista, E., Martinez-Balleste, A., & Solanas, A. (2021). Privacy and security in cognitive cities: A systematic review. *Applied Sciences (Basel, Switzerland)*, *11*(10), 4471.

Machin, J., & Solanas, A. (2018, July). A review on the meaning of cognitive cities. In *2018 9th International Conference on Information, Intelligence, Systems and Applications (IISA)* (pp. 1-5). IEEE.

Mazzetto, S. (2024). A review of urban digital twins integration, challenges, and future directions in smart city development. *Sustainability*, *16*(19), 8337.

Nusir, M., Alshirah, M., & Alghsoon, R. (2023). Investigating smart city adoption from the citizen's insights: Empirical evidence from the Jordan context. *PeerJ. Computer Science*, *9*, e1289. PMID: 37346561

Omrany, H., Al-Obaidi, K. M., Hossain, M., Alduais, N. A., Al-Duais, H. S., & Ghaffarianhoseini, A. (2024). IoT-enabled smart cities: A hybrid systematic analysis of key research areas, challenges, and recommendations for future direction. *Discover Cities*, *1*(1), 2.

Pacheco Rocha, N., Dias, A., Santinha, G., Rodrigues, M., Rodrigues, C., Queirós, A., & Pavão, J. (2022). Systematic literature review of context-awareness applications supported by smart cities' infrastructures. *SN Applied Sciences*, *4*(4), 90.

Pasolini, G., Guerra, A., Guidi, F., Decarli, N., & Dardari, D. (2020). Crowd-based cognitive perception of the physical world: Towards the Internet of Senses. *Sensors (Basel)*, *20*(9), 2437. PMID: 32344795

Patterson, J. J., & Huitema, D. (2019). Institutional innovation in urban governance: The case of climate change adaptation. *Journal of Environmental Planning and Management, 62*(3), 374–398.

Peldon, D., Banihashemi, S., LeNguyen, K., & Derrible, S. (2024). Navigating urban complexity: The transformative role of digital twins in smart city development. *Sustainable Cities and Society.*

Pérez, J. G., & Toledo, D. G. (2017, June). Cognitive analytics of smart cities: a case of Málaga city. In *Proceedings of the 18th Annual International Conference on Digital Government Research* (pp. 616-617).

Pliatsios, A., Kotis, K., & Goumopoulos, C. (2023). A systematic review on semantic interoperability in the IoE-enabled smart cities. *Internet of Things : Engineering Cyber Physical Human Systems, 22*, 100754.

Prabowo, O. M., Mulyana, E., Nugraha, I. G. B. B., & Supangkat, S. H. (2023). Cognitive city platform as digital public infrastructure for developing a smart, sustainable and resilient city in Indonesia. *IEEE Access : Practical Innovations, Open Solutions, 11*, 120157–120178.

Romualdo-Suzuki, L., & Finkelstein, A. (2020). Data as Infrastructure for Smart Cities: Linking Data Platforms to Business Strategies. *arXiv preprint arXiv:2005.11414.*

Ruhlandt, R. W. S. (2018). The governance of smart cities: A systematic literature review. *Cities (London, England), 81*, 1–23.

Sanchez, T. W., Brenman, M., & Ye, X. (2025). The ethical concerns of artificial intelligence in urban planning. *Journal of the American Planning Association, 91*(2), 294–307.

Sha, K., & Taeihagh, A. (2024). Designing adaptive policy packages for inclusive smart cities: Lessons from Singapore's smart nation program. *Sustainable Cities and Society, 115*, 105868.

Sharifi, A., Allam, Z., Bibri, S. E., & Khavarian-Garmsir, A. R. (2024). Smart cities and sustainable development goals (SDGs): A systematic literature review of co-benefits and trade-offs. *Cities (London, England), 146*, 104659.

Shepelev, V., Glushkov, A., Slobodin, I., & Balfaqih, M. (2023). Studying the relationship between the traffic flow structure, the traffic capacity of intersections, and vehicle-related emissions. *Mathematics, 11*(16), 3591.

Singh, J., Cobbe, J., & Norval, C. (2018). Decision provenance: Harnessing data flow for accountable systems. *IEEE Access : Practical Innovations, Open Solutions, 7*, 6562–6574.

Therias, A., & Rafiee, A. (2023). City digital twins for urban resilience. *International Journal of Digital Earth, 16*(2), 4164–4190.

Trigka, M., & Dritsas, E. (2025). Edge and cloud computing in smart cities. *Future Internet, 17*(3), 118.

Vallejo-Correa, P., Monsalve-Pulido, J., & Tabares-Betancur, M. (2021). A systematic mapping review of context-aware analysis and its approach to mobile learning and ubiquitous learning processes. *Computer Science Review, 39*, 100335.

Van Twist, A., Ruijer, E., & Meijer, A. (2023). Smart cities & citizen discontent: A systematic review of the literature. *Government Information Quarterly, 40*(2), 101799.

Wu, K., Ding, J., Lin, J., Zheng, G., Sun, Y., Fang, J., & Gu, B. (2025). Big-data empowered traffic signal control could reduce urban carbon emission. *Nature Communications, 16*(1), 2013. PMID: 40016199

Xia, L., Semirumi, D. T., & Rezaei, R. (2023). A thorough examination of smart city applications: Exploring challenges and solutions throughout the life cycle with emphasis on safeguarding citizen privacy. *Sustainable Cities and Society, 98*, 104771.

Yang, L., Luo, Z., Zhang, S., Teng, F., & Li, T. (2024). Continual learning for smart City: A survey. *IEEE Transactions on Knowledge and Data Engineering*.

Yao, H., Liu, Y., Wei, Y., Tang, X., & Li, Z. (2019, May). Learning from multiple cities: A meta-learning approach for spatial-temporal prediction. In *The world wide web conference* (pp. 2181-2191).

Yao, S., Ardabili, B. R., Pazho, A. D., Noghre, G. A., Neff, C., Bourque, L., & Tabkhi, H. (2025). From lab to field: Real-world evaluation of an AI-driven Smart Video Solution to enhance community safety. *Internet of Things : Engineering Cyber Physical Human Systems*, 101716.

Zakzak, L. (2019, June). Citizen-centric smart city development: The case of smart Dubai's "Happiness Agenda". In *Proceedings of the 20th Annual International Conference on Digital Government Research* (pp. 141-147).

Zaman, M., Puryear, N., Abdelwahed, S., & Zohrabi, N. (2024). A review of IoT-based smart city development and management. *Smart Cities, 7*(3), 1462–1501.

Zhang, F., Zhou, B., Liu, L., Liu, Y., Fung, H. H., Lin, H., & Ratti, C. (2018). Measuring human perceptions of a large-scale urban region using machine learning. *Landscape and Urban Planning*, *180*, 148–160.

Zhu, M., & Jin, J. (2025). Data-Driven Urban Digital Twins and Critical Infrastructure Under Climate Change: A Review of Frameworks and Applications. *Urban Planning*, 10.

Compilation of References

Abadía, J. J. P., Walther, C., Osman, A., & Smarsly, K. (2022). A systematic survey of Internet of Things frameworks for smart city applications. *Sustainable Cities and Society*, *83*, 103949.

Abbas, A. B., Almohammedi, A. A., Balfaqih, M., & Darshi, S. (2023, September). Conceptual Design of Wireless Smart Grid for the Optimization of Electric Transmission in Iraq. In *2023 3rd International Conference on Computing and Information Technology (ICCIT)* (pp. 577-580). IEEE.

Abdulmalek, S., Nasir, A., Jabbar, W. A., Almuhaya, M. A., Bairagi, A. K., Khan, M. A. M., & Kee, S. H. (2022, October). IoT-based healthcare-monitoring system towards improving quality of life: A review. In Healthcare (Vol. 10, No. 10, p. 1993). MDPI.

Abdulmanov, R., Miftakhov, I., Ishbulatov, M., Galeev, E., & Shafeeva, E. (2021). Comparison of the effectiveness of GIS-based interpolation methods for estimating the spatial distribution of agrochemical soil properties. *Environmental Technology & Innovation*, *24*, 101970. DOI: 10.1016/j.eti.2021.101970

Aboualola, M., Abualsaud, K., Khattab, T., Zorba, N., & Hassanein, H. S. (2023). Edge technologies for disaster management: A survey of social media and artificial intelligence integration. *IEEE Access : Practical Innovations, Open Solutions*, *11*, 73782–73802.

Aburukba, R., & El Fakih, K. (2025). Wireless Sensor Networks for Urban Development: A Study of Applications, Challenges, and Performance Metrics. *Smart Cities*, *8*(3), 89.

Acilar, A., & Sæbø, Ø. (2023). Towards understanding the gender digital divide: A systematic literature review. *Global Knowledge, Memory and Communication*, *72*(3), 233–249.

Adreani, L., Bellini, P., Fanfani, M., Nesi, P., & Pantaleo, G. (2024). Smart city digital twin framework for real-time multi-data integration and wide public distribution. *IEEE Access : Practical Innovations, Open Solutions*, *12*, 76277–76303.

Agarwal, S., Satyanarain, R., Sing, T. F., & Vollmer, D. (2016). Effects of construction activities on residential electricity consumption: Evidence from Singapore's public housing estates. *Energy Economics*, *55*, 101–111.

Ahad, M. A., Paiva, S., Tripathi, G., & Feroz, N. (2020). Enabling technologies and sustainable smart cities. *Sustainable Cities and Society*, *61*, 102301.

Ahamed, N. N., Karthikeyan, P., Velliangiri, S., & Gunjan, V. K. (2023). Blockchain Security for the Internet of Things. In *Modern Approaches in IoT and Machine Learning for Cyber Security: Latest Trends in AI* (pp. 75–92). Springer International Publishing.

Ahmad, K., Maabreh, M., Ghaly, M., Khan, K., Qadir, J., & Al-Fuqaha, A. (2022). Developing future human-centered smart cities: Critical analysis of smart city security, Data management, and Ethical challenges. *Computer Science Review*, *43*, 100452.

Akhtar, N., Rahman, S., Sadia, H., & Perwej, Y. (2021). A holistic analysis of Medical Internet of Things (MIoT). *Journal of Information and Computational Science*, *11*(4), 209–222.

Akintayo, T. A., Olobo, N. A., & Atinuke, A. T., & AbdulKareem, I. O. (2024). Computing Performance Optimization Through Parallelization: Techniques and Evaluation. *International Journal of Education, Management, and Technology*, *2*(3), 418–439. DOI: 10.58578/ijemt.v2i3.4210

Akroot, A., & Refaei, M. (2025). Thermodynamic and Exergoeconomic Assessment of a Solar-Assisted Combined Cooling, Heating, and Power System in Antalya, Turkey. *Gazi Üniversitesi Fen Bilimleri Dergisi Part C: Tasarım ve Teknoloji*, 1.

Akroot, A., & Al Shammre, A. S. (2024). Techno-Economic and Environmental Impact Analysis of a 50 MW Solar-Powered Rankine Cycle System. *Processes (Basel, Switzerland)*, *12*(6), 1059. DOI: 10.3390/pr12061059

Al Ghamdi, M. A. (2022). An Optimized and Secure Energy-Efficient Blockchain-Based Framework in IoT. *IEEE Access : Practical Innovations, Open Solutions*.

Alagirisamy, B., & Ramesh, P. (2022). Smart sustainable cities: Principles and future trends. In *Sustainable cities and resilience: Select proceedings of VCDRR 2021* (pp. 301–316). Springer Singapore.

Al-Ali, A. R., Gupta, R., Zualkernan, I., & Das, S. K. (2024). Role of IoT technologies in big data management systems: A review and Smart Grid case study. *Pervasive and Mobile Computing*, 101905.

Alam, T. (2021). Blockchain cities: The futuristic cities driven by Blockchain, big data and the internet of things. *GeoJournal*.

Alcaraz, C., & Zeadally, S. (2015). Critical infrastructure protection: Requirements and challenges for the 21st century. *International Journal of Critical Infrastructure Protection*, *8*, 53–66.

Alfaris, A., Akroot, A., Alqaed, S., & Almehmadi, F. A. (2025). Performance analysis of integrated solar and natural gas combined cycle power plants in high solar potential regions. *Scientific Reports*, *15*(1), 9181. PMID: 40097525

Alfaris, A., Akroot, A., & Deniz, E. (2024). The Exergo-Economic and Environmental Evaluation of a Hybrid Solar–Natural Gas Power System in Kirkuk. *Applied Sciences (Basel, Switzerland)*, *14*(22), 10113. DOI: 10.3390/app142210113

Al-Fuqaha, A., Guizani, M., Mohammadi, M., Aledhari, M., & Ayyash, M. (2015). Internet of things: A survey on enabling technologies, protocols, and applications. *IEEE Communications Surveys and Tutorials*, *17*(4), 2347–2376.

Algarni, M., & Mishra, S. (2024). A Secure and Reliable Framework for Explainable Artificial Intelligence (XAI) in Smart City Applications. *Engineering, Technology &. Applied Scientific Research*, *14*(4), 15291–15296. DOI: 10.48084/etasr.7676

Ali, Y., Shah, S. W., & Khan, W. A. (2023). Security at the Internet of Things. In Machine Tools: An Industry 4.0 Perspective (pp. 31–48). CRC Press.

Aljabr, A. A., Sharma, A., & Kumar, K. (2019). Mining Process in Cryptocurrency Using Blockchain Technology: Bitcoin as a Case Study. *Journal of Computational and Theoretical Nanoscience*, *16*(10), 4293–4298.

Al-khatib, A. A., Mohammed, B., & Abdelmajid, K. (2020). *A survey on outlier detection in Internet of Things big data. Big Data-Enabled Internet of Things*. IET.

Allam, Z., & Allam, Z. (2020). Revisiting energy policy and planning in future living cities. *Biotechnology and Future Cities: Towards Sustainability, Resilience and Living Urban Organisms*, 35-52.

Allwinkle, S., & Cruickshank, P. (2011). Creating smart-er cities: An overview. *Journal of Urban Technology*, *18*(2), 1–16.

Allwinkle, S., & Cruickshank, P. (2011). Creating Smart-er Cities: An Overview. *Journal of Urban Technology*, *18*(2), 1–16. DOI: 10.1080/10630732.2011.601103

Almanifi, O. R. A., Chow, C.-O., Tham, M.-L., Chuah, J. H., & Kanesan, J. (2023). Communication and computation efficiency in Federated Learning: A survey. *Internet of Things : Engineering Cyber Physical Human Systems*, *22*, 100742. DOI: 10.1016/j.iot.2023.100742

Almeida, F. (2023). Prospects of cybersecurity in smart cities. *Future Internet*, *15*(9), 285.

Almohammedi, A. A., & Yusof, M. H. M. bin Tuan Muda, S. R., Shepelev, V., Balfaqih, M., & Shamsuddin, S. (2023, October). Resistors Over Digital Analog Converter on PRBS Threat Generator with PLL Synchronizer: Comparison Between R2R & Weighted. In *2023 3rd International Conference on Emerging Smart Technologies and Applications (eSmarTA)* (pp. 1-6). IEEE.

Almohammedi, A. A., Balfaqih, M., Nahas, S., Bokhari, A., & Alqudsi, A. (2023). Design and Implementation of IoT-Enabled Intelligent Fire Detection System Using Neural Networks. In *International Conference on AI and Mobile Services* (pp. 63-70). Cham: Springer Nature Switzerland.

Almohammedi, A. A., Balfaqih, M., Nahas, S., Bokhari, A., & Alqudsi, A. (2023b). Design and Implementation of IoT-Enabled Intelligent Fire Detection System Using Neural Networks. In *International Conference on AI and Mobile Services* (pp. 63-70). Cham: Springer Nature Switzerland.

Almohammedi, A. A., Raad, A., Darshi, S., Shepelev, V., Yusof, M. H. M., & Balfaqih, M. (2023a). High Gain and Efficiency Triple U-shaped Slots Microstrip Patch Antenna for 5G Applications. In *2023 3rd International Conference on Computing and Information Technology (ICCIT)* (pp. 573-576). IEEE.

Almulhim, A. I. (2025). Building Urban Resilience Through Smart City Planning: A Systematic Literature Review. *Smart Cities (2624-6511), 8*(1).

Alnahari, M. S., & Ariaratnam, S. T. (2022). The Application of Blockchain Technology to Smart City Infrastructure. *Smart Cities*, *5*(3), 979–993. DOI: 10.3390/smartcities5030049

Aloini, D., Benevento, E., Stefanini, A., & Zerbino, P. (2023). Transforming healthcare ecosystems through blockchain: Opportunities and capabilities for business process innovation. *Technovation*, *119*, 102557.

Alshibli, F., Alqarni, K., & Balfaqih, H. (2024). Analyzing the causes and impact of essential medicines and supplies shortages in the supply chain of the Ministry of health in Saudi Arabia: A quantitative survey study. *Informatics in Medicine Unlocked*, *45*, 101457.

Alskaif, T., Crespo-Vazquez, J. L., Sekuloski, M., Leeuwen, G. V., & Catalao, J. P. S. (2022). Blockchain-based fully peer-to-peer energy trading strategies for residential energy systems. *IEEE Transactions on Industrial Informatics, 18*, 231–241. DOI: 10.1109/TII.2021.3050018

Alsunousi, M., & Kayabasi, E. (2024). Techno-economic assessment of a floating photovoltaic power plant assisted methanol production by hydrogenation of CO2 captured from Zawiya oil refinery. *International Journal of Hydrogen Energy, 57*, 589–600. DOI: 10.1016/j.ijhydene.2024.01.055

Al-Yadumi, S., Xion, T. E., Wei, S. G. W., & Boursier, P. (2021). Review on Integrating Geospatial Big Datasets and Open Research Issues. *IEEE Access : Practical Innovations, Open Solutions, 9*, 10604–10620. DOI: 10.1109/ACCESS.2021.3051084

Alzhrani, A. A., Balfaqih, M., Alsenani, F., Alharthi, M., Alshehri, A., & Balfagih, Z. (2024). Design and Implementation of an IoT-Integrated Smart Locker System utilizing Facial Recognition Technology. *Engineering, Technology &. Applied Scientific Research, 14*(4), 16000–16010.

Amadi-Echendu, A. P. (2021). Using blockchain technology to facilitate property transactions. *South African Journal of Information Management, 23*(1), a1421.

Andersen, B., & Skrede, J. (2017). Planning for a sustainable Oslo: The challenge of turning urban theory into practice. *Local Environment, 22*(5), 581–594.

Andreani, S., Kalchschmidt, M., Pinto, R., & Sayegh, A. (2019). Reframing technologically enhanced urban scenarios: A design research model towards human centered smart cities. *Technological Forecasting and Social Change, 142*, 15–25.

Angelidou, M. (2014). Smart city policies: A spatial approach. *Cities (London, England), 41*(1), S3–S11. DOI: 10.1016/j.cities.2014.06.007

Anikwe, C. V., Nweke, H. F., Ikegwu, A. C., Egwuonwu, C. A., Onu, F. U., Alo, U. R., & Teh, Y. W. (2022). Mobile and wearable sensors for data-driven health monitoring system: State-of-the-art and future prospect. *Expert Systems with Applications, 202*, 117362.

ANORBOYEVICH. A. R. (2024). Exploring the evolutionary path: A historical analysis of Internet of Things (IoT) technologies. Global Book Publishing Services, 1-70.

Antunes, R. S., Seewald, L. A., Rodrigues, V. F., & Costa, C. A. D.Jr, L. G., Righi, R. R. & Campanatti, G. (2018). A survey of sensors in healthcare workflow monitoring. *ACM Computing Surveys, 51*(2), 1–37.

Anupama, B. S., & Sunitha, N. R. (2023). Survey on blockchain scalability addressing techniques. In S. J. Patel, N. K. Chaudhary, B. N. Gohil, & S. S. Iyengar (Eds.), *Information Security, Privacy and Digital Forensics. ICISPD 2022.* (Lecture Notes in Electrical Engineering, Vol. 1075). Springer. https://doi.org/DOI: 10.1007/978-981-99-5091-1_28

Anvari, S., Mahian, O., Taghavifar, H., Wongwises, S., & Desideri, U. (2020). 4E analysis of a modified multigeneration system designed for power, heating/cooling, and water desalination. *Applied Energy*, *270*(February), 115107. DOI: 10.1016/j.apenergy.2020.115107

Arcila, B. B. (2024). How the Law Makes Smart Cities Unaccountable, and How to Start Making it Better: Lessons from Sidewalk Toronto. [-pages.]. *Vanderbilt Journal of Entertainment and Technology Law*, 58.

Armstrong, M. E., Jones, K. S., Namin, A. S., & Newton, D. C. (2020). Knowledge, skills, and abilities for specialized curricula in cyber defense: Results from interviews with cyber professionals. [TOCE]. *ACM Transactions on Computing Education*, *20*(4), 1–25.

Arundel, S. T., Li, W., & Campbell, B. B. (2023). Reimagining standardization and geospatial interoperability in today's GeoAI culture. *Proceedings of the 6th ACM SIGSPATIAL International Workshop on AI for Geographic Knowledge Discovery*, 83–84. https://doi.org/DOI: 10.1145/3615886.3627744

Asadi, A., Wang, Q., & Mancuso, V. (2014). A survey on device-to-device communication in cellular networks. *IEEE Communications Surveys and Tutorials*, *16*(4), 1801–1819.

Asadi, Z., Abdekhoda, M., & Nadrian, H. (2020). Cloud computing services adoption among higher education faculties: Development of a standardized questionnaire. *Education and Information Technologies*, *25*(1), 175–191. DOI: 10.1007/s10639-019-09932-0

Asif, R., & Hassan, S. R. (2023). Shaping the future of Ethereum: Exploring energy consumption in Proof-of-Work and Proof-of-Stake consensus. *Frontiers in Blockchain*, *6*, 1151724. Advance online publication. DOI: 10.3389/fbloc.2023.1151724

Astill, J., Dara, R. A., Fraser, E. D., Roberts, B., & Sharif, S. (2020). Smart poultry management: Smart sensors, big data, and the internet of things. *Computers and Electronics in Agriculture*, *170*, 105291.

Atieh, A. T. (2021). The next generation cloud technologies: A review on distributed cloud, fog and edge computing and their opportunities and challenges. *ResearchBerg Review of Science and Technology, 1*(1), 1–15. https://researchberg.com/

Atlanta Ransomware Attack https://coverlink.com/case-study/city-of-atlanta-ransomware/ [Accessed on 26th December, 2025].

Atzori, L., Iera, A., Morabito, G., & Nitti, M. (2012). The social internet of things (siot)–when social networks meet the internet of things: Concept, architecture and network characterization. *Computer Networks*, *56*(16), 3594–3608.

Awotunde, J. B., Adeniyi, A. E., Ogundokun, R. O., Ajamu, G. J., & Adebayo, P. O. (2021). MIoT-based big data analytics architecture, opportunities and challenges for enhanced telemedicine systems. Enhanced telemedicine and e-health: advanced IoT enabled soft computing framework, 199-220.

Babu, P. (2024), Future Trends in Remote Patient Monitoring: Innovations and Possibilities, *Firminiq*, Online: https://firminiq.com/future-trends-in-remote-patient -monitoring-innovations-and-possibilities/ [28/09/2025]

Bachiega, J., Holanda, M., & Araujo, A. P. F. (2021). A proposal to minimize the cost of processing big geospatial data in public cloud providers. *Transactions in GIS*, *25*(3), 1599–1624. DOI: 10.1111/tgis.12754

Bai, X., Blanco, H., Gurney, K. R., Kılkış, Ş., Lucon, O., Murakami, J., & White, L. (2022). Urban systems and other settlements. In *Climate Change 2022-Mitigation of Climate Change* (pp. 861–952). Cambridge University Press.

Bala, K., & Kaur, P. D. (2022, January). Changing trends of blockchain in IoT: Benefits and challenges. In *2022 12th International Conference on Cloud Computing, Data Science & Engineering (Confluence)* (pp. 324-329). IEEE.

Balfaqih, M. (2023). Design and Implementation of IoT-enabled Intelligent Irrigation System Using Machine Learning. In *2023 28th Asia Pacific Conference on Communications (APCC)* (pp. 436-441). IEEE.

Balfaqih, M. (2023, November). Design and Implementation of IoT-enabled Intelligent Irrigation System Using Machine Learning. In *2023 28th Asia Pacific Conference on Communications (APCC)* (pp. 436-441). IEEE.

Balfaqih, M. (2024). Enhancing security and flexibility in smart locker systems: A multi-authentication approach with IoT integration. In *2024 21st Learning and Technology Conference (L&T)* (pp. 325-329). IEEE.

Balfaqih, M. (2024). Enhancing security and flexibility in smart locker systems: A multi-authentication approach with IoT integration. In *2024 21st Learning and Technology Conference (L&T)* (pp. 325–329). IEEE.

Balfaqih, M., Alharbi, S. A., Alfarsi, F., Alharbi, O., & Alsalmi, F. (2021). An Efficient Method for Rapid Fabrication Using Low Cost 3D Printer. In *Research and Innovation Forum 2020: Disruptive Technologies in Times of Change* (pp. 537-544). Springer International Publishing.

Balfaqih, M., Altwaim, A., Almohammedi, A. A., & Yusof, M. H. M. (2023b). An intelligent movies recommendation system based facial attributes using machine learning. In *2023 3rd International Conference on Emerging Smart Technologies and Applications (eSmarTA)* (pp. 1-6). IEEE.

Balfaqih, M., Balfagih, Z., Almohammedi, A. A., & Alfawaz, K. M. (2023). A Smart and Privacy-Preserving Logistics System Based on IoT and Blockchain Technologies. In *2023 1st International Conference on Advanced Innovations in Smart Cities (ICAISC)* (pp. 1-5). IEEE.

Balfaqih, M., Balfagih, Z., Almohammedi, A. A., & Alfawaz, K. M. (2023a). A Smart and Privacy-Preserving Logistics System Based on IoT and Blockchain Technologies. In *2023 1st International Conference on Advanced Innovations in Smart Cities (ICAISC)* (pp. 1-5). IEEE.

Balfaqih, M., Balfagih, Z., Almohammedi, A. A., & Alfawaz, K. M. (2023b). A Smart and Privacy-Preserving Logistics System Based on IoT and Blockchain Technologies. In *2023 1st International Conference on Advanced Innovations in Smart Cities (ICAISC)* (pp. 1-5). IEEE.

Balfaqih, H. (2022, March). Artificial intelligence in logistics and supply chain management: A perspective on research trends and challenges. In *International Conference on Business and Technology* (pp. 1241-1247). Cham: Springer International Publishing.

Balfaqih, M. (2023). A Hybrid Movies Recommendation System Based on Demographics and Facial Expression Analysis using Machine Learning. *International Journal of Advanced Computer Science and Applications, 14*(11).

Balfaqih, M., & Alharbi, S. A. (2022). Associated information and communication technologies challenges of smart city development. *Sustainability, 14*(23), 16240.

Balfaqih, M., & Alharbi, S. A. (2022). Associated Information and Communication Technologies Challenges of Smart City Development. *Sustainability, 14*(23), 16240. DOI: 10.3390/su142316240

Balfaqih, M., Alharbi, S. A., Alzain, M., Alqurashi, F., & Almilad, S. (2021). An accident detection and classification system using internet of things and machine learning towards smart city. *Sustainability, 14*(1), 210.

Balfaqih, M., Balfagih, Z., Lytras, M. D., Alfawaz, K. M., Alshdadi, A. A., & Al-solami, E. (2023). A blockchain- enabled iot logistics system for efficient tracking and management of high-price shipments: A resilient, scalable and sustainable approach to smart cities. *Sustainability, 15*(18), 13971.

Balfaqih, M., Balfagih, Z., Lytras, M. D., Alfawaz, K. M., Alshdadi, A. A., & Al-solami, E. (2023). A blockchain-enabled IoT logistics system for efficient tracking and management of high-price shipments: A resilient, scalable and sustainable approach to smart cities. *Sustainability, 15*(18), 13971.

Balfaqih, M., Balfagih, Z., Lytras, M. D., Alfawaz, K. M., Alshdadi, A. A., & Alsol-ami, E. (2023). A blockchain-enabled n nIoT logistics system for efficient tracking and management of high-price shipments: A resilient, scalable and sustainable approach to smart cities. *Sustainability, 15*(18), 13971.

Balfaqih, M., Balfaqih, Z., Shepelev, V., Alharbi, S. A., & Jabbar, W. A. (2022). An analytical framework for distributed and centralized mobility management protocols. *Journal of Ambient Intelligence and Humanized Computing, 13*(7), 3393–3405.

Balfaqih, M., Ismail, M., Nordin, R., & Balfaqih, Z. A. (2017). 802.21-assisted distributed mobility management solution in vehicular networks. *IEEE Access : Practical Innovations, Open Solutions, 5*, 9518–9532.

Balfaqih, M., Ismail, M., Nordin, R., Rahem, A. A., & Balfaqih, Z. (2017). Fast handover solution for network-based distributed mobility management in intelligent transportation systems. *Telecommunication Systems, 64*, 325–346.

Balfaqih, M., Jabbar, W., Khayyat, M., & Hassan, R. (2021). Design and development of smart parking system based on fog computing and internet of things. *Electronics (Basel), 10*(24), 3184. DOI: 10.3390/electronics10243184

Bamiah, M. A., & Brohi, S. N. (2011). Cloud deployment and service delivery models cloud service models. *International Journal of Research and Reviews in Information Sciences, 1*(3), 77–80.

Barata, J., Rupino Da Cunha, P., & Stal, J. (2018). Mobile supply chain management in the Industry 4.0 era: An annotated bibliography and guide for future research. *Journal of Enterprise Information Management, 31*(1), 173–192.

Bardhan, R., Debnath, R., Gama, J., & Vijay, U. (2020). REST framework: A modelling approach towards cooling energy stress mitigation plans for future cities in warming Global South. *Sustainable Cities and Society*, *61*, 102315. PMID: 33014694

Barns, S. (2019). *Platform urbanism: Negotiating platform ecosystems in connected cities*. Springer Nature.

Bastos, D., Fernández-Caballero, A., Pereira, A., & Rocha, N. P. (2022, October). Smart city applications to promote citizen participation in city management and governance: A systematic review. [). MDPI.]. *Informatics (MDPI)*, *9*(4), 89.

Batra, R., Shrivastava, V. K., & Goel, A. K. (2021). Anomaly Detection over SDN Using Machine Learning and Deep Learning for Securing Smart City. In *Green Internet of Things for Smart Cities* (pp. 191–204). CRC Press.

Batty, M., Axhausen, K. W., Giannotti, F., Pozdnoukhov, A., Bazzani, A., Wachowicz, M., & Portugali, Y. (2012). Smart cities of the future. *The European Physical Journal. Special Topics*, *214*, 481–518.

Baudier, P., Ammi, C., & Deboeuf-Rouchon, M. (2020). Smart home: Highly-educated students' acceptance. *Technological Forecasting and Social Change*, *153*, 119355.

Bayih, A. Z., Morales, J., Assabie, Y., & de By, R. A. (2022). Utilization of internet of things and wireless sensor networks for sustainable smallholder agriculture. *Sensors (Basel)*, *22*(9), 3273. PMID: 35590963

Beard, C., & Stallings, W. (2002). *Wireless communication networks and systems* (2nd ed.). Pearson.

Bebortta, S., Das, S. K., Kandpal, M., Barik, R. K., & Dubey, H. (2020). Geospatial Serverless Computing: Architectures, Tools and Future Directions. *ISPRS International Journal of Geo-Information*, *9*(5), 311. DOI: 10.3390/ijgi9050311

Bécue, A., Praça, I., & Gama, J. (2021). Artificial intelligence, cyber-threats and Industry 4.0: Challenges and opportunities. *Artificial Intelligence Review*, *54*(5), 3849–3886.

Belhadi, A., Kamble, S., Jabbour, C. J. C., Gunasekaran, A., Ndubisi, N. O., & Venkatesh, M. (2021). Manufacturing and service supply chain resilience to the COVID-19 outbreak: Lessons learned from the automobile and airline industries. *Technological Forecasting and Social Change*, *163*, 120447. PMID: 33518818

Benfeldt, O., Persson, J. S., & Madsen, S. (2020). Data Governance as a Collective Action Problem. *Information Systems Frontiers*, *22*(2), 299–313. DOI: 10.1007/s10796-019-09923-z

Bera, S., Dey, T., Mukherjee, A., & Buyya, R. (2023). E-CropReco: A dew-edge-based multi-parametric crop recommendation framework for internet of agricultural things. *The Journal of Supercomputing*, 1–35.

Bertot, J. C., Jaeger, P. T., & Hansen, D. (2012). The impact of polices on government social media usage: Issues, challenges, and recommendations. *Government Information Quarterly*, *29*(1), 30–40.

Besevli, B., Kayabasi, E., Akroot, A., Talal, W., Alfaris, A., Assaf, Y. H., Nawaf, M. Y., Bdaiwi, M., & Khudhur, J. (2024). Technoeconomic Analysis of Oxygen-Supported Combined Systems for Recovering Waste Heat in an Iron-Steel Facility. *Applied Sciences (Basel, Switzerland)*, *14*(6), 2563. DOI: 10.3390/app14062563

Bhardwaj, A., Bachu, S., Rajendran, S. P., Rastogi, S., Ramakrishna, K., Balalakshmi, C., & Sharma, P. (2024). The Future of Territorial Management: AI, IoT, and Big Data Synergies. In Alla, L., Alj, B., & Bentalha, B. (Eds.), (pp. 239–272). Advances in Electronic Government, Digital Divide, and Regional Development. IGI Global., DOI: 10.4018/979-8-3693-6854-1.ch008

Bhati, A., Hansen, M., & Chan, C. M. (2017). Energy conservation through smart homes in a smart city: A lesson for Singapore households. *Energy Policy*, *104*, 230–239.

Bhattacharyya, S., Darshi, S., Kumbhani, B., Pan, J., & Almohammedi, A. A. (2024). RSMUC: An alternative to coded cooperation for QoS enhancements in next generation vehicular network. In 2024 *IEEE Middle East Conference on Communications and Networking (MECOM) (IEEE MECOM 2024)*, pp. (1–6), Abu Dhabi, United Arab Emirates.

Bhattacharyya, S., Kumar, P., Darshi, S., Burnwal, O., & Singhal, K. (2021, December). Full duplex relay assisted coded cooperation for next generation wireless networks. In *2021 IEEE 18th India Council International Conference (INDICON)* (pp. 1-6). IEEE.

Bhattacharyya, S., Darshi, S., & Lin, Z. (2025). Outage analysis of downlink rsma in vehicular cluster: A copula- based approach. *IEEE Wireless Communications Letters*, 1–5.

Bhattacharyya, S., Kumar, P., Darshi, S., Agarwal, S., & Shailendra, S. (2023). Cross layer MAC protocol for a peer conscious opportunistic network coded cooperation system. *IEEE Transactions on Mobile Computing*, *22*(12), 7014–7026.

Bhattacharyya, S., Kumar, P., Darshi, S., Majhi, S., Kumbhani, B., & Almohammedi, A. A. (2024). Hybrid relaying based cross layer MAC protocol using variable beacon for cooperative vehicles. *IEEE Transactions on Vehicular Technology*.

Bhattacharyya, S., Kumar, P., Sharma, S., Darshi, S., & Almohammedi, A. A. (2021, December). A hybrid-combine-and-forward relaying scheme for network coded cooperative systems. In *2021 IEEE International Conference on Advanced Networks and Telecommunications Systems (ANTS)* (pp. 426-431). IEEE.

Bhushan, B., Khamparia, A., Martin Sagayam, K., Sharma, S. K., Ahad, M. A., & Debnath, N. C. (2020). Blockchain for smart cities: A review of architectures, integration trends and future research directions. *Sustainable Cities and Society*.

Bhushan, B., Sinha, P., Sagayam, K. M., & Andrew, J. (2021). Untangling blockchain technology: A survey on state of the art, security threats, privacy services, applications and future research directions. *Computers & Electrical Engineering*, *90*, 106897.

Bibri, S. E. (2023). Data-driven smart eco-cities of the future: An empirically informed integrated model for strategic sustainable urban development. *World Futures*, *79*(7-8), 703–746.

Bibri, S. E., & Krogstie, J. (2020). Environmentally data-driven smart sustainable cities: Applied innovative solutions for energy efficiency, pollution reduction, and urban metabolism. *Energy Informatics*, *3*(1), 29.

Bin Tajuddin, A. F., Yusof, M. H. M., Almohammedi, A. A., Dael, F. A., Balfaqih, M., & Qaid, N. (2025, May). Developing Secure MFA through QR code, SHA-256 OTP & Device Verification. In *2025 International Conference on Innovation in Artificial Intelligence and Internet of Things (AIIT)* (pp. 1-6). IEEE.

Birkmann, J., Garschagen, M., & Setiadi, N. (2014). New challenges for adaptive urban governance in highly dynamic environments: Revisiting planning systems and tools for adaptive and strategic planning. *Urban Climate*, *7*, 115–133.

Biswas, S., Yao, Z., Yan, L., Alqhatani, A., Bairagi, A. K., Asiri, F., & Masud, M. (2023). Interoperability benefits and challenges in smart city services: Blockchain as a solution. *Electronics (Basel)*, *12*(1036). Advance online publication. DOI: 10.3390/electronics12051036

Blasi, S., Ganzaroli, A., & De Noni, I. (2022). Smartening sustainable development in cities: Strengthening the theoretical linkage between smart cities and SDGs. *Sustainable Cities and Society*, *80*, 103793.

Blot, S., Ruppé, E., Harbarth, S., Asehnoune, K., Poulakou, G., Luyt, C. E., & Zahar, J. R. (2022). Healthcare-associated infections in adult intensive care unit patients: Changes in epidemiology, diagnosis, prevention and contributions of new technologies. *Intensive & Critical Care Nursing*, *70*, 103227. PMID: 35249794

Bodkhe, U., Tanwar, S., Parekh, K., Khanpara, P., Tyagi, S., Kumar, N., & Alazab, M. (2020). Blockchain for industry 4.0: A comprehensive review. *IEEE Access : Practical Innovations, Open Solutions*, *8*, 79764–79800.

Bokolo, A. J. (2023). Data driven approaches for smart city planning and design: A case scenario on urban data management. *Digital Policy. Regulation & Governance*, *25*(4), 351–367.

Bradley, S., Mahmoud, I. H., & Arlati, A. (2022). Integrated collaborative governance approaches towards urban transformation: Experiences from the CLEVER cities project. *Sustainability*, *14*(23), 15566.

Braun, T., Fung, B., Iqbal, F., & Shah, B. (2018). Security and privacy challenges in smart cities. *Sustainable Cities and Society*, *39*, 499–507.

Bruno, S. da M. (2024). *Processing Geospatial Data from Public Datasets to Assess Cycling Infrastructure in Smart Cities* [Master in Electrical and Computer Engineering Thesis, Faculdade De Engenharia, Universidade Do Porto]. https://repositorio-aberto.up.pt/bitstream/10216/162384/2/694027.pdf

Buzási, A., & Csizovszky, A. (2023). Urban sustainability and resilience: What the literature tells us about "lock-ins"? *Ambio*, *52*(3), 616–630. PMID: 36520412

Calder, K. E. (2016). *Singapore: Smart city, smart state*. Brookings Institution Press.

Calzada, I. (2021). Data co-operatives through data sovereignty. *Smart Cities*, *4*(3), 1158–1172.

Calzati, S., & Van Loenen, B. (2023). An ethics assessment list for geoinformation ecosystems: Revisiting the integrated geospatial information framework of the United Nations. *International Journal of Digital Earth*, *16*(1), 1418–1438. DOI: 10.1080/17538947.2023.2200041

Cao, F., Liu, B., He, J., Xu, J., & Xiao, Y. (2024). Privacy preservation-based federated learning with uncertain data. *Information Sciences*, *678*, 121024. DOI: 10.1016/j.ins.2024.121024

Caragliu, A., Del Bo, C., & Nijkamp, P. (2011). Smart cities in Europe. *Journal of Urban Technology*, *18*(2), 65–82.

Case, D. U. (2016). Analysis of the cyber attack on the Ukrainian power grid. *Electricity information sharing and analysis center (E-ISAC), 388*(1-29), 3.

Catarinucci, L., De Donno, D., Mainetti, L., Palano, L., Patrono, L., Stefanizzi, M. L., & Tarricone, L. (2015). An IoT-aware architecture for smart healthcare systems. *IEEE Internet of Things Journal, 2*(6), 515–526.

Cavalcanti, E. J. C., Ferreira, J. V. M., & Carvalho, M. (2021). On the consideration of different dead states in the exergy assessment of a solar-assisted combined cooling, heat and power system. *Sustainable Energy Technologies and Assessments, 47*, 101361. DOI: 10.1016/J.SETA.2021.101361

Cavoukian, A. (2021). *Privacy by design: The seven foundational principles*. IAPP Resource Center.

Chandran, M. S., Chandran, R., Das, D., & Vinodkumar, K. (2025). Examining the role of smart cities and sustainability research in achieving SDGs through bibliometric lens. *Discover Sustainability, 6*(1), 679.

Chang, S. C., Lu, M. T., Pan, T. H., & Chen, C. S. (2021). Evaluating the e-health cloud computing systems adoption in Taiwan's healthcare industry. *Life (Chicago, Ill.), 11*(4), 310. DOI: 10.3390/life11040310 PMID: 33918246

Chang, S., & Smith, M. (2023). Residents' Quality of Life in Smart Cities: A Systematic Literature Review. *Land (Basel)*.

Chanthaweethip, W., Han, X., Crespi, N., Chen, Y., Farahbakhsh, R., & Cuevas, A. (2013). "Current city" prediction for coarse location based applications on Facebook. In *GLOBECOM - IEEE Global Telecommunications Conference* (pp. 3188–3193). IEEE. https://doi.org/DOI: 10.1109/GLOCOM.2013.6831562

Chatterjee, S., & Kar, A. K. (2018). Effects of successful adoption of information technology enabled services in proposed smart cities of India: From user experience perspective. *Journal of Science and Technology Policy Management, 9*(2), 189–209.

Chattopadhyay, P., Patel, H. P., & Parmar, V. (2022, August). Internet of Things (IoT) in Smart Agriculture. In *2022 3rd International Conference on Electronics and Sustainable Communication Systems (ICESC)* (pp. 536-540). IEEE.

Chauhan, N., & Dwivedi, R. K. (2022, April). Designing a secure smart healthcare system with blockchain: a review. In *2022 1st International Conference on Informatics (ICI)* (pp. 214-219). IEEE.

Chen, E., Zhao, Y., Wang, M., Bian, M., Cai, W., Li, B., & Dai, Y. (2023). Experimental investigation of a solar-assisted absorption-compression system for heating and cooling. *Solar Energy*, *257*, 18–33. DOI: 10.1016/J.SOLENER.2023.04.025

Chen, T., Barbarossa, S., Wang, X., Giannakis, G. B., & Zhang, Z. L. (2019). Learning and management for Internet of Things: Accounting for adaptivity and scalability. *Proceedings of the IEEE*, *107*(4), 778–796.

Chen, Z., Ding, Z., Dai, X., & Karagiannidis, G. K. (2016). On the application of quasi-degradation to MISO-NOMA downlink. *IEEE Transactions on Signal Processing*, *64*(23), 6174–6189.

Chik, W. B. (2013). The Singapore Personal Data Protection Act and an assessment of future trends in data privacy reform. *Computer Law & Security Review*, *29*(5), 554–575.

Chintala, S. (2024). Harnessing AI and BI for Smart Cities: Transforming Urban Life with Data Driven Solutions. [IJSR]. *International Journal of Scientific Research*, *13*(9), 337–342.

Chiu, M. C., Huang, J. H., Gupta, S., & Akman, G. (2021). Developing a personalized recommendation system in a smart product service system based on unsupervised learning model. *Computers in Industry*, *128*, 103421. DOI: 10.1016/j.compind.2021.103421

Choi, J., Park, J., & Evans, B. L. (2017). Spectral efficiency bounds for interference-limited SVD-MIMO cellular communication systems. *IEEE Wireless Communications Letters*, *6*(1), 46–49.

Choi, T. M., Kumar, S., Yue, X., & Chan, H. L. (2022). Disruptive technologies and operations management in the Industry 4.0 era and beyond. *Production and Operations Management*, *31*(1), 9–31.

Chourabi, H., Nam, T., Walker, S., Gil-Garcia, J. R., Mellouli, S., Nahon, K., Pardo, T. A., & Scholl, H. J. (2012, January). Understanding smart cities: An integrative framework. In *2012 45th Hawaii international conference on system sciences* (pp. 2289-2297). IEEE.

Chu, X., Yang, D., & Li, J. (2019). Sustainability assessment of combined cooling, heating, and power systems under carbon emission regulations. *Sustainability*, *11*(21), 5917.

Chu, Z., Cheng, M., & Yu, N. (2021). A smart city is a less polluted city. *Technological Forecasting and Social Change*, *172*, 121037.

Clement, J., Manjon, M., & Crutzen, N. (2022). Factors for collaboration amongst smart city stakeholders: A local government perspective. *Government Information Quarterly*, *39*(4), 101746.

Clerckx, B., Mao, Y., Jorswieck, E. A., Yuan, J., Love, D. J., Erkip, E., & Niyato, D. (2023). A primer on rate-splitting multiple access: Tutorial, myths, and frequently asked questions. *IEEE Journal on Selected Areas in Communications*, *41*(5), 1265–1308.

Codina Barberà, M. (2024). Gait-Analysis MIoT Platform for Health Assessment.

Coetzee, S., Camboim, S. P., Griffin, A. L., Kubicek, P., Harvey, F., & Behr, F.-J. (2023). "Cartography and Open Standards": A Proposal for an Open Geospatial Consortium Technical Paper. *Abstracts of the ICA, 6*, 1–2. https://doi.org/DOI: 10.5194/ica-abs-6-39-2023

Cole, J., McDonald, J., Wen, X., & Kramer, R. (2018). Marketing energy efficiency: Perceived benefits and barriers to home energy efficiency. *Energy Efficiency*, *11*, 1811–1824.

Connolly, D., Nam, S., & Goodman, K. (2022). Solving old problems or making new ones? Blockchain technology for the protection of refugees and migrants. *Journal of Human Rights*, *22*(02), 1–26. DOI: 10.1080/14754835.2022.2100984

Culot, G., Nassimbeni, G., Podrecca, M., & Sartor, M. (2021). The ISO/IEC 27001 information security management standard: Literature review and theory-based research agenda. *The TQM Journal*, *33*(7), 76–105.

Cybersecurity, C. I. (2018). Framework for improving critical infrastructure cybersecurity. *URL:* https://nvlpubs. nist. gov/nistpubs/CSWP/NIST. *CSWP, 4162018*, 7 [Accessed on 24th December, 2025].

Dadayan, P. (2019). *City of Surveillance? The Implications of Sidewalk Labs' Resolution to Build a Smart City in Toronto* (Doctoral dissertation, Toronto Metropolitan University).

Daimon, H. (2022). A Fundamental Study on the Influences of Smart City on Residential Location Choice. *Toshi Keikaku Rombunshu.*

Dallasega, P., Rauch, E., & Linder, C. (2018). Industry 4.0 as an enabler of proximity for construction supply chains: A systematic literature review. *Computers in Industry*, *99*, 205–225.

Das, J., Ghosh, S. K., & Buyya, R. (2021). Geospatial Edge-Fog Computing: A Systematic Review, Taxonomy, and Future Directions. In Mukherjee, A., De, D., Ghosh, S. K., & Buyya, R. (Eds.), *Mobile Edge Computing* (pp. 47–69). Springer International Publishing., DOI: 10.1007/978-3-030-69893-5_3

Davoli, L., Belli, L., Dall'Olio, A., Di Nocera, F., Adorni, P., Cantelli, A., & Ferrari, G. (2024). Data Integration in a Smart City: A Real Case. In Menozzi, R. (Ed.), *Information and Communications Technologies for Smart Cities and Societies* (Vol. 5, pp. 11–24). Springer Nature Switzerland., DOI: 10.1007/978-3-031-39446-1_2

De Las Heras, A., Luque-Sendra, A., & Zamora-Polo, F. (2020). Machine learning technologies for sustainability in smart cities in the post-covid era. *Sustainability, 12*(22), 9320.

Degirmenci, K., Desouza, K. C., & Watson, R. T. (2024). Urban Resilience Through Cognitive Computing Systems. *Journal of Urban Technology, 31*(2), 117–124.

Delmastro, F., Arnaboldi, V., & Conti, M. (2016). People-centric computing and communications in smart cities. *IEEE Communications Magazine, 54*(7), 122–128.

Dener, M. (2019). The role of cloud computing in smart cities. *The Eurasia Proceedings of Science Technology Engineering and Mathematics, 7*, 39–43.

Dhunny, A. Z., Allam, Z., Lobine, D., & Lollchund, M. R. (2019). Sustainable renewable energy planning and wind farming optimization from a biodiversity perspective. *Energy, 185*, 1282–1297.

Dias, D., & Paulo Silva Cunha, J. (2018). Wearable health devices—Vital sign monitoring, systems and technologies. *Sensors (Basel), 18*(8), 2414. PMID: 30044415

Dimitrov, D. V. (2016). Medical internet of things and big data in healthcare. *Healthcare Informatics Research, 22*(3), 156–163. PMID: 27525156

Dinc, E., Kuscu, M., Bilgin, B. A., & Akan, O. B. (2019). Internet of everything: A unifying framework beyond internet of things. In *Harnessing the Internet of Everything (IoE) for Accelerated Innovation Opportunities* (pp. 1–30). IGI Global.

Ding, Z., Adachi, F., & Poor, H. V. (2015). The application of MIMO to non-orthogonal multiple access. *IEEE Transactions on Wireless Communications, 15*(1), 537–552.

Dittakavi, R. S. S. (2021). Deep learning-based prediction of CPU and memory consumption for cost-efficient cloud resource allocation. *Sage Science Review of Applied Machine Learning, 4*(1), 45–58.

Dogheim, G. M., & Hussain, A. (2023). Patient care through AI-driven remote monitoring: Analyzing the role of predictive models and intelligent alerts in preventive medicine. *Journal of Contemporary Healthcare Analytics*, *7*(1), 94–110.

Dogra, N., & Adil, M. (2024). Should we or should we not? Examining travelers' perceived privacy, perceived security and actual behavior in online travel purchases. *Journal of Vacation Marketing*, *30*(1), 123–142.

Dong, L., Liu, Z., & Bian, Y. (2021). Match circular economy and urban sustainability: Re-investigating circular economy under sustainable development goals (SDGs). *Circular Economy and Sustainability*, *1*, 243–256.

Dong, Z., Luo, F., & Liang, G. (2018). Blockchain: A secure, decentralized, trusted cyber infrastructure solution for future energy systems. *Journal of Modern Power Systems and Clean Energy*, *6*, 958–967.

Dorduncu, H., & Oruç, Z. S. (2021). Industry 4.0, digitalization, and big data: perspective of logistics and supply chain management. *Çukurova Üniversitesi Sosyal Bilimler Enstitüsü Dergisi, 30*(3), 170-180.

Douaioui, K., Fri, M., Mabroukki, C., & Semma, E. A. (2018, April). The interaction between industry 4.0 and smart logistics: concepts and perspectives. In *2018 international colloquium on logistics and supply chain management (LOGISTIQUA)* (pp. 128-132). IEEE.

Dritsas, E., & Trigka, M. (2024). Machine learning for blockchain and IoT systems in smart cities: A survey. *Future Internet*, *16*(9), 324. DOI: 10.3390/fi16090324

Duygan, M., Fischer, M., Pärli, R., & Ingold, K. (2022). Where do Smart Cities grow? The spatial and socio-economic configurations of smart city development. *Sustainable Cities and Society*, *77*, 103578.

Dwivedi, A. D., Singh, R., Kaushik, K., Mukkamala, R. R., & Alnumay, W. S. (2021). Blockchain and AI for 5G-enabled IoT: Challenges, opportunities and solutions. *Transactions on Emerging Telecommunications Technologies (Online)*, Article e4329. https://doi.org/DOI: 10.1002/ett.4329

Dwivedi, S. K., Roy, P., Karda, C., Agrawal, S., & Amin, R. (2021). Blockchain-based internet of things and industrial IoT: A comprehensive survey. *Security and Communication Networks*, *2021*, 1–21.

Dwork, C. (2006, July). Differential privacy. In *International colloquium on automata, languages, and programming* (pp. 1-12). Berlin, Heidelberg: Springer Berlin Heidelberg.

Eisavi, B., Khalilarya, S., Chitsaz, A., & Rosen, M. A. (2018). Thermodynamic analysis of a novel combined cooling, heating and power system driven by solar energy. *Applied Thermal Engineering*, *129*, 1219–1229.

El Hamdani, S., Benamar, N., & Younis, M. (2020). Pedestrian support in intelligent transportation systems: Challenges, solutions and open issues. *Transportation Research Part C, Emerging Technologies*, *121*, 102856.

Elnagar, S., Thomas, M. A., & Osei-Bryson, K. M. (2023). What is Cognitive Computing? An Architecture and State of The Art. *arXiv preprint arXiv:2301.00882*.

Elsafi, A., Almohammedi, A. A., Balfaqih, M., Balfagih, Z., & Sabri, S. (2025). Comparative analysis of maximum power point tracking methods for power optimization in grid tied photovoltaic solar systems. *Discover Applied Sciences*, *7*(9), 976.

Erskine, M. A., Pick, J. B., Satpathy, A., Díaz López, A., Sarkar, A., & Shin, N. (2024). Location Analytics in Information Systems: Opportunities for Research and Teaching. *Communications of the Association for Information Systems*, *55*. https://aisel.aisnet.org/cais/vol55/iss1/21

Eskhita, R., Manda, V. K., & Hlali, A. (2021). Dubai and Barcelona as Smart Cities: Some Reflections on Data Protection Law and Privacy. *Environmental Policy and Law*, *51*(6), 403–407. DOI: 10.3233/EPL-210023

Esposito, M., Palma, L., Belli, A., Sabbatini, L., & Pierleoni, P. (2022). Recent advances in internet of things solutions for early warning systems: A review. *Sensors (Basel)*, *22*(6), 2124.

Etkin, R. H., David, N. C., & Wang, H. (2008). Gaussian interference channel capacity to within one bit. *IEEE Transactions on Information Theory*, *54*(12), 5534–5562.

Eubanks, V. (2018). *Automating inequality: How high-tech tools profile, police, and punish the poor*. St. Martin's Press.

Evtodieva, T. E. (2019). *Sustainable Growth and Development of Economic Systems. Contributions to Economics*. Springer., DOI: 10.1007/978-3-030-11754-2_16

Excel, D. S. (2024). Privacy-preserving data analysis. *World Journal of Advanced Research and Reviews*, *23*(3), 597–609. DOI: 10.30574/wjarr.2024.23.3.2724

Fani, M., & Sadreddin, A. (2017). Solar assisted CCHP system, energetic, economic and environmental analysis, case study: Educational office buildings. *Energy and Building*, *136*, 100–109. DOI: 10.1016/j.enbuild.2016.11.052

Farahani, B., Firouzi, F., Chang, V., Badaroglu, M., Constant, N., & Mankodiya, K. (2018). Towards fog-driven IoT eHealth: Promises and challenges of IoT in medicine and healthcare. *Future Generation Computer Systems*, *78*, 659–676.

Fatema, N., Hua, G., Xiang, Y., Peng, D., & Natgunanathan, I. (2018). Massive MIMO linear precoding: A survey. *IEEE Systems Journal*, *12*(4), 3920–3931.

Ferdous, M. S., Biswas, K., Chowdhury, M. J. M., Chowdhury, N., & Muthukkumarasamy, V. (2019). Chapter Two—Integrated platforms for blockchain enablement. In Kim, S., Deka, G. C., & Zhang, P. (Eds.), *Role of Blockchain Technology in IoT Applications* (Vol. 115, pp. 41–72). Academic Press.

Ferrag, M. A., & Shu, L. (2021). The performance evaluation of blockchain-based security and privacy systems for the Internet of Things: A tutorial. *IEEE Internet of Things Journal*, *8*(24), 17236–17260.

Ferre, M., Batista, E., Solanas, A., & Martinez-Balleste, A. (2021). Smart health-enhanced early mobilisation in intensive care units. *Sensors (Basel)*, *21*(16), 5408. PMID: 34450850

Ferro-Diez, L. E., Villegas, N. M., & Diaz-Cely, J. (2020). Location Data Analytics in the Business Value Chain: A Systematic Literature Review. *IEEE Access : Practical Innovations, Open Solutions*, *8*, 204639–204659. DOI: 10.1109/ACCESS.2020.3036835

Ferro-Escobar, R., Vacca-González, H., & Gómez-Castillo, H. (2022). Smart and sustainable cities in collaboration with IoT: The Singapore success case. In *Machine Learning for Smart Environments/Cities: An IoT Approach* (pp. 213–243). Springer International Publishing.

Ferro, M., Silva, E., & Fidalgo, R. (2023). AStar: A modeling language for document-oriented geospatial data warehouses. *Data & Knowledge Engineering*, *145*, 102174. DOI: 10.1016/j.datak.2023.102174

Fieuw, W., Foth, M., & Caldwell, G. A. (2022). Towards a more-than-human approach to smart and sustainable urban development: Designing for multispecies justice. *Sustainability*, *14*(2), 948.

Flores, A., Morales, A., Campos, G., & Gelso, J. (2021, December). Energy Efficiency Using IOTA Tangle for Greenhouse Agriculture. In *Annual International Conference on Information Management and Big Data* (pp. 122-138). Cham: Springer International Publishing.

Fodor, G., Parkvall, S., Sorrentino, S., Wallentin, P., Lu, Q., & Brahmi, N. (2014). Device-to-device communications for national security and public safety. *IEEE Access : Practical Innovations, Open Solutions*, *2*, 1510–1520.

Fortino, G., Savaglio, C., Spezzano, G., & Zhou, M. (2020). Internet of things as system of systems: A review of methodologies, frameworks, platforms, and tools. *IEEE Transactions on Systems, Man, and Cybernetics. Systems*, *51*(1), 223–236.

Fouad, K. M., Ismail, M. M., Azar, A. T., & Arafa, M. M. (2021). Advanced methods for missing values imputation based on similarity learning. *PeerJ. Computer Science*, *7*, e619. DOI: 10.7717/peerj-cs.619 PMID: 34395861

Frascella, A., Brutti, A., Gessa, N., De Sabbata, P., Novelli, C., Burns, M., & He, L. (2018). A minimum set of common principles for enabling Smart City Interoperability. *Techne : Journal of Technology for Architecture and Environment*, 56–61. PMID: 39474009

Fu, B., Zhang, J., Wu, X., & Meadows, M. E. (2025). Geography's hotspots and frontiers: Diverse, systematic, and intelligent trends. *Geography and Sustainability*, *100285*. Advance online publication. DOI: 10.1016/j.geosus.2025.100285

Gambazzi, L., Schaller, P., Mermoud, A., & Lenders, V. (2021). Blockchain in Cyberdefence: A Technology Review from a Swiss Perspective. *The Swiss Technology Observatory*. https://doi.org//arXiv.2103.02606DOI: 10.48550

Gangwani, P., Perez-Pons, A., Joshi, S., Upadhyay, H., & Lagos, L. (2023). *Integration of Data Science and IoT with Blockchain for Industry 4.0. In S. Namasudra & K. Akkaya (Eds.), Blockchain and its Applications in Industry 4.0* (Studies in Big Data, vol 119, pp. 6). Springer. https://doi.org/DOI: 10.1007/978-981-19-8730-4_6

Gao, Y., Yao, W., Wang, J., & Cui, Z. (2023). Thermodynamic analysis of solid oxide fuel cell based combined cooling, heating, and power system integrated with solar-assisted electrolytic cell. *Journal of Thermal Science*, *32*(1), 93–108.

Garber, M. D., Watkins, K. E., & Kramer, M. R. (2019). Comparing bicyclists who use smartphone apps to record rides with those who do not: Implications for representativeness and selection bias. *Journal of Transport & Health*, *15*, 100661. PMID: 31938687

Gassmann, O., Böhm, J., & Palmié, M. (2019). *Smart cities: Introducing digital innovation to cities*. Emerald Publishing Limited.

Gawazah, L., Rondla, A., & Balhareth, M. S. A. (2024). To pay or not to pay: the us colonial pipeline ransomware attack.

Gellman, R. (2009, February). Privacy in the clouds: Risks to privacy and confidentiality from cloud computing. In *World privacy forum* (Vol. 23).

Geneve.ch. (2024). GenevaSustainable city: Environmental commitments. Available online: https://www.geneve.ch/geneve-ville-durable-engagements-lien-environnement [08 April 2025]

Gentry, C. (2009). *A fully homomorphic encryption scheme*. Stanford university.

Ghadami, N., Gheibi, M., Kian, Z., Faramarz, M. G., Naghedi, R., Eftekhari, M., & Tian, G. (2021). Implementation of solar energy in smart cities using an integration of artificial neural network, photovoltaic system and classical Delphi methods. *Sustainable Cities and Society*, *74*, 103149.

Ghandour, O., El Kafhali, S., & Hanini, M. (2023). Computing resources scalability performance analysis in cloud computing data center. *Journal of Grid Computing*, *21*(4), 96. Advance online publication. DOI: 10.1007/s10723-023-09696-5

Ghosh, S., & Eldawy, A. (2022). AID*: A Spatial Index for Visual Exploration of Geo-Spatial Data. *IEEE Transactions on Knowledge and Data Engineering*, *34*(8), 3569–3582. DOI: 10.1109/TKDE.2020.3026657

Giampietro, M., & Mayumi, K. (2018). Unraveling the complexity of the Jevons Paradox: The link between innovation, efficiency, and sustainability. *Frontiers in Energy Research*, *6*, 349753.

Gibson, D. V., Kozmetsky, G., & Smilor, R. W. (Eds.). (1992). *The technopolis phenomenon: Smart cities, fast systems, global networks*. Rowman & Littlefield.

Giffinger, R., & Gudrun, H. (2010). Smart cities ranking: an effective instrument for the positioning of the cities? *ACE: Architecture. City and Environment*, *4*(12), 7–26.

Gillanders, R., & Whelan, K. (2014). Open For Business? Institutions, Business Environment and Economic Development. *Kyklos*, *67*(4), 535–558. DOI: 10.1111/kykl.12067

Gohari, A., Ahmad, A. B., Rahim, R. B. A., Supa'at, A. S. M., Abd Razak, S., & Gismalla, M. S. M. (2022). Involvement of surveillance drones in smart cities: A systematic review. *IEEE Access : Practical Innovations, Open Solutions*, *10*, 56611–56628.

Goldsmith, S., & Crawford, S. (2014). *The responsive city: Engaging communities through data-smart governance*. John Wiley & Sons.

González, A., Kelly, C., & Rymszewicz, A. (2020). Advancements in web-mapping tools for land use and marine spatial planning. *Transactions in GIS*, *24*(2), 253–267. DOI: 10.1111/tgis.12603

Görçün. (2016). *Ömer Faruk. "Dördüncü endüstri devrimi endüstri 4.0."*. Beta Yayıncılık.

Government of Telangana. (2024). *T-chits*. Retrieved February 01, 2024 from Registration and Stamps Department: https://t-chits.telangana.gov.in

Govindan, K., Mina, H., & Alavi, B. (2020). A decision support system for demand management in healthcare supply chains considering the epidemic outbreaks: A case study of coronavirus disease 2019 (COVID-19). *Transportation Research Part E, Logistics and Transportation Review*, *138*, 101967. PMID: 32382249

Goyal, P., Sahoo, A. K., & Sharma, T. K. (2021). Internet of things: Architecture and enabling technologies. *Materials Today: Proceedings*, *34*, 719–735.

Greenberg, A. (2021). A hacker tried to poison a Florida city's water supply, officials say. *Wired. com, 2*.

Grindsted, T. S., Christensen, T. H., Freudendal-Pedersen, M., Friis, F., & Hartmann-Petersen, K. (2022). The urban governance of autonomous vehicles–In love with AVs or critical sustainability risks to future mobility transitions. *Cities (London, England)*, *120*, 103504.

Gritsenko, S., Karpun, O., & Kolisnichenko, A. (2022). Digitalization of logistics company business process in conditions of industry 4.0. *Electronic scientific and practical publication in economic sciences*.

Grosse, E. H., Glock, C. H., & Neumann, W. P. (2017). Human factors in order picking: A content analysis of the literature. *International Journal of Production Research*, *55*(5), 1260–1276.

Guarascio-Howard, L. (2011). Examination of wireless technology to improve nurse communication, response time to bed alarms, and patient safety. *HERD: Health Environments Research & Design Journal*, *4*(2), 109–120. PMID: 21465438

Guelzim, T., Obaidat, M. S., & Sadoun, B. (2016). Introduction and overview of key enabling technologies for smart cities and homes. In *Smart cities and homes* (pp. 1–16). Morgan Kaufmann.

Gupta, R., & Sharma, P. (2024). Protecting Privacy in the Smart City: Novel Approaches for Urban Informatics and Geospatial Data Management. *Emerging Trends in Machine Intelligence and Big Data*, 17.

Gupta, S., & Loukissas, Y. A. (2023). Making Smart Cities Explainable: What XAI Can Learn from the "Ghost Map." *Extended Abstracts of the 2023 CHI Conference on Human Factors in Computing Systems*, 1–8. https://doi.org/DOI: 10.1145/3544549.3585847

Gupta, D., Bhatt, P., & Bhatt, S. (2020, December). A game theoretic analysis for cooperative smart farming. In *2020 IEEE International Conference on Big Data (Big Data)* (pp. 2350-2359). IEEE.

Gupta, R., Kumari, A., & Tanwar, S. (2021). A taxonomy of blockchain envisioned edge-as-a-connected autonomous vehicles. *Transactions on Emerging Telecommunications Technologies, 32*(6), e4009. DOI: 10.1002/ett.4009

Guven, I., & Parlak, M. (2022, November). Blockchain, AI and IoT Empowered Swarm Drones for Precision Agriculture Applications. In *2022 IEEE 1st Global Emerging Technology Blockchain Forum: Blockchain & Beyond (iGETblockchain)* (pp. 1-6). IEEE.

Habib, A., Alsmadi, D., & Prybutok, V. R. (2020). Factors that determine residents' acceptance of smart city technologies. *Behaviour & Information Technology, 39*(6), 610–623.

Hafiz, M. Y. M., Balfaqih, M., Khan, S., Almohammedi, A., & Shepelev, V. (2024). Review on Comparative Analysis of Univariate and Multivariate Input Performance Over Time Series Ids Prediction Models in Healthcare Infrastructure. Available at *SSRN*: https://ssrn.com/abstract=4931271

Haghani, N. (2023). Reduction of Blood Pressure Among Adult Hypertensive Patients Using Remote Patient Monitoring Followed by Telehealth (Doctoral dissertation, University of Massachusetts Global).

Hair, J. F., Risher, J. J., Sarstedt, M., & Ringle, C. M. (2019). When to use and how to report the results of PLS-SEM. *European Business Review, 31*(1), 2–24.

Hameed, K., Barika, M., Garg, S., Amin, M. B., & Kang, B. (2022). A taxonomy study on securing Blockchain-based Industrial applications: An overview, application perspectives, requirements, attacks, countermeasures, and open issues. *Journal of Industrial Information Integration, 26*, 100312. DOI: 10.1016/j.jii.2021.100312

Hamid, B., Jhanjhi, N. Z., Humayun, M., Khan, A., & Alsayat, A. (2019, December). Cyber security issues and challenges for smart cities: A survey. In *2019 13th International Conference on Mathematics, Actuarial Science, Computer Science and Statistics (MACS)* (pp. 1-7). IEEE.

Hammouti, I., Addaim, A., & Guennoun, Z. (2023). Decentralized Energy Trading in Smart Grids: A Blockchain-based Framework for Efficiency and Transparency. *IEEE Access : Practical Innovations, Open Solutions.*

Han, L., & Han, X. (2023). The influence of price value on purchase intention among patients with chronic diseases in medical e-commerce during the COVID-19 pandemic in China. *Frontiers in Public Health, 11*, 1081196. PMID: 36817906

Hannan, M. A., Begum, R. A., Al-Shetwi, A. Q., Ker, P. J., Al Mamun, M. A., Hussain, A., & Mahlia, T. M. I. (2020). Waste collection route optimisation model for linking cost saving and emission reduction to achieve sustainable development goals. *Sustainable Cities and Society, 62*, 102393.

Hansen, C., Hansen, C., Alstrup, S., & Lioma, C. (2017, November). Smart city analytics: Ensemble-learned prediction of citizen home care. In *Proceedings of the 2017 ACM on Conference on Information and Knowledge Management* (pp. 2095-2098).

Hansen, M. M., & Dahiya, B. (2025). Traffy Fondue: A smart city citizen engagement. *Frontiers in Sustainable Cities, 7*, 1491621.

Haque, A. K. M. B., Bhushan, B., & Dhiman, G. (2021). Conceptualizing smart city applications: Requirements, architecture, security issues, and emerging trends. *Expert Systems: International Journal of Knowledge Engineering and Neural Networks.*

Hardi, R., Nurmandi, A., Purwaningsih, T., & Manaf, H. A. (2025). Smart city governance and interoperability: Enhancing human security in Yogyakarta and Makassar, Indonesia. *Frontiers in Political Science, 7*, 1553177.

Hartzog, W. (2018). *PrivacyOs Blueprint: The Battle to Control the Design of New Technologies.* Harvard University Press.

Hasan, H. F., Najwadi, M., Yusoff, B., & Ali, S. M. A. (2023). Bitcoin Layer Two Scaling Solutions: Lightening Payment Channels Network Comprehensive Review, Mechanisms, Challenges, Open Issues and Future Research Directions. *Iraqi Journal for Computer Science and Mathematics, 5*(1), 25–59.

Hasan, M. K., Habib, A. A., Islam, S., Balfaqih, M., Alfawaz, K. M., & Singh, D. (2023). Smart grid communication networks for electric vehicles empowering distributed energy generation: Constraints, challenges, and recommendations. *Energies, 16*(3), 1140.

Hassan, M. M., Song, B., & Huh, E. N. (2009, February). A framework of sensor-cloud integration opportunities and challenges. In *Proceedings of the 3rd international conference on Ubiquitous information management and communication* (pp. 618-626).

Hassebo, A., & Tealab, M. (2023). Global models of smart cities and potential IoT applications: A review. *IoT*, *4*(3), 366–411.

Hayat, P. (2016). Smart cities: A global perspective. *India Quarterly*, *72*(2), 177–191.

He, H. Z., & Kua, H. W. (2013). Lessons for integrated household energy conservation policy from Singapore's southwest Eco-living Program. *Energy Policy*, *55*, 105–116.

Henseler, J., Ringle, C. M., & Sarstedt, M. (2015). A new criterion for assessing discriminant validity in variance-based structural equation modeling. *Journal of the Academy of Marketing Science*, *43*, 115–135.

Herath, H. M. K. K. M. B., & Mittal, M. (2022). Adoption of artificial intelligence in smart cities: A comprehensive review. *International Journal of Information Management Data Insights*, *2*(1), 100076.

Hidayat, M. N. F., Baharun, H., Aisyah, E. N., Zaini, A. W., Sanjani, M. A. F., & Hasanah, R. (2024, October). Bridging the Digital Divide: The Role of Public Relations in Enhancing Digital Inclusivity. In *2024 10th International Conference on Education and Technology (ICET)* (pp. 59-66). IEEE.

Hierink, F., Yaghmaei, N., Bakker, M. I., Ray, N., & Van Den Homberg, M. (2024). Geospatial tools and data for health service delivery during disasters: Opportunities and challenges across the management cycle. *Geospatial Health*, *19*(2). Advance online publication. DOI: 10.4081/gh.2024.1284 PMID: 39470294

Ho, E. (2015). Bound by ethical complexities and socio-material histories: An exploration of household energy consumption in Singapore. *Energy Research & Social Science*, *10*, 150–164.

Hofmann, E., & Rüsch, M. (2017). Industry 4.0 and the current status as well as future prospects on logistics. *Computers in Industry*, *89*, 23–34.

Hofstad, H., & Torfing, J. (2017). Towards a climate-resilient city: Collaborative innovation for a 'green shift'in Oslo. In *Carbon Footprint and the Industrial Life Cycle: From Urban Planning to Recycling* (pp. 221–242). Springer International Publishing.

Hollands, R. G. (2020). Will the real smart city please stand up?: Intelligent, progressive or entrepreneurial? In *The Routledge companion to smart cities* (pp. 179–199). Routledge.

Hong, J. H., & Shi, Y. T. (2023). Integration of heterogeneous sensor systems for disaster responses in smart cities: Flooding as an example. *ISPRS International Journal of Geo-Information*, *12*(7), 279.

Horenberg, D. (2017). *Applications within Logistics 4.0: A research conducted on the visions of 3PL service providers* (Bachelor's thesis, University of Twente).

Hossain, M. M., Fotouhi, M., & Hasan, R. (2015, June). *Towards an analysis of security issues, challenges, and open problems in the internet of things. In 2015 ieee world congress on services*. IEEE.

Hou, H., Wu, J., Ding, Z., Yang, B., & Hu, E. (2021). Performance analysis of a solar-assisted combined cooling, heating and power system with an improved operation strategy. *Energy, 227*, 120516. DOI: 10.1016/J.ENERGY.2021.120516

Houssein, E. H., Othman, M. A., Mohamed, W. M., & Younan, M. (2024). Internet of Things in Smart Cities: Comprehensive Review, Open Issues and Challenges. *IEEE Internet of Things Journal*.

Huang, B., Thomas, T., Groenewolt, B., & van Berkum, E. C. (2022). An investigation of the motivators and barriers of smartphone app incentives for encouraging cycling. *Decision Analytics Journal, 5*, 100127.

Hua, W., Chen, Y., Qadrdan, M., Jiang, J., Sun, H., & Wu, J. (2022). Applications of blockchain and artificial intelligence technologies for enabling prosumers in smart grids: A review. *Renewable & Sustainable Energy Reviews, 161*, 112308.

Hughes, L., Dwivedi, Y. K., Misra, S. K., Rana, N. P., Raghavan, V., & Akella, V. (2019). Blockchain research, practice and policy: Applications, benefits, limitations, emerging research themes and research agenda. *International Journal of Information Management, 49*, 114–129. DOI: 10.1016/j.ijinfomgt.2019.02.00

Huseien, G. F., & Shah, K. W. (2021). Potential applications of 5G network technology for climate change control: A scoping review of singapore. *Sustainability, 13*(17), 9720.

Hussain, A., Wenbi, R., Da Silva, A. L., Nadher, M., & Mudhish, M. (2015). Health and emergency-care platform for the elderly and disabled people in the Smart City. *Journal of Systems and Software, 110*, 253–263.

Imbault, F., Swiatek, M., de Beaufort, R., & Plana, R. (2017). The green blockchain: Managing decentralized energy production and consumption. In *IEEE International Conference on Environment and Electrical Engineering and 2017 IEEE Industrial and Commercial Power Systems Europe (EEEIC / I&CPS Europe).*

International Institute for Management Development. (2019). Singapore tops new 'citizen-centric' global smart city index. International Institute for Management Development. Available online: https://www.imd.org/news/competitiveness/updates-singapore-tops-new-citizen-centric-global-smart-city-index/ [08 April 2025]

Ioanid, A., & Andrei, N. (2024). *Artificial Intelligence and Geospatial Technologies for Sustainable Maritime Logistics. Case Study: Port Of Constanta, Romania*. 15. https://marlog.aast.edu/files/marlog13/MARLOG13_paper_112.pdf

Irajifar, L., Chen, H., Lak, A., Sharifi, A., & Cheshmehzangi, A. (2023). The nexus between digitalization and sustainability: A scientometrics analysis. *Heliyon*, *9*(5). PMID: 37153424

Ishikawa, T., & Kryvoi, Y. (Eds.). (2023). *Public and Private Governance of Cybersecurity: Challenges and Potential*. Cambridge University Press.

Islam, M. M., Hasan, M. K., Islam, S., Balfaqih, M., Alzahrani, A. I., Alalwan, N., & Ghazal, T. M. (2024). Enabling pandemic-resilient healthcare: Narrowband Internet of Things and edge intelligence for real-time monitoring. *CAAI Transactions on Intelligence Technology*.

Ismagilova, E., Hughes, D., Dwivedi, Y., & Raman, K. (2019). Smart cities: Advances in research - An information systems perspective. *International Journal of Information Management*, *47*, 88–100.

Ismagilova, E., Hughes, L., Dwivedi, Y. K., & Raman, K. R. (2019). Smart cities: Advances in research—An information systems perspective. *International Journal of Information Management*, *47*, 88–100.

ISO/IEC 17067:2013, Conformity assessment — Fundamentals of product certification and guidelines for product certification schemes

ISO/SAE 21434:2021 for Road Vehicles—Cybersecurity Engineering

Ivanov, D., Das, A., & Choi, T. M. (2018). New flexibility drivers for manufacturing, supply chain and service operations. *International Journal of Production Research*, *56*(10), 3359–3368.

Jabbar, W. A., Kian, T. K., Ramli, R. M., Zubir, S. N., Zamrizaman, N. S., Balfaqih, M., & Alharbi, S. (2019). Design and fabrication of smart home with internet of things enabled automation system. *IEEE Access : Practical Innovations, Open Solutions*, *7*, 144059–144074.

Jacobs, N., Edwards, P., Markovic, M., Cottrill, C. D., & Salt, K. (2020). Who trusts in the smart city? Transparency, governance, and the internet of things. *Data & Policy*, *2*, e11.

Jain, A., Gue, I. H., & Jain, P. (2023). Research trends, themes, and insights on artificial neural networks for smart cities towards SDG-11. *Journal of Cleaner Production*, *412*, 137300.

Jang, K. M., & Kim, Y. (2019). Crowd-sourced cognitive mapping: A new way of displaying people's cognitive perception of urban space. *PLoS One*, *14*(6), e0218590. PMID: 31220148

Jan, M. A., Zhang, W., Usman, M., Tan, Z., Khan, F., & Luo, E. (2019). SmartEdge: An end-to-end encryption framework for an edge-enabled smart city application. *Journal of Network and Computer Applications*, *137*, 1–10.

Jararweh, Y., Otoum, S., & Al Ridhawi, I. (2020). Trustworthy and sustainable smart city services at the edge. *Sustainable Cities and Society*, *62*, 102394.

Javadi, M. A., Jafari Najafi, N., Khalili Abhari, M., Jabery, R., & Pourtaba, H. (2021). 4E analysis of three different configurations of a combined cycle power plant integrated with a solar power tower system. *Sustainable Energy Technologies and Assessments*, *48*. Advance online publication. DOI: 10.1016/j.seta.2021.101599

Javaid, M., Haleem, A., Singh, R. P., Rab, S., Suman, R., & Khan, I. H. (2022). Evolutionary trends in progressive cloud computing based healthcare: Ideas, enablers, and barriers. *International Journal of Cognitive Computing in Engineering*, *3*, 124–135.

Jiang, L., Jun, M., & Yang, Z. (2016). Customer-perceived value and loyalty: How do key service quality dimensions matter in the context of B2C e-commerce? *Service Business*, *10*(2), 301–317.

Jitta, A. S., Manda, V. K., & Tarnanidis, T. (2024). Geospatial Technologies for Smart Cities. In Darwish, D., & Chemingui, H. (Eds.), (pp. 103–136). Advances in Geospatial Technologies. IGI Global., DOI: 10.4018/979-8-3693-8054-3.ch004

Joia, L. A., & Vieira, J. A. P. (2021). Going beyond cryptocurrencies: The social representation of blockchain. *Sustainability*, *13*, 14054.

Joo, Y. M. (2023). Developmentalist smart cities? The cases of Singapore and Seoul. *International Journal of Urban Sciences, 27*(sup1), 164-182.

Julsrud, T. E., & Krogstad, J. R. (2020). Is there enough trust for the smart city? exploring acceptance for use of mobile phone data in oslo and tallinn. *Technological Forecasting and Social Change*, *161*, 120314. PMID: 32981976

Junge, A. L., Verhoeven, P., Reipert, J., & Mansfeld, M. (2019). *Pathway of digital transformation in logistics: best practice concepts and future developments* (Vol. 8). Universitätsverlag der TU Berlin.

Kache, F., & Seuring, S. (2017). Challenges and opportunities of digital information at the intersection of Big Data Analytics and supply chain management. *International Journal of Operations & Production Management*, *37*(1), 10–36.

Kai, C., Zhou, H., Yi, Y., & Huang, W. (2021). Collaborative cloud-edge-end task offloading in mobile-edge computing networks with limited communication capability. *IEEE Transactions on Cognitive Communications and Networking*, *7*(2), 624–634. DOI: 10.1109/TCCN.2020.3018159

Kalfas, D., Kalogiannidis, S., Spinthiropoulos, K., Chatzitheodoridis, F., & Ziouziou, E. (2025). Enhancing Predictive Urban Planning in European Smart Cities Through AI-Driven Digital Twin Technology: A Case Study of Greece. *Urban Science (Basel, Switzerland)*, *9*(7), 267.

Kamal, R. (2025), Sleep Better with AI and IoT-Enabled Smart Beds, *Intuz*, Online: https://www.intuz.com/blog/sleep-smarter-with-multifunctional-smart-bed [28/09/2025]

Kamal, H. (2024). IoT-enabled Smart Hospital Infrastructure for Efficient Resource Management: Designing IoT-enabled systems to optimize resource allocation and management within hospitals. *Journal of AI in Healthcare and Medicine*, *4*(2), 45–56.

Kamble, S. S., Gunasekaran, A., & Gawankar, S. A. (2018). Sustainable Industry 4.0 framework: A systematic literature review identifying the current trends and future perspectives. *Process Safety and Environmental Protection*, *117*, 408–425.

Karandikar, N., Chakravorty, A., & Rong, C. (2019). Transactive Energy on Hyperledger Fabric. In 2019 *Sixth international Conference on Internet of Things Systems, Management and Security (*IOTSMS), pp.539-546.

Karunarathne, G. G. K. W. M. S. I. R., Kulawansa, K. A. D. T., & Firdhous, M. F. M. (2018, December). Wireless communication technologies in internet of things: A critical evaluation. In *2018 International conference on intelligent and innovative computing applications (ICONIC)* (pp. 1-5). IEEE.

Karvounis, E., Polymeni, S., Tsipouras, M., Koritsoglou, K., & Tzovaras, D. (2021, September). Smart beds and bedding surfaces for personalized patient care: a review. In 2021 6th South-East Europe Design Automation, Computer Engineering, Computer Networks and Social Media Conference (SEEDA-CECNSM) (pp. 1-8). IEEE.

Katz, J., & Lindell, Y. (2020). Introduction to modern cryptography crc press.

Kaur, U., Kaur, M., & Khang, A. (2023). Futuristic Technologies in Agriculture Challenges and Future Prospects. *Handbook of Research on AI-Equipped IoT Applications in High-Tech Agriculture*, 347-364.

Kayikci, Y. (2018). Sustainability impact of digitization in logistics. *Procedia Manufacturing, 21*, 782–789.

Khalifeh, A., Rajendiran, K., Darabkh, K. A., Khasawneh, A. M., AlMomani, O., & Zinonos, Z. (2019). On the potential of fuzzy logic for solving the challenges of cooperative multi-robotic wireless sensor networks. *Electronics (Basel), 8*(12), 1513.

Khanna, A., Sah, A., Bolshev, V., Jasinski, M., Vinogradov, A., Leonowicz, Z., & Jasiński, M. (2021). Blockchain: Future of e-Governance in Smart Cities. *Sustainability, 13*(21), 11840. DOI: 10.3390/su132111840

Khan, S., Shael, M., Majdalawieh, M., Nizamuddin, N., & Nicho, M. (2022). Blockchain for governments: The case of the Dubai Government. *Sustainability, 14*(11), 6576. DOI: 10.3390/su14116576

Khan, Y., & Singh, P. K. (2024). Thermo-Economic and Environmental Assessment of a Novel SOFC-Based Hybrid Energy Generation System for Combined Cooling Heating and Power Generation. *Journal of Energy Resources Technology, Part A: Sustainable and Renewable Energy, 1*(1). Advance online publication. DOI: 10.1115/1.4066110

Khan, Z., Ludlow, D., Loibl, W., & Soomro, K. (2014). ICT enabled participatory urban planning and policy development: The UrbanAPI project. *Transforming Government: People. Process and Policy, 8*(2), 205–229.

Khan, Z., Pervez, Z., & Abbasi, A. G. (2017). Towards a secure service provisioning framework in a smart city environment. *Future Generation Computer Systems, 77*, 112–135.

Khaqqia, K. N., Sikorskib, J. J., Hadinotoa, K., & Kraft, M. (2018). Incorporating seller/buyer reputation-based system in blockchain-enabled emission trading application. *Applied Energy, 209*, 8–19.

Khayyat, M. M., Jambi, J., Kutbi, N., Garout, R., Al-Noumani, L., Asiri, R., & Balfaqih, M. (2025). Logistic caravan: A smart transportation system for green logistics and sustainable practices.

Khayyat, M., Balfaqih, M., Balfaqih, H., & Ismail, M. (2024). Challenges and factors influencing the implementation of green logistics: A case study of Saudi Arabia. *Sustainability, 16*(13), 5617.

Khayyat, M., Balfaqih, M., Balfaqih, H., & Ismail, M. (2024). Challenges and Factors Influencing the Implementation of Green Logistics: A Case Study of Saudi Arabia. *Sustainability, 16*(13), 5617.

Khayyat, M., Jambi, J., Kutbi, N., Garout, R., Al-Noumani, L., Asiri, R., & Balfaqih, M. (2025). Logistic caravan: A smart transportation system for green logistics and sustainable practices. *International Journal of Advanced and Applied Sciences, 12*(6), 154–168.

Khoshgoftar Manesh, M. H., Mousavi Rabeti, S. A., Nourpour, M., & Said, Z. (2022). Energy, exergy, exergoeconomic, and exergoenvironmental analysis of an innovative solar-geothermal-gas driven polygeneration system for combined power, hydrogen, hot water, and freshwater production. *Sustainable Energy Technologies and Assessments, 51*(April 2021). https://doi.org/DOI: 10.1016/j.seta.2021.101861

Kim, H. W., Chan, H. C., & Gupta, S. (2007). Value-based adoption of mobile internet: An empirical investigation. *Decision Support Systems, 43*(1), 111–126.

Kim, J., Hong, S., Jeong, S., Park, S., & Yu, K. (2024). SGIR-Tree: Integrating R-Tree Spatial Indexing as Subgraphs in Graph Database Management Systems. *ISPRS International Journal of Geo-Information, 13*(10), 346. DOI: 10.3390/ijgi13100346

Kitchin, R. (2014). The data revolution: Big data, open data, data infrastructures and their consequences. *Sage (Atlanta, Ga.).*

Kitchin, R. (2016). The ethics of smart cities and urban science. *Philosophical Transactions. Series A, Mathematical, Physical, and Engineering Sciences, 374*(2083), 20160115. PMID: 28336794

Klumpp, M. (2018). Automation and artificial intelligence in business logistics systems: Human reactions and collaboration requirements. *International Journal of Logistics Research and Applications, 21*(3), 224–242.

Kock, N. (2015). Common method bias in PLS-SEM: A full collinearity assessment approach. [ijec]. *International Journal of e-Collaboration, 11*(4), 1–10.

Kolajo, T., Daramola, O., & Adebiyi, A. (2019). Big data stream analysis: A systematic literature review. *Journal of Big Data, 6*(1), 1–30.

Kolotouchkina, O., Barroso, C. L., & Sánchez, J. L. M. (2022). Smart cities, the digital divide, and people with disabilities. *Cities (London, England), 123*, 103613.

Komninos, N. (2008). *Intelligent Cities and Globalisation of Innovation Networks.* Routledge.

Komninos, N. (2011). Intelligent Cities: Towards Interactive and Global Innovation Environments. *International Journal of Innovative Research and Development, 1*, 337–355. DOI: 10.1504/IJIRD.2009.022726

Koppel, R. (2022). Estimating the United States' cost of healthcare information technology. In *Healthcare Information Management Systems: Cases, Strategies, and Solutions* (pp. 3–38). Springer International Publishing.

Kshetri, N. (2017). The economics of the Internet of Things in the Global South. *Third World Quarterly, 38*(2), 311–339.

Kumar, A., Nanthaamornphong, A., Selvi, R., Venkatesh, J., Alsharif, M. H., Uthansakul, P., & Uthansakul, M. (2023). Evaluation of 5G techniques affecting the deployment of smart hospital infrastructure: Understanding 5G, AI and IoT role in smart hospital. *Alexandria Engineering Journal, 83*, 335–354.

Kumari, A., & Sharma, I. (2023, October). Augmentation of Internet of Things Security and Privacy with IOTA Tangle Network. In *2023 4th IEEE Global Conference for Advancement in Technology (GCAT)* (pp. 1-5). IEEE.

Kumar, M., Kumar, A., Verma, S., Bhattacharya, P., Ghimire, D., Kim, S. H., & Hosen, A. S. (2023). Healthcare Internet of Things (H-IoT): Current trends, future prospects, applications, challenges, and security issues. *Electronics (Basel), 12*(9), 2050.

Kumar, P., Bhattacharyya, S., Darshi, S., Sharma, A., Almohammedi, A. A., Shepelev, V., & Shailendra, S. (2022). Drone assisted network coded cooperation with energy harvesting: Strengthening the lifespan of the wireless networks. *IEEE Access : Practical Innovations, Open Solutions, 10*, 43055–43070.

Kumar, P., Darshi, S., & Shailendra, S. (2021). Drone assisted device to device cooperative communication for critical environments. *IET Communications, 15*(7), 957–972.

Kumar, P., & Majhi, S. (2024). Uav-assisted network coded cooperation by using height-dependency shaping parameters in nakagami-m faded channel. *IEEE Access : Practical Innovations, Open Solutions*.

Kumar, T. V., & Dahiya, B. (2017). Smart economy in smart cities. In *Smart Economy in Smart Cities* (pp. 3–76). Springer.

Kunduru, A. R. (2023). Cloud BPM application (Appian) robotic process automation capabilities. *Asian Journal of Research in Computer Science, 16*(3), 267–280. DOI: 10.9734/ajrcos/2023/v16i3361

Kurt, A., Saputro, N., Akkaya, K., & Uluagac, A. S. (2021). Distributed connectivity maintenance in swarm of drones during post-disaster transportation applications. *IEEE Transactions on Intelligent Transportation Systems, 22*(9), 6061–6073.

Kutty, A. A., Abdella, G. M., Kucukvar, M., Onat, N. C., & Bulu, M. (2020). A system thinking approach for harmonizing smart and sustainable city initiatives with United Nations sustainable development goals. *Sustainable Development*, *28*(5), 1347–1365.

Lager, H., Virgillito, A., & Buchberger, T. P. (2021). Digitalization of logistics work: Ergonomic improvements versus work intensification. In *Digital supply chains and the human factor* (pp. 33–53). Springer International Publishing.

Lai, C. M. T., Cole, A., & Stivas, D. (2025). Understanding Trust in the Smart City Through Focus Groups in Hong Kong. *Urban Science (Basel, Switzerland)*, *9*(4), 123.

Lamba, K., & Singh, S. P. (2017). Big data in operations and supply chain management: Current trends and future perspectives. *Production Planning and Control*, *28*(11-12), 877–890.

Lamtzidis, O., & Gialelis, J. (2018, December). An IOTA based distributed sensor node system. In *2018 IEEE Globecom Workshops (GC Wkshps)* (pp. 1-6). IEEE.

Lapointe, C., & Fishbane, L. (2019). The Blockchain Ethical Design Framework. *Innovations: Technology, Governance, Globalization*, *12*, 50–71.

Lark, J. (2020). Uneven innovation: The work of smart cities. Columbia University Press.

Lartey, D., & Law, K. M. (2025). Artificial intelligence adoption in urban planning governance: A systematic review of advancements in decision-making, and policy making. Landscape and Urban Planning, 258, 105337.

Law, K. H., & Lynch, J. P. (2019). Smart city: Technologies and challenges. *IT Professional*, *21*(6), 46–51.

Lazaroiu, C., & Roscia, M. (2018, October). Smart resilient city and IoT towards sustainability of Africa. In *2018 7th International Conference on Renewable Energy Research and Applications (ICRERA)* (pp. 1292-1298). IEEE.

Lea, P. (2020). *IoT and Edge Computing for Architects: Implementing edge and IoT systems from sensors to clouds with communication systems, analytics, and security.* Packt Publishing Ltd.

Lee, I., & Lee, K. (2018). The Internet of Things (IoT): Applications, investments, and challenges for enterprises. *Business Horizons*, *58*(4), 431–440.

Lee, J. (2023). Beyond Geospatial Inquiry—How Can We Integrate the Latest Technological Advances into Geography Education? *Education Sciences*, *13*(11), 1128. DOI: 10.3390/educsci13111128

Lee, S. K., Bae, M., & Kim, H. (2017). Future of IoT networks: A survey. *Applied Sciences (Basel, Switzerland)*, *7*(10), 1072.

Leow, N. X., & Krishnaswamy, J. (2021). Smart cities need environmental consciousness and more social responsibilities as an outcome of COVID-19 – reflections from urban road commuters. *Foresight, ahead-of-print*(ahead-of-print). https://doi.org/DOI: 10.1108/fs-02-2021-0035

Lewandowski, J. (2014). Mobile application of artificial intelligence to vital signs monitoring: multi parametric, user adaptable model for ubiquitous well-being monitoring (Doctoral dissertation, Coventry University).

Liao, Y., Deschamps, F., Loures, E. D. F. R., & Ramos, L. F. P. (2017). Past, present and future of Industry 4.0-a systematic literature review and research agenda proposal. *International Journal of Production Research*, *55*(12), 3609–3629.

Liaqat, R., Sajjad, I. A., Waseem, M., & Alhelou, H. H. (2021). Appliance level energy characterization of residential electricity demand: Prospects, challenges and recommendations. *IEEE Access : Practical Innovations, Open Solutions*, *9*, 148676–148697.

Li, H., Han, D., & Tang, M. (2021). A privacy-preserving storage scheme for logistics data with assistance of blockchain. *IEEE Internet of Things Journal*, *9*(6), 4704–4720.

Li, J., Greenwood, D., & Kassem, M. (2019). Blockchain in the built environment and construction industry: A systematic review, conceptual models and practical use cases. *Automation in Construction*, *102*, 288–307.

Li, L. L., Ren, X. Y., Tseng, M. L., Wu, D. S., & Lim, M. K. (2022). Performance evaluation of solar hybrid combined cooling, heating and power systems: A multi-objective arithmetic optimization algorithm. *Energy Conversion and Management*, *258*, 115541. DOI: 10.1016/J.ENCONMAN.2022.115541

Lim, M. K., Xiong, W., & Lei, Z. (2020). Theory, supporting technology and application analysis of cloud manufacturing: A systematic and comprehensive literature review. *Industrial Management & Data Systems*, *120*(8), 1585–1614. DOI: 10.1108/IMDS-10-2019-0570

Lim, Y., Edelenbos, J., & Gianoli, A. (2019). Identifying the results of smart city development: Findings from systematic literature review. *Cities (London, England)*, *95*, 102397.

Lin, C., Zhao, G., Yu, C., & Wu, Y. J. (2019). Smart city development and residents' well-being. *Sustainability*, *11*(3), 676.

Lin, J., Yu, W., Zhang, N., Yang, X., Zhang, H., & Zhao, W. (2017). A survey on internet of things: Architecture, enabling technologies, security and privacy, and applications. *IEEE Internet of Things Journal*, *4*(5), 1125–1142.

Liu, C.. (2020). Smart Energy Systems: A Comprehensive Review of IoT-Enabled Solutions for Sustainable and Optimized Energy Usage. *Energy Reports*, *6*, 1342–1355.

Liu, Y., Hao, T., Gong, X., Kong, D., & Wang, J. (2021). Research on Hybrid Index Based on 3D Multi-Level Adaptive Grid and R+ Tree. *IEEE Access : Practical Innovations, Open Solutions*, *9*, 146010–146022. DOI: 10.1109/ACCESS.2021.3115510

Liu, Y., Zhang, S., Mu, X., Ding, Z., Schober, R., Al-Dhahir, N., & Shen, X. (2022). Evolution of NOMA toward next generation multiple access (NGMA) for 6G. *IEEE Journal on Selected Areas in Communications*, *40*(4), 1037–1071.

Liu, Z., Yang, Z., & Osmani, M. (2021). The relationship between sustainable built environment, art therapy and therapeutic design in promoting health and well-being. *International Journal of Environmental Research and Public Health*, *18*(20), 10906. PMID: 34682646

Lnenicka, M., Hervert, P., & Horak, O. (2024). Understanding big data and data protection measures in smart city strategies: An analysis of 28 cities. *Urban Governance*, *4*(4), 255–273.

Loh, S., Foth, M., Caldwell, G. A., Garcia-Hansen, V., & Thomson, M. (2020). A more-than-human perspective on understanding the performance of the built environment. *Architectural Science Review*, *63*(3-4), 372–383.

Lopes de Sousa Jabbour, A. B., Jabbour, C. J. C., Godinho Filho, M., & Roubaud, D. (2018). Industry 4.0 and the circular economy: A proposed research agenda and original roadmap for sustainable operations. *Annals of Operations Research*, *270*(1), 273–286.

Luca, M., Lepri, B., Gallotti, R., Paolazzi, S., Bigi, M., & Pistore, M. (2024). Towards civic digital twins: Co-design the citizen-centric future of bologna. *arXiv preprint arXiv:2412.06328*.

Luusua, A., Ylipulli, J., Foth, M., & Aurigi, A. (2023). Urban AI: Understanding the emerging role of artificial intelligence in smart cities. *AI & Society*, *38*(3), 1039–1044. PMID: 36196366

Ma, C. (2021). Smart city and cyber-security; technologies used, leading challenges and future recommendations. *Energy Reports*, *7*, 7999–8012.

Machin, J., & Solanas, A. (2018, July). A review on the meaning of cognitive cities. In *2018 9th International Conference on Information, Intelligence, Systems and Applications (IISA)* (pp. 1-5). IEEE.

Machin, J., Batista, E., Martinez-Balleste, A., & Solanas, A. (2021). Privacy and security in cognitive cities: A systematic review. *Applied Sciences (Basel, Switzerland)*, *11*(10), 4471.

Madanian, S., Chinbat, T., Subasinghage, M., Airehrour, D., Hassandoust, F., & Yongchareon, S. (2024). Health IoT Threats: Survey of Risks and Vulnerabilities. *Future Internet*, *16*(11), 389.

Mahmmod, B. M., Naser, M. A., Al-Sudani, A. H. S., Alsabah, M., Mohammed, H. J., Alaskar, H., & Abdulhussain, S. H. (2024). Patient Monitoring System Based on Internet of Things: A Review and Related Challenges with Open Research Issues. *IEEE Access : Practical Innovations, Open Solutions*.

Mahmood, M. (2018). *The three regularities in development: growth, jobs and macro policy in developing countries*. Palgrave Macmillan.

Makani, S., Pittala, R., Alsayed, E., Aloqaily, M., & Jararweh, Y. (2022). A survey of blockchain applications in sustainable and smart cities. *Cluster Computing*, *25*(6), 3915–3936.

Maleki Varnosfaderani, S., & Forouzanfar, M. (2024). The role of AI in hospitals and clinics: Transforming healthcare in the 21st century. *Bioengineering (Basel, Switzerland)*, *11*(4), 337. PMID: 38671759

Malik, H., Yong, H., & Mohamad, Z., CharlesRamendranS.P., R., Yip, C., Fong, C., Rehman, M., & Zailani, S. (2022). Sustainable Smart Homes and Community Happiness in the Malaysian Context. *International Journal of Asian Business and Information Management*, *13*, 1–18.

Malik, P. K., Wadhwa, D. S., & Khinda, J. S. (2020). A survey of device to device and cooperative communication for the future cellular networks. *International Journal of Wireless Information Networks*, *27*(3), 411–432.

Manavi, T., Zafar, H., & Sharif, F. (2024). An era of digital healthcare—A comprehensive review of sensor technologies and telehealth advancements in chronic heart failure management. *Sensors (Basel)*, *24*(8), 2546. PMID: 38676163

Manda, V. K., Christy, V., & Hlali, A. (2024). Current Trends, Opportunities, and Futures Research Directions in Geospatial Technologies for Smart Cities. In Darwish, D., & Chemingui, H. (Eds.), (pp. 239–270). Advances in Geospatial Technologies. IGI Global., DOI: 10.4018/979-8-3693-8054-3.ch009

Mann, M., Mitchell, P., & Foth, M. (2022). Between surveillance and technological solutionism: A critique of privacy-preserving apps for COVID-19 contact-tracing. *New Media & Society*, *26*(7), 4099–4117.

Mao, Y., Clerckx, B., & Li, V. O. (2018). Rate-splitting multiple access for downlink communication systems: Bridging, generalizing, and outperforming SDMA and NOMA. *EURASIP Journal on Wireless Communications and Networking*, *2018*, 1–54. PMID: 30996723

Mao, Y., Dizdar, O., Clerckx, B., Schober, R., Popovski, P., & Poor, H. V. (2022). Rate-splitting multiple access: Fundamentals, survey, and future research trends. *IEEE Communications Surveys and Tutorials*, *24*(4), 2073–2126.

Margret, M. K., & Julie, E. G. (2024). Smarter and resilient smart contracts applications for smart cities environment using blockchain technology. *Automatika (Zagreb)*, *65*(2), 572–583. DOI: 10.1080/00051144.2024.2307228

Marsh-Hunn, D., Trilles, S., Gonzalez-Perez, A., Torres-Sospedra, J., & Ramos, F. (2021). A Comparative Study in the Standardization of IoT Devices Using Geospatial Web Standards. *IEEE Sensors Journal*, *21*(4), 5512–5528. DOI: 10.1109/JSEN.2020.3031315

Marune, A. E. M. S., & Hartanto, B. (2022). Strengthening personal data protection, cyber security, and improving public awareness in Indonesia: Progressive legal perspective. *International Journal of Business, Economics, and Social Development*, *2*(4), 143–152.

Masik, G., Sagan, I., & Scott, J. W. (2021). Smart City strategies and new urban development policies in the Polish context. *Cities (London, England)*, *108*, 102970. DOI: 10.1016/j.cities.2020.102970

Mauladdawilah, H., Balfaqih, M., Balfagih, Z., Gago, E. J., & Pegalajar, M. C. (2025, January). Optimization of Photovoltaic Power Forecasting: A Comparative Study of Deep Learning Architectures, Optimization Techniques, and Evaluation Metrics. In *2025 22nd International Learning and Technology Conference (L&T)* (Vol. 22, pp. 109-114). IEEE.

Mauladdawilah, H., Balfaqih, M., Balfagih, Z., Gago, E. J., & Pegalajar, M. C. (2025b). Optimization of Photovoltaic Power Forecasting: A Comparative Study of Deep Learning Architectures, Optimization Techniques, and Evaluation Metrics. In *2025 22nd International Learning and Technology Conference (L&T)* (Vol. 22, pp. 109-114). IEEE.

Mauladdawilah, H., Gago, E. J., Pegalajar, M. C., & Balfaqih, M. (2025c). An Evaluation of Meteorological Variables Impact on Photovoltaic Power Generation Estimation Based on Deep Learning Model. In *2025 4th International Conference on Computing and Information Technology (ICCIT)* (pp. 638-642). IEEE.

Mauladdawilah, H., Balfaqih, M., Balfagih, Z., Pegalajar, M. D. C., & Gago, E. J. (2025). Deep Feature Selection of Meteorological Variables for LSTM-Based PV Power Forecasting in High-Dimensional Time-Series Data. *Algorithms, 18*(8), 496.

Mazhar, T., Irfan, H. M., Haq, I., Ullah, I., Ashraf, M., Shloul, T. A., Ghadi, Y. Y., Imran, & Elkamchouchi, D. H. 2023. Analysis of Challenges and Solutions of IoT in Smart Grids Using AI and Machine Learning Techniques:Review, A. . *Electronics (Basel), 12*(1), 242.

Mazzetto, S. (2024). A review of urban digital twins integration, challenges, and future directions in smart city development. *Sustainability, 16*(19), 8337.

Menendez, M., & Ambühl, L. (2022). Implementing design and operational measures for sustainable mobility: Lessons from Zurich. *Sustainability, 14*(2), 625.

Menezes, A. J., Van Oorschot, P. C., & Vanstone, S. A. (2018). *Handbook of applied cryptography*. CRC press.

Merhi, M., Hone, K., & Tarhini, A. (2019). A cross-cultural study of the intention to use mobile banking between Lebanese and British consumers: Extending UTAUT2 with security, privacy and trust. *Technology in Society, 59*, 101–151.

Merodio Gómez, P., Ramírez Santiago, A., García Seco, G., Casanova, R., MacKenzie, D., & Tucker, C. (2022). Ethics in the use of geospatial information in the Americas. *Technology in Society, 69*, 101964. DOI: 10.1016/j.techsoc.2022.101964

Micheli, M., Ponti, M., Craglia, M., & Berti Suman, A. (2020). Emerging models of data governance in the age of datafication. *Big Data & Society, 7*(2), 2053951720948087. DOI: 10.1177/2053951720948087

Miglani, A., Kumar, N., Chamola, V., & Zeadally, S. (2020). Blockchain for Internet of Energy management: Review, solutions, and challenges. *Computer Communications, 151*, 395–418. DOI: 10.1016/j.comcom.2020.01.006

Mithat Delibaş, H., Kayabaşı, E., Mithat Delibas, H., & Kayabasi, E. (2021). Environment and Economy Assessment of Waste Heat Recovery Technologies in Marine Industry. *The International Journal of Materials and Engineering Technology, 004*(2), 133–146. http://dergipark.gov.tr/tijmet

Mohamad, Z., Nee, A., Yang, F., Rehman, M., & Yin, Y. (2019). *Achieving Community Happiness through Affordable Eco-Friendly Smart Houses*. KnE Social Sciences.

Monzon, A. (2015). Smart cities concept and challenges: Bases for the assessment of smart city projects. In *2015 international conference on smart cities and green ICT systems (SMARTGREENS)* (pp. 1-11). IEEE.

Mora, H., Mendoza-Tello, J. C., Varela-Guzmán, E. G., & Szymanski, J. (2021). Blockchain technologies to address smart city and society challenges. *Computers in Human Behavior*, *122*, 106854.

Moreno, C., Allam, Z., Chabaud, D., Gall, C., & Pratlong, F. (2021). Introducing the "15-Minute City": Sustainability, resilience and place identity in future post-pandemic cities. *Smart cities, 4*(1), 93-111.

Moreno-Álvarez, S., Paoletti, M. E., Sanchez-Fernandez, A. J., Rico-Gallego, J. A., Han, L., & Haut, J. M. (2024). Federated learning meets remote sensing. *Expert Systems with Applications*, *255*, 124583. DOI: 10.1016/j.eswa.2024.124583

Morozov, V., Petrov, A. I., Shepelev, V., & Balfaqih, M. (2024). Ideology of urban road transport chaos and accident risk management for sustainable transport systems. *Sustainability*, *16*(6), 2596.

Mostafa, M. M., & El-Masry, A. A. (2013). Citizens as consumers: Profiling e-government services' users in Egypt via data mining techniques. *International Journal of Information Management*, *33*, 627–641.

Mukherjee, A., De, D., Dey, N., Crespo, R. G., & Herrera-Viedma, E. (2023). DisastDrone: A disaster-aware consumer Internet of Drone Things system in ultra-low latent 6G network. *IEEE Transactions on Consumer Electronics*, *69*(1), 38–48.

Munawar, H. S., Mojtahedi, M., Hammad, A. W., Kouzani, A., & Mahmud, M. P. (2022). Disruptive technologies as a solution for disaster risk management: A review. *The Science of the Total Environment*, *806*, 151351.

Muthunagai, S. U., Girija, M. S., Kumar, B. P., & Anitha, R. (2024). Storage Management Techniques for Medical Internet of Things (MIoT). In Intelligent Systems and Sustainable Computational Models (pp. 34-47). Auerbach Publications.

Nain, G., Pattanaik, K. K., & Sharma, G. K. (2022). Towards edge computing in intelligent manufacturing: Past, present and future. *Journal of Manufacturing Systems*, *62*, 588–611.

Namasudra, S. (2021). Data access control in the cloud computing environment for bioinformatics. *International Journal of Applied Research in Bioinformatics, 11*(1), 40–50. DOI: 10.4018/ijarb.2021010105

Nam, T., & Pardo, T. A. (2020). Conceptualizing smart city with dimensions of technology, people, and institutions. *Proceedings of the 12th Annual International Digital Government Research Conference on Digital Government Innovation in Challenging Times - Dg.o '11*. https://doi.org/DOI: 10.1145/2037556.2037602

Naphade, M., Banavar, G., Harrison, C., Paraszczak, J., & Morris, R. (2011). Smarter cities and their innovation challenges. *Computer, 44*(6), 32–39.

Naqash, M. T., Syed, T. A., Alqahtani, S. S., Siddiqui, M. S., Alzahrani, A., & Nauman, M. (2023). A Blockchain Based Framework for Efficient Water Management and Leakage Detection in Urban Areas. *Urban Science (Basel, Switzerland), 7*(4), 99. DOI: 10.3390/urbansci7040099

Narayanan, A., & Shmatikov, V. (2008, May). Robust de-anonymization of large sparse datasets. In *2008 IEEE Symposium on Security and Privacy (sp 2008)* (pp. 111-125). IEEE.

Naser, S., Sofotasios, P. C., Bariah, L., Jaafar, W., Muhaidat, S., Al-Qutayri, M., & Dobre, O. A. (2020). Rate-splitting multiple access: Unifying NOMA and SDMA in MISO VLC channels. *IEEE Open Journal of Vehicular Technology, 1*, 393–413.

Nasir, T., Bukhari, S. S. H., Raza, S., Munir, H. M., Abrar, M., Muqeet, H. A. U., & Masroor, R. (2021). Recent Challenges and Methodologies in Smart Grid Demand Side Management: State-of-the-Art Literature Review. *Mathematical Problems in Engineering, 2021*(1), 5821301.

Nathali Silva, B., Khan, M., & Han, K. (2017). Big data analytics embedded smart city architecture for performance enhancement through real-time data processing and decision-making. *Wireless Communications and Mobile Computing, 2017*(1), 9429676.

Nguyen, G., Dlugolinsky, S., Tran, V., & Garcia, A. L. (2020). Deep learning for proactive network monitoring and security protection. *ieee. Access : the Newsmagazine of the American Dental Hygienists' Association, 8*, 19696–19716.

Nikolov, N. (2024, June). Smart cities as a tool for environmental sustainability: opportunities and challenges. In *ENVIRONMENT. TECHNOLOGIES. RESOURCES. Proceedings of the International Scientific and Practical Conference* (Vol. 1, pp. 261-266).

Ning, X., Ramirez, R., & Khuntia, J. (2021). Blockchain-enabled government efficiency and impartiality: Using blockchain for targeted poverty alleviation in a city in China. *Information Technology for Development, 27*(3), 599–616. DOI: 10.1080/02681102.2021.1925619

Nirandjan, S., Koks, E. E., Ward, P. J., & Aerts, J. C. J. H. (2022). A Spatially Explicit Harmonized Global Dataset of Critical Infrastructure. *Scientific Data, 9*(1), 150. DOI: 10.1038/s41597-022-01218-4 PMID: 35365664

Nitsche, B., Straube, F., & Wirth, M. (2021, July). Application areas and antecedents of automation in logistics and supply chain management: A conceptual framework. In *Supply Chain Forum* []. Taylor & Francis.]. *International Journal (Toronto, Ont.), 22*(3), 223–239.

Nitzl, C., Roldan, J. L., & Cepeda, G. (2016). Mediation analysis in partial least squares path modeling: Helping researchers discuss more sophisticated models. *Industrial Management & Data Systems, 116*(9), 1849–1864.

Noaman, M., Khan, M. S., Abrar, M. F., Ali, S., Alvi, A., & Saleem, M. A. (2022). Challenges in integration of heterogeneous internet of things. *Scientific Programming, 2022*(1), 8626882.

Nusir, M., Alshirah, M., & Alghsoon, R. (2023). Investigating smart city adoption from the citizen's insights: Empirical evidence from the Jordan context. *PeerJ. Computer Science, 9*, e1289. PMID: 37346561

Ochoa, S. F., & Santos, R. (2015). Human-centric wireless sensor networks to improve information availability during urban search and rescue activities. *Information Fusion, 22*, 71–84.

Ogala, J. O., Ahmad, S., Shakeel, I., Ahmad, J., & Mehfuz, S. (2023). Strengthening KMS Security with Advanced Cryptography, Machine Learning, Deep Learning, and IoT Technologies. *SN Computer Science, 4*(5), 530.

Ogiemwonyi, O., Alam, M. N., Alshareef, R., Alsolamy, M., Azizan, N. A., & Mat, N. (2023). Environmental factors affecting green purchase behaviors of the consumers: Mediating role of environmental attitude. *Cleaner Environmental Systems, 10*, 100130.

Ohm, P. (2009). Broken promises of privacy: Responding to the surprising failure of anonymization. *UCLA l. Rev., 57*, 1701.

Oliveira, A. C., Afonso, C., Matos, J., Riffat, S., Nguyen, M., & Doherty, P. (2002). A combined heat and power system for buildings driven by solar energy and gas. *Applied Thermal Engineering, 22*(6), 587–593. DOI: 10.1016/S1359-4311(01)00110-7

Omrany, H., Al-Obaidi, K. M., Hossain, M., Alduais, N. A., Al-Duais, H. S., & Ghaffarianhoseini, A. (2024). IoT-enabled smart cities: A hybrid systematic analysis of key research areas, challenges, and recommendations for future direction. *Discover Cities*, *1*(1), 2.

Ordonez García, B. J., & Martos Nunez, M. V. (2023). Blockchain in Agriculture: A PESTELS Analysis JAVIER.

Osama, M., Ateya, A. A., Sayed, M. S., Hammad, M., Pławiak, P., Abd El-Latif, A. A., & Elsayed, R. A. (2023). Internet of medical things and healthcare 4.0: Trends, requirements, challenges, and research directions. *Sensors (Basel)*, *23*(17), 7435. PMID: 37687891

Otjacques, B., Hitzelberger, P., Naumann, S., & Wohlgemuth, V. (2018). *From Science to Society: New Trends in Environmental Informatics*. Springer.

Öztuna, B. (2021). Logistics 4.0 and technologic applications. In *Logistics 4.0 and Future of Supply Chains* (pp. 9–27). Springer Nature Singapore.

Pacheco Rocha, N., Dias, A., Santinha, G., Rodrigues, M., Rodrigues, C., Queirós, A., & Pavão, J. (2022). Systematic literature review of context-awareness applications supported by smart cities' infrastructures. *SN Applied Sciences*, *4*(4), 90.

Palmisano, S. J. (2008). A smarter planet: the next leadership agenda. *IBM. November, 6*, 1-8.

Pandey, S., & Sen, C. (2022). Blockchain Technology in Real-Time Governance: An Indian Scenario. *The Indian Journal of Public Administration*, *68*(3), 397–413. DOI: 10.1177/00195561221105241

Paprocki, W. (2016, May). How transport and logistics operators can implement the solutions of "Industry 4.0". In *TranSopot Conference* (pp. 185–196). Springer International Publishing.

Park, J. S., Kim, Y. S., Choi, C. H., & Shim, J. (2018). How to define value on data under blockchain driven open data system for e-government. In *Proceedings of the 11th International Conference on Theory and Practice of Electronic Governance* (pp. 670–672). Galway, Ireland, April 4–6, 2018.

Park, L. W., Lee, S., & Chang, H. (2018). A sustainable home energy prosumer-chain methodology with energy tags over the blockchain. *Sustainability*, *10*(3), 658. DOI: 10.3390/su10030658

Parmentola, A., Petrillo, A., Tutore, I., & De Felice, F. (2022). Is blockchain able to enhance environmental sustainability? A systematic review and research agenda from the perspective of Sustainable Development Goals (SDGs). *Business Strategy and the Environment*, *31*(1), 194–217.

Pasolini, G., Guerra, A., Guidi, F., Decarli, N., & Dardari, D. (2020). Crowd-based cognitive perception of the physical world: Towards the Internet of Senses. *Sensors (Basel)*, *20*(9), 2437. PMID: 32344795

Patterson, J. J., & Huitema, D. (2019). Institutional innovation in urban governance: The case of climate change adaptation. *Journal of Environmental Planning and Management*, *62*(3), 374–398.

Pawar, N. S., & Sharma, K. V. (2025). Comprehensive review of remote sensing integration with deep learning in landslide forecasting and future directions. *Natural Hazards*. Advance online publication. DOI: 10.1007/s11069-025-07171-w

Peldon, D., Banihashemi, S., LeNguyen, K., & Derrible, S. (2024). Navigating urban complexity: The transformative role of digital twins in smart city development. *Sustainable Cities and Society*.

Pereira, P. J., Costa, N., Barros, M., Cortez, P., Durães, D., Silva, A., & Machado, J. (2022). A Comparison of Automated Time Series Forecasting Tools for Smart Cities. In Marreiros, G., Martins, B., Paiva, A., Ribeiro, B., & Sardinha, A. (Eds.), *Progress in Artificial Intelligence* (Vol. 13566, pp. 551–562). Springer International Publishing., DOI: 10.1007/978-3-031-16474-3_45

Perera, S., Nanayakkara, S., Rodrigo, M. N. N., Senaratne, S., & Weinand, R. (2020). Blockchain technology: Is it hype or real in the construction industry? *Journal of Industrial Information Integration*, *17*, 100125.

Pérez, J. G., & Toledo, D. G. (2017, June). Cognitive analytics of smart cities: a case of Málaga city. In *Proceedings of the 18th Annual International Conference on Digital Government Research* (pp. 616-617).

Peric Momcilovic, A., De Blust, S., Jiang, Y., Guariento, N., & Wälty, S. (2022). Urban Strategies for Dense and Green Zurich: From Healthy Neighbourhoods towards Healthy Communities? In *Proceedings of the 58th ISOCARP World Planning Congress* (pp. 1676-1691). ISOCARP.

Perić, A., Jiang, Y., Menz, S., & Ricci, L. (2023). Green Cities: Utopia or Reality? Evidence from Zurich, Switzerland. *Sustainability*, *15*(15), 12079.

Peters, M. A., Jandrić, P., & Hayes, S. (2022). Biodigital philosophy, technological convergence, and postdigital knowledge ecologies. In *Bioinformational philosophy and postdigital knowledge ecologies* (pp. 3–22). Springer International Publishing.

Petrolo, R., Loscri, V., & Mitton, N. (2017). Towards a smart city based on cloud of things, a survey on the smart city vision and paradigms. *Transactions on Emerging Telecommunications Technologies*, *28*(1), e2931.

Phang, S. K., Chiang, T. H. A., Happonen, A., & Chang, M. M. L. (2023). From Satellite to UAV-Based Remote Sensing: A Review on Precision Agriculture. *IEEE Access : Practical Innovations, Open Solutions*.

Phunchongharn, P., Hossain, E., & Kim, D. I. (2013). Resource allocation for device-to-device communications underlaying LTE-advanced networks. *IEEE Wireless Communications*, *20*(4), 91–100.

Phuttharak, J., & Loke, S. W. (2023). An event-driven architectural model for integrating heterogeneous data and developing smart city applications. *Journal of Sensor and Actuator Networks*, *12*(1), 12.

Piaggio, D., Zarro, M., Pagliara, S., Andellini, M., Almuhini, A., Maccaro, A., & Pecchia, L. (2023). The use of smart environments and robots for infection prevention control: A systematic literature review. *American Journal of Infection Control*, *51*(10), 1175–1181. PMID: 36924997

Planning Strategy, A. C. T. (2024). Sustainable and resilient territory. Available online: https://www.planning.act.gov.au/act-planning-strategy/strategic-directions/sustainable-and-resilient-territory [08 April 2025]

Pliatsios, A., Kotis, K., & Goumopoulos, C. (2023). A systematic review on semantic interoperability in the IoE-enabled smart cities. *Internet of Things : Engineering Cyber Physical Human Systems*, *22*, 100754.

Podder, A. K., Al Bukhari, A., Islam, S., Mia, S., Mohammed, M. A., Kumar, N. M., & Abdulkareem, K. H. (2021). IoT based smart agrotech system for verification of Urban farming parameters. *Microprocessors and Microsystems*, *82*, 104025.

Poell, T., Nieborg, D., & Van Dijck, J. (2019). Platformisation. *Internet Policy Review*, *8*(4), 1–13.

Poyyamozhi, M., Murugesan, B., Rajamanickam, N., Shorfuzzaman, M., & Aboelmagd, Y. (2024). IoT—A Promising Solution to Energy Management in Smart Buildings: A Systematic Review, Applications, Barriers, and Future Scope. *Buildings (Basel, Switzerland)*, *14*(11), 3446.

Prabowo, O. M., Mulyana, E., Nugraha, I. G. B. B., & Supangkat, S. H. (2023). Cognitive city platform as digital public infrastructure for developing a smart, sustainable and resilient city in Indonesia. *IEEE Access : Practical Innovations, Open Solutions*, *11*, 120157–120178.

Pratama, A. B., & Imawan, S. A. (2019). A scale for measuring perceived bureaucratic readiness for smart cities in Indonesia. *Public Administration and Policy*, *22*(1), 25–39.

Priyashani, N., Kankanamge, N., & Yigitcanlar, T. (2023). Multisource Open Geospatial Big Data Fusion: Application of the Method to Demarcate Urban Agglomeration Footprints. *Land (Basel)*, *12*(2), 407. DOI: 10.3390/land12020407

Protection, F. D. (2018). General data protection regulation (GDPR). *Intersoft Consulting, Accessed in October, 24*(1).

Puliafito, A., Tricomi, G., Zafeiropoulos, A., & Papavassiliou, S. (2021). Smart cities of the future as cyber physical systems: Challenges and enabling technologies. *Sensors (Basel)*, *21*(10), 3349. PMID: 34066019

Puron-Cid, G., & Gil-García, J. (2022). Are Smart Cities Too Expensive in the Long Term? Analyzing the Effects of ICT Infrastructure on Municipal Financial Sustainability. *Sustainability*.

Qonita, M., Rachmawati, R., & Rijanta, R. (2025). Assessing Amsterdam's urban mobility through the lens of smart mobility goals. *Case Studies on Transport Policy*, *20*, 101443.

Qu, C., Sorbelli, F. B., Singh, R., Calyam, P., & Das, S. K. (2023). Environmentally-aware and energy-efficient multi- drone coordination and networking for disaster response. *IEEE Transactions on Network and Service Management*, *20*(2), 1093–1109.

Rahman, M. A., Rashid, M. M., Shamim Hossain, M., Hassanain, E., Alhamid, M. F., & Guizani, M. (2019). Blockchain and IoT-based cognitive edge framework for sharing economy services in a smart city. *IEEE Access : Practical Innovations, Open Solutions*, *7*, 18611–18621.

Raisa, F. F., Orthy, N. T., Nibir, M. D., & Faysal, H. M. (2023). IoT based medicine dispenser for improving medication adherence.

Rajković, T., Vasiljević, D., & Lečić-Cvetković, D. (2022, June). Logistics 4.0–Smart Transformation of Logistics and Supply Chain Management. In *International Symposium SymOrg* (pp. 386-402). Cham: Springer International Publishing.

Rakhee, M., & Elayidom, S. (2023, January). Blockchain Technology as a Solution for Vulnerabilities in Internet of Things-A Survey. In *2023 Advanced Computing and Communication Technologies for High Performance Applications (ACCTHPA)* (pp. 1-6). IEEE.

Ramu, S. P., Boopalan, P., Pham, Q.-V., Maddikunta, P. K. R., Huynh-The, T., Alazab, M., Nguyen, T. T., & Gadekallu, T. R. (2022). Federated learning enabled digital twins for smart cities: Concepts, recent advances, and future directions. *Sustainable Cities and Society*, *79*, 103663. DOI: 10.1016/j.scs.2021.103663

Rao, P. M., & Deebak, B. D. (2023). Security and privacy issues in smart cities/ industries: Technologies, applications, and challenges. *Journal of Ambient Intelligence and Humanized Computing*, *14*(8), 10517–10553.

Rathore, S. P. S., Dalabhai, C. V., Patel, C. K. B., Sharma, R., Mathur, A., & Yadav, A. (2024, December). Big Data Analytics for Smart Cities. In *2024 1st International Conference on Advances in Computing, Communication and Networking (ICAC2N)* (pp. 1289-1294). IEEE.

Rathore, M. M., Paul, A., Hong, W. H., Seo, H., Awan, I., & Saeed, S. (2018). Exploiting IoT and big data analytics: Defining smart digital city using real-time urban data. *Sustainable Cities and Society*, *40*, 600–610.

Ravindran, M. (2022). Energy Efficient Clustering for Precision Agriculture using LEACH Protocol. *NeuroQuantology : An Interdisciplinary Journal of Neuroscience and Quantum Physics*, *20*(7), 2702.

Rawindaran, N. (2023). Legal Considerations and Ethical Challenges of Artificial Intelligence on Internet of Things and Smart Cities. In *Data Protection in a Post-Pandemic Society: Laws, Regulations, Best Practices and Recent Solutions* (pp. 217–239). Springer International Publishing.

Razavi-Termeh, S. V., Sadeghi-Niaraki, A., Abba, S. I., Ali, F., & Choi, S.-M. (2024). Enhancing spatial prediction of groundwater-prone areas through optimization of a boosting algorithm with bio-inspired metaheuristic algorithms. *Applied Water Science*, *14*(11), 244. DOI: 10.1007/s13201-024-02301-4

Reddy, A., Ghantasala, G. P., Kurra, M., & Ayyappa, R. M. K. (2025a). PATIENT EMPOWERMENT THROUGH SECURE DATA MANAGEMENT. Using Blockchain Technology in Healthcare Settings: Empowering Patients with Trustworthy Data, 153.

Reddy, K. V., Saketh, G., Priyanshu, S. S., Nitesh, M., Hussain, S. K., & Sharma, K. V. (2025b). AI-Driven Healthcare Management Platform: Enhancing Accessibility, Efficiency, and Security in Digital Health Systems. Synthesis: A Multidisciplinary Research Journal, 3(1), 1-14.

Regulation, P. (2016). Regulation (EU) 2016/679 of the European Parliament and of the Council. [eu]. *Regulation*, *679*, 2016.

Rejeb, A., Rejeb, K., Simske, S., & Treiblmaier, H. (2021). Blockchain Technologies in Logistics and Supply Chain Management: A Bibliometric Review. *Logistics (Basel)*, *5*(72). Advance online publication. DOI: 10.3390/logistics5040072

Rennung, F., Luminosu, C. T., & Draghici, A. (2016). Service provision in the framework of Industry 4.0. *Procedia: Social and Behavioral Sciences*, *221*, 372–377.

Ridha, S., & Kamil, P. A. (2021). The Problems of Teaching Geospatial Technology in Developing Countries: Concepts, Curriculum, and Implementation in Indonesia. *The Journal of Geography*, *120*(2), 72–82. DOI: 10.1080/00221341.2021.1872681

Rihar, M., Hrovatin, N., & Zorič, J. (2015). Household valuation of smart-home functionalities in Slovenia. *Utilities Policy*, *33*, 42–53.

Rittenbruch, M., Foth, M., Mitchell, P., Chitrakar, R., Christensen, B., & Pettit, C. (2022). Co-designing planning support systems in urban science: The questions they answer and the questions they raise. *Journal of Urban Technology*, *29*(2), 7–32.

Robinson, T., & Ji, M. (2022). Smart City Seoul: Solving the Urban Puzzle. In *Sustainable, Smart and Solidary Seoul: Transforming an Asian Megacity* (pp. 99–134). Springer International Publishing.

Roche, S., Nabian, N., Kloeckl, K., & Ratti, C. (2012, May). Are 'smart cities' smart enough. In *Global geospatial conference* (pp. 215-235).

Rodrigues, M., & Franco, M. (2019). Measuring the urban sustainable development in cities through a Composite Index: The case of Portugal. *Sustainable Development*, *28*(4), 507–520.

Roman, R., Zhou, J., & Lopez, J. (2018). On the features and challenges of security and privacy in distributed internet of things. *Computer Networks*, *57*(10), 2266–2279.

Romualdo-Suzuki, L., & Finkelstein, A. (2020). Data as Infrastructure for Smart Cities: Linking Data Platforms to Business Strategies. *arXiv preprint arXiv:2005.11414*.

Rudrakar, S., & Rughani, P. (2022). *IoT based Agriculture (IoTA)*. Architecture, Cyber Attack, Cyber Crime and Digital Forensics Challenges.

Ruggieri, R., Ruggeri, M., Vinci, G., & Poponi, S. (2021). Electric mobility in a smart city: European overview. *Energies*, *14*(2), 315.

Ruhlandt, R. W. S. (2018). The governance of smart cities: A systematic literature review. *Cities (London, England)*, *81*, 1–23.

Sachdeva, S., Bhatia, S., Al Harrasi, A., Shah, Y. A., Anwer, K., Philip, A. K., & Halim, S. A. (2024). Unraveling the role of cloud computing in health care system and biomedical sciences. *Heliyon*. PMID: 38601602

Saleem, J., Hammoudeh, M., Raza, U., Adebisi, B., & Ande, R. (2018, June). IoT standardisation: Challenges, perspectives and solution. In *Proceedings of the 2nd international conference on future networks and distributed systems* (pp. 1-9).

Salimi, M., Hosseinpour, M., Mansouri, S., & Borhani, N., T. (2022). Environmental aspects of the combined cooling, heating, and power (CCHP) systems: A review. *Processes (Basel, Switzerland)*, *10*(4), 711.

Salman, M. Y., & Hasar, H. (2023). Review on environmental aspects in smart city concept: Water, waste, air pollution and transportation smart applications using IoT techniques. *Sustainable Cities and Society*, *94*, 104567.

Sami, M., Noordin, N. K., Khabazian, M., Hashim, F., & Subramaniam, S. (2016). A survey and taxonomy on medium access control strategies for cooperative communication in wireless networks: Research issues and challenges. *IEEE Communications Surveys and Tutorials*, *18*(4), 2493–2521.

Sanchez, T. W., Brenman, M., & Ye, X. (2025). The ethical concerns of artificial intelligence in urban planning. *Journal of the American Planning Association*, *91*(2), 294–307.

Santos, T., Deus, R., Rocha, J., & Tenedório, J. A. (2021). Assessing sustainable urban development trends in a dynamic tourist coastal area using 3D spatial indicators. *Energies*, *14*(16), 5044.

Sargiotis, D. (2024). Data Security and Privacy: Protecting Sensitive Information. In *Data Governance: A Guide* (pp. 217–245). Springer Nature Switzerland.

Sarker, I. H. (2021). Data science and analytics: An overview from data-driven smart computing, decision-making and applications perspective. *SN Computer Science*, *2*(5), 377. PMID: 34278328

Savchenko, A. B., & Borodina, T. L. (2020). Green and digital economy for sustainable development of urban areas. *Regional Research of Russia*, *10*, 583–592.

Savithramma, R. M., Sumathi, R., & Sudhira, S. H. (2022). Variability Analysis of Isolated Intersections Through Case Study. *International Journal of Engineering Trends and Technology*, *70*(8), 359–374. DOI: 10.14445/22315381/ijett-v70i8p237

Schaffers, H., Komninos, N., Pallot, M., Trousse, B., Nilsson, M., & Oliveira, A. (2011). Smart Cities and the Future Internet: Towards Cooperation Frameworks for Open Innovation. *Future Internet*, *6656*, 431–446. DOI: 10.1007/978-3-642-20898-0_31

Schreck, J., Jung, P., & Stańczak, S. (2018). Compressive rate estimation with applications to device-to-device communications. *IEEE Transactions on Wireless Communications*, *17*(10), 7001–7012.

Schrotter, G., & Hürzeler, C. (2020). The digital twin of the city of Zurich for urban planning. *PFG–Journal of Photogrammetry. Remote Sensing and Geoinformation Science*, *88*(1), 99–112.

Semeniuk, V. (2024). Development of Optimization Algorithms for Geospatial Search in Large Volumes of Data. *Наука і Техніка Сьогодні, 5(33).* https://doi .org/DOI: 10.52058/2786-6025-2024-5(33)-1083-1094

Senior, C., Jowkar, M., Temeljotov-Salaj, A., & Johansen, A. (2021, March). Empowering citizens in a smart city project one step at a time: A Norwegian case study. In *2021 IEEE European Technology and Engineering Management Summit (E-TEMS)* (pp. 10-15). IEEE.

SERVICE CANTONAL DE GESTION DES DECHETS – GESDEC. (2006). *Le recyclage des matériaux de construction à Genève*. Etat de Genève.

Sha, K., & Taeihagh, A. (2024). Designing adaptive policy packages for inclusive smart cities: Lessons from Singapore's smart nation program. *Sustainable Cities and Society*, *115*, 105868.

Shamsuzzoha, A., Nieminen, J., Piya, S., & Rutledge, K. (2021). Smart city for sustainable environment: A comparison of participatory strategies from Helsinki, Singapore and London. *Cities (London, England)*, *114*, 103194.

Shanmathi, N., & Jagannath, M. (2018). Computerised decision support system for remote health monitoring: A systematic review. *Ingénierie et Recherche Biomédicale : IRBM = Biomedical Engineering and Research*, *39*(5), 359–367.

Shao, M., Zhou, Z., Bin, G., Bai, Y., & Wu, S. (2020). A wearable electrocardiogram telemonitoring system for atrial fibrillation detection. *Sensors (Basel)*, *20*(3), 606. DOI: 10.3390/s20030606 PMID: 31979184

Sharifi, A., Allam, Z., Bibri, S. E., & Khavarian-Garmsir, A. R. (2024). Smart cities and sustainable development goals (SDGs): A systematic literature review of co-benefits and trade-offs. *Cities (London, England)*, *146*, 104659.

Sharma, M., Gupta, R., & Acharya, P. (2021). Analysing the adoption of cloud computing service: A systematic literature review. *Global Knowledge, Memory and Communication*, *70*(1–2), 114–153. DOI: 10.1108/GKMC-10-2019-0126

Sharma, N., Anand, A., & Husain, A. (2020). Cloud based healthcare services for telemedicine practices using internet of things. *Journal of Critical Reviews*, *7*(14), 2605–2611. DOI: 10.31838/jcr.07.14.510

Shatz, S., & Chylik, S. E. (2019). The California consumer privacy act of 2018: A sea change in the protection of California consumers' personal information. *Bus. LAw.*, *75*, 1917.

Sheikh, H., Mitchell, P., & Foth, M. (2023). More-than-human smart urban governance: A research agenda. *Digital Geography and Society*, *4*, 100045.

Shepelev, V., Glushkov, A., Slobodin, I., & Balfaqih, M. (2023). Studying the relationship between the traffic flow structure, the traffic capacity of intersections, and vehicle-related emissions. *Mathematics*, *11*(16), 3591.

Shim, K., Nguyen, T. V., & An, B. (2019). Exploiting opportunistic scheduling schemes to improve physical-layer security in MU-MISO NOMA systems. *IEEE Access : Practical Innovations, Open Solutions*, *7*, 180867–180886.

Shuaib, M., Hassan, N. H., Usman, S., Alam, S., Bhatia, S., Koundal, D., Mashat, A., & Belay, A. (2022). Identity model for blockchain-based land registry system: A comparison. *Wireless Communications and Mobile Computing*, *2022*. Advance online publication. DOI: 10.1155/2022/5670714

Sicari, S., Rizzardi, A., Grieco, L. A., & Coen-Porisini, A. (2015). Security, privacy and trust in Internet of Things: The road ahead. *Computer Networks*, *76*, 146–164.

Siddiqui, A., & Pandit, R. K. (2021). Smart Cities in India: Linkages with circular economy. In *Sustainability in the built environment in the 21st century: Lessons learned from India and the region* (pp. 185-200). Springer International Publishing.

Siddiqui, S., Hameed, S., Shah, S. A., Khan, A. K., & Aneiba, A. (2023). Smart contract-based security architecture for collaborative services in municipal smart cities. *Journal of Systems Architecture*, *135*, 102802.

Silva, B. N., Khan, M., & Han, K. (2018). Towards sustainable smart cities: A review of trends, architectures, components, and open challenges in smart cities. *Sustainable Cities and Society, 38*, 697–713. DOI: 10.1016/j.scs.2018.01.053

Silvestri, S., Tricomi, G., Damiano, E., Sicuranza, M., & Ciampi, M. (2025). A Data Platform for the Integration of Smart City Subsystems. In L. Barolli (Ed.), *Advances on P2P, Parallel, Grid, Cloud and Internet Computing* (Vol. 232, pp. 187–198). Springer Nature Switzerland. https://doi.org/DOI: 10.1007/978-3-031-76462-2_17

Sindhuja, M., Alice, K., Nancy, A. M., & Vinoth, N. A. S. (2025). Cybersecurity Strategies for Smart Cities. In *Harnessing AI in Geospatial Technology for Environmental Monitoring and Management* (pp. 77–98). IGI Global Scientific Publishing.

Singh, B. (2023). Federated Learning for Envision Future Trajectory Smart Transport System for Climate Preservation and Smart Green Planet: Insights into Global Governance and SDG-9 (Industry, Innovation and Infrastructure). *National Journal of Environmental Law, 6*(2), 6–17.

Singh, J., Cobbe, J., & Norval, C. (2018). Decision provenance: Harnessing data flow for accountable systems. *IEEE Access : Practical Innovations, Open Solutions, 7*, 6562–6574.

Sinha, A., Kumar, P., Rana, N. P., Islam, R., & Dwivedi, Y. K. (2019). Impact of internet of things (IoT) in disaster management: A task-technology fit perspective. *Annals of Operations Research, 283*, 759–794.

Siswipraptini, P. C., Aziza, R. N., & Sangadji, I., Indrianto, Siregar, R. R. A., & Sondakh, G. (2021). IoT for smart home system. *Indonesian Journal of Electrical Engineering and Computer Science, 23*(2), 733–739. DOI: 10.11591/ijeecs.v23.i2.pp733-739

Siwach, V., Sehrawat, H., & Dahiya, O. (2022, October). Hybrid Blockchain Security Mechanism for protection of data related to Precision Agriculture. In *2022 10th International Conference on Reliability, Infocom Technologies and Optimization (Trends and Future Directions)(ICRITO)* (pp. 1-7). IEEE.

Sokolov, B., & Ivanov, D. (2015). Integrated scheduling of material flows and information services in industry 4.0 supply networks. *IFAC-PapersOnLine, 48*(3), 1533–1538.

Solove, D. J. (2005). A taxonomy of privacy. *U. Pa. l. Rev., 154*, 477.

Sookhak, M., Tang, H., He, Y., & Yu, F. R. (2018). Security and privacy of smart cities: A survey, research issues and challenges. *IEEE Communications Surveys and Tutorials, 21*(2), 1718–1743.

Sowmya, T., & Anita, E. M. (2024). A comprehensive review of AI based intrusion detection system. *Measurement. Sensors, 28*, 100827.

Sripan, M., Lin, X., Petchlorlean, P., & Ketcham, M. (2012). Research and thinking of smart home technology. In *Proceedings of the International Conference on Systems and Electronic Engineering (ICSEE'2012)* (pp. 61–63). Phuket, Thailand.

Stamp, M. (2011). *Information security: principles and practice*. John Wiley & Sons.

Strobel, G. (2020). Agriculture in the Era of the Internet of Things: An Information System Architecture for Smart Agriculture.

Su, K., Li, J., & Fu, H. (2011). Smart city and the applications. In *2011 international conference on electronics, communications and control (ICECC)* (pp. 1028-1031). IEEE.

Sukawan, H. A. R., & Rachmawati, R. (2021, March). MyTransport. SG as a new communication platform in implementing smart mobility in Singapore. []. IOP Publishing.]. *Journal of Physics: Conference Series, 1834*(1), 012006.

Su, M., Wang, G., & Choo, K. K. R. (2022). Prediction-based resource deployment and task scheduling in edge-cloud collaborative computing. *Wireless Communications and Mobile Computing, 2022*, 2568503. Advance online publication. DOI: 10.1155/2022/2568503

Sumi, L., & Ranga, V. (2016). Sensor enabled Internet of Things for smart cities. In *2016 fourth international conference on parallel, distributed and grid computing (PDGC)* (pp. 295-300). IEEE.

Sundaramurthy, S. K., Ravichandran, N., Inaganti, A. C., & Muppalaneni, R. (2022). The future of enterprise automation: Integrating AI in cybersecurity, cloud operations, and workforce analytics. *Artificial Intelligence and Machine Learning Review, 3*(2), 1–15.

Sun, J., Yan, J., & Zhang, K. Z. K. (2016). Blockchain-based sharing services: What blockchain technology can contribute to smart cities. *Financial Innovation, 2*(26), 1–9. DOI: 10.1186/s40854-016-0040-y

Sun, Y., Chen, C., Xu, Y., Xie, S., Blum, R. S., & Venkitasubramaniam, P. (2024). On the generalization discrepancy of spatiotemporal dynamics-informed graph convolutional networks. *Frontiers of Mechanical Engineering, 10*, 1397131. DOI: 10.3389/fmech.2024.1397131

Su, Z., Yang, L., Wang, H., Liu, S., Song, J., & Jin, X. (2023). Solar-assisted combined cooling and power system integrating energy storage and desulfurization for coal-fired power plants. *Thermal Science and Engineering Progress*, *45*, 102110. DOI: 10.1016/J.TSEP.2023.102110

Sweeney, L. (2002). k-anonymity: A model for protecting privacy. *International Journal of Uncertainty, Fuzziness and Knowledge-based Systems*, *10*(05), 557–570.

Syed, A. S., Sierra-Sosa, D., Kumar, A., & Elmaghraby, A. (2021). IoT in smart cities: A survey of technologies, practices and challenges. *Smart Cities*, *4*(2), 429–475.

Tajuddin, A. F. B., Yusof, M. H. M., Balfaqih, M., Almohammedi, A. A., Darshi, S., & Qaid, N. (2025). Comparative analysis of AES, RC and Twofish algorithms for data encryption and decryption. In *2025 4th International Conference on Computing and Information Technology (ICCIT)* (pp. 14–19). IEEE.

Talal, W., & Akroot, A. (2023). Exergoeconomic Analysis of an Integrated Solar Combined Cycle in the Al-Qayara Power Plant in Iraq. *Processes (Basel, Switzerland)*, *11*(3). Advance online publication. DOI: 10.3390/pr11030656

Tan, L., & Wang, N. (2010, August). Future internet: The internet of things. In *2010 3rd international conference on advanced computer theory and engineering (ICACTE)* (Vol. 5, pp. V5-376). IEEE.

Tantray, J., Kosey, S., Prajapati, J. B., & Prajapati, B. (2024). The Future of the Healthcare System: A Meta-Analysis of Remote Patient Monitoring. In Handbook on Augmenting Telehealth Services (pp. 105-126). CRC Press.

Tariq, M. A. U. R., Faumatu, A., Hussein, M., Shahid, M. L. U. R., & Muttil, N. (2020). Smart city-ranking of major Australian cities to achieve a smarter future. *Sustainability*, *12*(7), 2797.

Tavares, P., Ingi, D., Araújo, L., Pinho, P., & Bhusal, P. (2021). Reviewing the role of outdoor lighting in achieving sustainable development goals. *Sustainability*, *13*(22), 12657.

Tehrani, M. N., Uysal, M., & Yanikomeroglu, H. (2014). Device-to-device communication in 5G cellular networks: Challenges, solutions, and future directions. *IEEE Communications Magazine*, *52*(5), 86–92.

Thakkalapelli, D. (2023). Cost analysis of cloud migration for small businesses. *Journal Name*, *44*(4), 4778–4784.

Thakur, H., & Rao, R. (2024). Emphasis of infection prevention and control: A review. *Journal of Population Therapeutics and Clinical Pharmacology = Journal de la Thérapeutique des Populations et de la Pharmacologie Clinique*, *31*, 2238–2249.

Thangavelu, M. (2020). Outer space activities and city evolution. In *ASCEND 2020* (p. 4059).

Therias, A., & Rafiee, A. (2023). City digital twins for urban resilience. *International Journal of Digital Earth*, *16*(2), 4164–4190.

Thomas, T., & Rajabi, E. (2021). A systematic review of machine learning-based missing value imputation techniques. *Data Technologies and Applications*, *55*(4), 558–585. DOI: 10.1108/DTA-12-2020-0298

Tiwari, S. (2021). Supply chain integration and Industry 4.0: A systematic literature review. *Benchmarking*, *28*(3), 990–1030. DOI: 10.1108/BIJ-08-2020-0428

Tiwari, S., Wee, H. M., & Daryanto, Y. (2018). Big data analytics in supply chain management between 2010 and 2016: Insights to industries. *Computers & Industrial Engineering*, *115*, 319–330.

Toh, C. K. (2022). Smart city indexes, criteria, indicators and rankings: An in-depth investigation and analysis. *IET Smart Cities*, *4*(3), 211–228.

Toli, A. M., & Murtagh, N. (2020). The concept of sustainability in smart city definitions. *Frontiers in Built Environment*, *6*, 77.

Tomatsu, Y., & Han, W. (2023). Bitcoin and Renewable Energy Mining: A Survey. *Blockchains*, *1*(2), 90–110. DOI: 10.3390/blockchains1020007

Tranos, E., & Gertner, D. (2012). Smart networked cities? *Innovation (Abingdon)*, *25*(2), 175–190. DOI: 10.1080/13511610.2012.660327

Trigka, M., & Dritsas, E. (2025). Edge and cloud computing in smart cities. *Future Internet*, *17*(3), 118.

Truby, J. (2020). Governing artificial intelligence to benefit the UN sustainable development goals. *Sustainable Development*, *28*(4), 946–959.

Turečková, K., & Nevima, J. (2020). The cost benefit analysis for the concept of a smart city: How to measure the efficiency of smart solutions? *Sustainability*, *12*(7), 2663.

Türetken, O., Grefen, P., Gilsing, R., Adali, E., & Ozkan, B. (2019). Business model innovation in the smart mobility domain. In Lu, M. (Ed.), *Cooperative Intelligent Transport Systems* (Vol. 25, pp. 63–86). IET Transportation Series., DOI: 10.1049/PBTR025E_ch4

Tzanou, M. (2020). *Health Data Privacy under theGDPR - Big Data Challenges and RegulatoryResponses*. Routledge.

Udeh, E. O., Amajuoyi, P., Adeusi, K. B., & Scott, A. O. (2024). The role of IoT in boosting supply chain transparency and efficiency. *Magna Scientia Advanced Research and Reviews, 11*(01), 178–197.

Ullo, S. L., & Sinha, G. R. (2020). Advances in smart environment monitoring systems using IoT and sensors. *Sensors (Basel), 20*(11), 3113. PMID: 32486411

UN. (2022). The Sustainable Development Goals Report, United Nations. Available online: https://unstats.un.org/sdgs/report/2022/ [09/04/2025].

Ünal, F., Temir, G., & Köten, H. (2018). Energy, exergy and exergoeconomic analysis of solar-assisted vertical ground source heat pump system for heating season. *Journal of Mechanical Science and Technology, 32*, 3929–3942.

UN-IGIF. (2024). *United Nations Integrated Geospatial Information Framework (UN-IGIF)*. https://ggim.un.org/UN-IGIF/

United Nations, Department of Economic and Social Affairs, Population Division. (2018). World Urbanization Prospects (WUP): The 2018 Revision, Methodology (Working Paper No. ESA/P/WP.252). New York, NY: United Nations.

Uslu, B. Ç., Okay, E., & Dursun, E. (2020). Analysis of factors affecting IoT-based smart hospital design. *Journal of Cloud Computing (Heidelberg, Germany), 9*(1), 67. PMID: 33532168

Vahidi, V., & Saberinia, E. (2019). MIMO–OFDM communication systems for traffic data transmission in 5G drone small cells. *IET Communications, 13*(20), 3565–3574.

Vallejo-Correa, P., Monsalve-Pulido, J., & Tabares-Betancur, M. (2021). A systematic mapping review of context-aware analysis and its approach to mobile learning and ubiquitous learning processes. *Computer Science Review, 39*, 100335.

Van Twist, A., Ruijer, E., & Meijer, A. (2023). Smart cities & citizen discontent: A systematic review of the literature. *Government Information Quarterly, 40*(2), 101799.

Vangipuram, S. L., Mohanty, S. P., Kougianos, E., & Ray, C. (2022). G-DaM: A Distributed Data Storage with Blockchain Framework for Management of Groundwater Quality Data. *Sensors (Basel), 22*(22), 8725. PMID: 36433322

Varesano, D., Amos, E., Domer, B., Laferrière, F., Redaelli, D., & Viviani, M. (2018). Towards a" comprehensive" and smart platform for coordinated waste management in construction in Geneva?: a case study. In *Proceedings of 20. Status-Seminar" Forschen für den Bau im Kontext von Energie und Umwelt", Zurich, Switzerland, 6-7 September 2018*. 6-7 September 2018.

Venkatesh, V., Thong, J. Y., & Xu, X. (2012). Consumer acceptance and use of information technology: Extending the unified theory of acceptance and use of technology. *Management Information Systems Quarterly*, **•••**, 157–178.

Verma, A., Bhattacharya, P., Madhani, N., Trivedi, C., Bhushan, B., Tanwar, S., & Sharma, R. (2022). Blockchain for industry 5.0: Vision, opportunities, key enablers, and future directions. *IEEE Access : Practical Innovations, Open Solutions, 10*, 69160–69199.

Vermesan, O., & Friess, P. (Eds.). (2022). *Digitising the industry internet of things connecting the physical, digital and VirtualWorlds*. CRC Press.

Veuger, J. (2018). Trust in a viable real estate economy with disruption and blockchain. *Facilities*, *36*, 103–120.

Vimalkumar, M., Sharma, S. K., Singh, J. B., & Dwivedi, Y. K. (2021). 'Okay google, what about my privacy?': User's privacy perceptions and acceptance of voice based digital assistants. *Computers in Human Behavior*, *120*, 106763.

Vinodhini, V., Sathiyabhama, B., Sankar, S., Ramasubbareddy, S., Sundarakumar, M. R., & Balusamy, B. (2023). Enabling Technologies for Internet of Things (IoT)-Based Smart Cities. In *Convergence of IoT, Blockchain, and Computational Intelligence in Smart Cities* (pp. 41-57). CRC Press.

Vinuesa, R., Azizpour, H., Leite, I., Balaam, M., Dignum, V., Domisch, S., & Fuso Nerini, F. (2020). The role of artificial intelligence in achieving the Sustainable Development Goals. *Nature Communications*, *11*(1), 233. PMID: 31932590

Vodák, J., Šulyová, D., & Kubina, M. (2021). Advanced technologies and their use in smart city management. *Sustainability*, *13*(10), 5746.

Voigt, P., & Von dem Bussche, A. (2017). The eu general data protection regulation (GDPR). *A Practical Guide, 1st Ed., Cham: Springer International Publishing*, *10*(3152676), 10-5555.

Vojković, G., & Katulić, T. (2021). Data protection and smart cities. In *Handbook of smart cities* (pp. 903–928). Springer International Publishing.

Voulgaris, S., Fotiou, N., Siris, V. A., Polyzos, G. C., Jaatinen, M., & Oikonomidis, Y. (2019). Blockchain technology for intelligent environments. *Future Internet, 11*(10), 213.

Vuong, Q. P. (2024). The potential for artificial intelligence and machine learning in healthcare: the future of healthcare through smart technologies.

Wahrstätter, A., Taudes, A., & Svetinovic, D. (2024). Reducing Privacy of Coin-Join Transactions: Quantitative Bitcoin Network Analysis. *IEEE Transactions on Dependable and Secure Computing*.

Wang, Y., Feng, L., Chang, H., & Wu, M. (2017). Research on the impact of big data on logistics. In *MATEC Web of Conferences* (Vol. 100, p. 02015). EDP Sciences.

Wang, G., Gunasekaran, A., Ngai, E. W., & Papadopoulos, T. (2016). Big data analytics in logistics and supply chain management: Certain investigations for research and applications. *International Journal of Production Economics, 176*, 98–110.

Wang, J., Han, Z., & Guan, Z. (2020). Hybrid solar-assisted combined cooling, heating, and power systems: A review. *Renewable & Sustainable Energy Reviews, 133*, 110256.

Wang, J., Ma, C., & Wu, J. (2019a). Thermodynamic analysis of a combined cooling, heating and power system based on solar thermal biomass gasification☆. *Applied Energy, 247*, 102–115. DOI: 10.1016/J.APENERGY.2019.04.039

Wang, K. (2020). Migration strategy of cloud collaborative computing for delay-sensitive industrial IoT applications in the context of intelligent manufacturing. *Computer Communications, 150*, 413–420. DOI: 10.1016/j.comcom.2019.12.014

Wang, L., Xu, S., Wang, X., & Zhu, Q. (2021). Addressing Class Imbalance in Federated Learning. *Proceedings of the AAAI Conference on Artificial Intelligence, 35*(11), 10165–10173. DOI: 10.1609/aaai.v35i11.17219

Wang, W., Hoang, D. T., Hu, P., Xiong, Z., Niyato, D., & Wang, P.. (2019). A survey on consensus mechanisms and mining strategy management in blockchain networks. *IEEE Access : Practical Innovations, Open Solutions, 7*, 22328–22370.

Wang, X. X., Cao, W. Q., Cao, M. S., & Yuan, J. (2020). Assembling nano–microarchitecture for electromagnetic absorbers and smart devices. *Advanced Materials, 32*(36), 2002112.

Wang, Y. P. E., Lin, X., Adhikary, A., Grovlen, A., Sui, Y., Blankenship, Y., & Razaghi, H. S. (2017). A primer on 3GPP narrowband Internet of Things. *IEEE Communications Magazine, 55*(3), 117–123.

Wang, Y., Lin, W., & Zhang, T. (2010, October). Study on security of wireless sensor networks in smart grid. In *2010 International Conference on Power System Technology* (pp. 1-7). IEEE.

Wenge, R., Zhang, X., Dave, C., Chao, L., & Hao, S. (2014). Smart city architecture: A technology guide for implementation and design challenges. *China Communications, 11*(3), 56–69.

Wen, M. H., Chen, P. Y., Lin, S., Lien, C. W., Tu, S. H., Chueh, C. Y., & Bai, D. (2024). Enhancing Patient Safety Through an Integrated Internet of Things Patient Care System: Large Quasi-Experimental Study on Fall Prevention. *Journal of Medical Internet Research, 26*, e58380. PMID: 39361417

Winkle, H., & Zinsmeister, L. (2019). Trends in digitalization of intralogistics and the critical success factors of its implementation. *Brazilian Journal of Operations & Production Management, 16*(3), 537–549.

Wirth, J., Maier, C., Laumer, S., Oehlhorn, C. E., & Weitzel, T. (2020, June). Benefits in privacy research: a literature review, status quo and future research directions. In *Proceedings of the 2020 on Computers and People Research Conference* (pp. 135-141).

Witkowski, K. (2017). Internet of things, big data, industry 4.0–innovative solutions in logistics and supply chains management. *Procedia Engineering, 182*, 763–769.

Wolniak, R., & Grebski, W. (2023). *Smart mobility in smart city–Singapore and Tokyo comparison.* Zeszyty Naukowe. Organizacja i Zarządzanie/Politechnika Śląska.

Wong, G., & Jie, W. J. (2019). Singapore in 2018: Between uncharted waters and old ghosts. *Southeast Asian Affairs, 2019*(1), 296–323.

Wood, G. (2014). Ethereum: A secure decentralised generalised transaction ledger. *Ethereum project yellow paper, 151*(2014), 1-32.

Wu, M., Lu, T. J., Ling, F. Y., Sun, J., & Du, H. Y. (2010, August). Research on the architecture of Internet of Things. In *2010 3rd international conference on advanced computer theory and engineering (ICACTE)* (Vol. 5, pp. V5-484). IEEE.

Wu, C., Zheng, P., Xu, X., Chen, S., Wang, N., & Hu, S. (2020). Discovery of the environmental factors affecting urban Dwellers' mental health: A data-driven approach. *International Journal of Environmental Research and Public Health, 17*(21), 8167. PMID: 33167348

Wu, H., Zhang, Z., Guan, C., Wolter, K., & Xu, M. (2020). Collaborate edge and cloud computing with distributed deep learning for smart city internet of things. *IEEE Internet of Things Journal*, *7*(9), 8099–8110. DOI: 10.1109/JIOT.2020.2996784

Wu, J., Gan, W., Chao, H.-C., & Yu, P. S. (2024). Geospatial Big Data: Survey and Challenges. *IEEE Journal of Selected Topics in Applied Earth Observations and Remote Sensing*, *17*, 17007–17020. DOI: 10.1109/JSTARS.2024.3438376

Wu, K.-C., Chen, C.-H., & Hsu, C.-H. (2022). Three-dimensional visualization of thermal environments in urban canyons. *Geocarto International*, *37*(27), 16261–16282. DOI: 10.1080/10106049.2022.2107713

Wu, K., Ding, J., Lin, J., Zheng, G., Sun, Y., Fang, J., & Gu, B. (2025). Big-data empowered traffic signal control could reduce urban carbon emission. *Nature Communications*, *16*(1), 2013. PMID: 40016199

Wüst, K., & Gervais, A. (2018). Do you need a blockchain? In *Proceedings of the 2018 Crypto Valley Conference on Blockchain Technology (CVCBT)* (pp. 45–54). Zug, Switzerland.

Wu, Z., & Dai, L. (2023). Multiple access for near-field communications: SDMA or LDMA? *IEEE Journal on Selected Areas in Communications*, *41*(6), 1918–1935.

Xia, L., Semirumi, D. T., & Rezaei, R. (2023). A thorough examination of smart city applications: Exploring challenges and solutions throughout the life cycle with emphasis on safeguarding citizen privacy. *Sustainable Cities and Society*, *98*, 104771.

Xue, M., Xiu, G., Saravanan, V., & Montenegro-Marin, C. E. (2021). Cloud computing with AI for banking and e-commerce applications. *The Electronic Library*, *39*(4), 539–552. DOI: 10.1108/EL-07-2020-0207

Yalli, J. S., Hasan, M. H., & Badawi, A. (2024). Internet of things (iot): Origin, embedded technologies, smart applications and its growth in the last decade. *IEEE Access : Practical Innovations, Open Solutions*.

Yang, K., He, Y., Du, N., Yan, P., Zhu, N., Chen, Y., Wang, J., & Lund, P. D. (2024). Exergy, exergoeconomic, and exergoenvironmental analyses of novel solar- and biomass-driven trigeneration system integrated with organic Rankine cycle. *Energy*, *301*. Advance online publication. DOI: 10.1016/j.energy.2024.131605

Yang, K., Zhu, N., Ding, Y., Chang, C., Wang, D., & Yuan, T. (2019). Exergy and exergoeconomic analyses of a combined cooling, heating, and power (CCHP) system based on dual-fuel of biomass and natural gas. *Journal of Cleaner Production*, *206*, 893–906. DOI: 10.1016/j.jclepro.2018.09.251

Yang, L., Luo, Z., Zhang, S., Teng, F., & Li, T. (2024). Continual learning for smart City: A survey. *IEEE Transactions on Knowledge and Data Engineering*.

Yang, R., Zhao, T., Yu, F. R., Li, M., Zhang, D., & Zhao, X. (2024). Blockchain-based federated learning with enhanced privacy and security using homomorphic encryption and reputation. *IEEE Internet of Things Journal, 11*, 21674–21688.

Yang, W., & Lam, P. (2021). Evaluating non-market costs of ICT involving data transmission in smart cities. *Building Research and Information, 49*, 715.

Yang, X., Yan, Z., Bu, S., Li, W., Yu, N., Su, C., Lu, S., & Wang, X. (2024). Energy, exergy, exergoeconomic, economic, and environmental analyses and multi-objective optimization of a novel combined cooling and power system with dual-pressure Kalina cycle-absorption refrigeration. *Energy Conversion and Management, 315*. Advance online publication. DOI: 10.1016/j.enconman.2024.118781

Yao, H., Liu, Y., Wei, Y., Tang, X., & Li, Z. (2019, May). Learning from multiple cities: A meta-learning approach for spatial-temporal prediction. In *The world wide web conference* (pp. 2181-2191).

Yao, S., Ardabili, B. R., Pazho, A. D., Noghre, G. A., Neff, C., Bourque, L., & Tabkhi, H. (2025). From lab to field: Real-world evaluation of an AI-driven Smart Video Solution to enhance community safety. *Internet of Things : Engineering Cyber Physical Human Systems*, 101716.

Yaqoob, I., Ahmed, E., Hashem, I. A. T., Ahmed, A. I. A., Gani, A., Imran, M., & Guizani, M. (2017). Internet of things architecture: Recent advances, taxonomy, requirements, and open challenges. *IEEE Wireless Communications, 24*(3), 10–16.

Yesmin, T., Carter, M. W., & Gladman, A. S. (2022). Internet of things in healthcare for patient safety: An empirical study. *BMC Health Services Research, 22*(1), 278. PMID: 35232433

Yigitcanlar, T., Foth, M., & Kamruzzaman, M. (2019). Towards post-anthropocentric cities: Reconceptualizing smart cities to evade urban ecocide. *Journal of Urban Technology, 26*(2), 147–152.

Younan, M., Elhoseny, M., Ali, A. A., & Houssein, E. H. (2021, December). Quantum chain of things (qcot): A new paradigm for integrating quantum computing, blockchain, and internet of things. In *2021 17th International computer engineering conference (ICENCO)* (pp. 101-106). IEEE.

Yu, R., Wu, C., Yan, B., Yu, B., Zhou, X., Yu, Y., & Chen, N. (2021). Analysis of the impact of big data on e-commerce in cloud computing environment. *Complexity, 2021*, 5613599. Advance online publication. DOI: 10.1155/2021/5613599

Zachariadis, M., Hileman, G., & Scott, S. V. (2019). Governance and control in distributed ledgers: Understanding the challenges facing blockchain technology in financial services. *Information and Organization*, *29*(2), 105–117. DOI: 10.1016/j.infoandorg.2019.03.001

Zaheer, A. (2020). Sustainability and resilience in megacities through energy diversification, land fragmentation and fiscal mechanisms. *Sustainable Cities and Society*, *53*, 101841.

Zakzak, L. (2019, June). Citizen-centric smart city development: The case of smart Dubai's "Happiness Agenda". In *Proceedings of the 20th Annual International Conference on Digital Government Research* (pp. 141-147).

Zaman, M., Puryear, N., Abdelwahed, S., & Zohrabi, N. (2024). A review of IoT-based smart city development and management. *Smart Cities*, *7*(3), 1462–1501.

Zavratnik, V., Podjed, D., Trilar, J., Hlebec, N., Kos, A., & Stojmenova Duh, E. (2020). Sustainable and Community-Centred Development of Smart Cities and Villages. *Sustainability*, *12*(10), 3961. DOI: 10.3390/su12103961

Zeng, F., Pang, C., & Tang, H. (2023). Sensors on the internet of things systems for urban disaster management: A systematic literature review. *Sensors (Basel)*, *23*(17), 7475.

Zhang, F., Zhou, B., Liu, L., Liu, Y., Fung, H. H., Lin, H., & Ratti, C. (2018). Measuring human perceptions of a large-scale urban region using machine learning. *Landscape and Urban Planning*, *180*, 148–160.

Zhang, K., Ni, J., Yang, K., Liang, X., Ren, J., & Shen, X. S. (2017). Security and privacy in smart city applications: Challenges and solutions. *IEEE Communications Magazine*, *55*(1), 122–129.

Zhang, L., Li, F., Sun, B., & Zhang, C. (2019a). Integrated optimization design of combined cooling, heating, and power system coupled with solar and biomass energy. *Energies*, *12*(4). Advance online publication. DOI: 10.3390/en12040687

Zhang, N., Wang, Z., Lior, N., & Han, W. (2018). Advancement of distributed energy methods by a novel high efficiency solar-assisted combined cooling, heating and power system. *Applied Energy*, *219*, 179–186. DOI: 10.1016/J.APENERGY.2018.03.050

Zhang, Y., Zeng, D., Luo, J., Fu, X., Chen, G., Xu, Z., & King, I. (2024). A Survey of Trustworthy Federated Learning: Issues, Solutions, and Challenges. *ACM Transactions on Intelligent Systems and Technology*, *15*(6), 1–47. DOI: 10.1145/3678181

Zhao, J., Zhang, S., Sun, Y., Zhou, N., Yu, H., Zhang, H., & Jia, D. (2022). Wearable optical sensing in the medical internet of things (MIoT) for pervasive medicine: Opportunities and challenges. *ACS Photonics*, *9*(8), 2579–2599.

Zhao, S., & Chen, L. (2021). Exploring Residents' Purchase Intention of Green Housings in China: An Extended Perspective of Perceived Value. *International Journal of Environmental Research and Public Health*, 18. PMID: 33924311

Zhao, W. (2022). On Blockchain: Design Principle, Building Blocks, Core Innovations, and Misconceptions. *IEEE Systems, Man, and Cybernetics Magazine*, *8*(4), 6–14.

Zhao, Y., Li, S., Chen, H., & Xu, Y. (2022). Application of Smart City Construction in a New Data Environment. *Frontiers in Energy Research*, *10*, 908338. DOI: 10.3389/fenrg.2022.908338

Zheng, Y., Capra, L., Wolfson, O., & Yang, H. (2014). Urban computing: Concepts, methodologies, and applications. [TIST]. *ACM Transactions on Intelligent Systems and Technology*, *5*(3), 1–55.

Zheng, Y., Hao, Q., Wang, J., Gao, C., Chen, J., Jin, D., & Li, Y. (2024). A survey of machine learning for urban decision making: Applications in planning, transportation, and healthcare. *ACM Computing Surveys*, *57*(4), 1–41.

Zheng, Z., Xie, S., Dai, H., Chen, X., & Wang, H. (2017, June). *An overview of blockchain technology: Architecture, consensus, and future trends. In 2017 IEEE international congress on big data (BigData congress)*. IEEEE.

Zhong, C., Batty, M., Manley, E., Wang, J., Wang, Z., Chen, F., & Schmitt, G. (2016). Variability in regularity: Mining temporal mobility patterns in London, Singapore and Beijing using smart-card data. *PLoS One*, *11*(2), e0149222. PMID: 26872333

Zhong, R. Y., Xu, X., Klotz, E., & Newman, S. T. (2017). Intelligent manufacturing in the context of industry 4.0: A review. *Engineering*, *3*(5), 616–630.

Zhou, Y., & Kankanhalli, A. (2021). AI Regulation for Smart Cities: Challenges and Principles. In Estevez, E., Pardo, T. A., & Scholl, H. J. (Eds.), *Smart Cities and Smart Governance* (Vol. 37, pp. 101–118). Springer International Publishing., DOI: 10.1007/978-3-030-61033-3_5

Zhu, G., & Chow, T. T. (2019). Design optimization and two-stage control strategy on combined cooling, heating and power system. *Energy Conversion and Management*, *199*, 111869. DOI: 10.1016/J.ENCONMAN.2019.111869

Zhu, M., & Jin, J. (2025). Data-Driven Urban Digital Twins and Critical Infrastructure Under Climate Change: A Review of Frameworks and Applications. *Urban Planning*, 10.

Ziosi, M., Hewitt, B., Juneja, P., Taddeo, M., & Floridi, L. (2022). Smart cities: Mapping their ethical implications. SSRN *Electronic Journal, 10*. https://doi.org/ DOI: 10.2139/ssrn.1234567

Zurbaran, M. A., Salazar, A., Brovelli, M. A., & Wightman, P. M. (2020). An Evaluation Framework for Assessing the Impact of Location Privacy on Geospatial Analysis. *IEEE Access : Practical Innovations, Open Solutions*, *8*, 158224–158236. DOI: 10.1109/ACCESS.2020.3019631

Zygiaris, S., Saleh, M. F., & Al-Imamy, S. Y. (2023). The Smart City Blockchain Governance: A Literature Review. *Central European Management Journal,* 31(1).

About the Contributors

Mohammed Balfaqih earned his PhD in Electrical, Electronic, and Systems Engineering and received the On-Time Postgraduate Award from the National University of Malaysia (UKM). He is currently an Associate Professor with Department of Computer and Network Engineering, University of Jeddah, Saudi Arabia. He worked as a Senior Researcher at South Ural State University in Russia from 2018 to 2019, as a Postgraduate Research Assistant at the National University of Malaysia from 2015 to 2017, and as a Research Assistant at MIMOS Research Center in Malaysia from 2011 to 2013. He is an experienced academic with a distinguished record in teaching, research, and supervision at the international level. He has led collaborative research projects that have resulted in high-quality outcomes. His research interests focus on wireless communication networks, Internet of Vehicles (IoV), cooperative communications, and O-RAN. He has published around 25 articles in peer-reviewed journals, 20 international conference papers, two book chapters, and holds two patents.

Akram A. Almohammedi earned his PhD in Wireless Communications and Networks Engineering from Universiti Putra Malaysia (UPM). He is currently an Assistant Professor in the Department of Electrical and Electronics Engineering at Karabuk University, Turkey. He worked as a Senior Researcher at South Ural State University in Russia from 2020 to 2023. Akram has a strong academic background with extensive experience in teaching, research, and supervision at the international level. In 2018, he received the Best Paper Award in Malaysia from IEEE ComSoc/VTS for the paper, "An Adaptive Multi-Channel Assignment and Coordination Scheme for IEEE 802.11 P/1609.4 in Vehicular Ad-Hoc Networks," published in IEEE Access. His research interests focus on wireless communication networks, Internet of Things (IoT), Vehicular Ad Hoc Networks (VANETs), and machine learning applications in smart cities. He has been a member of several international collaborative research projects and serves as a reviewer for numerous international journals and conferences. Akram has published over 30 articles in peer-reviewed journals and numerous international conference papers. His contributions to wireless networks research continue to impact the field globally.

Malek Alsmadi is a dedicated researcher and professor with a Ph.D. in Electrical and Computer Engineering from Lakehead University, Canada, and a Master's in Communication and Computer Engineering from UKM University, Malaysia. He is currently a full-time professor at the School of Engineering Technology, Confederation College, Thunder Bay, Ontario, and an adjunct professor in the Electrical and Computer Engineering department at Lakehead University. He teaches a range of courses, including Programming, Networking, and Operating Systems. Dr. Alsmadi's research focuses on wireless communications, network security, and the impact of hardware impairments on advanced communication systems, contributing to numerous IEEE journal and conference publications. He has received several prestigious awards, including the Dean's Scholar Award and the President's International Scholarship. In addition to his academic roles, Dr. Alsmadi has extensive experience in supervising Ph.D. students, leading research teams, and writing funding proposals for organizations like NSERC and NDRIO-Portage. Dr. Alsmadi is also a Senior IEEE Member, recognized for his significant contributions to the field of communication systems, and actively participates in various IEEE societies and technical program committees.

Arun Agrawal is an accomplished academic in the field of Computer Science & Engineering, Currently serving as an Assistant Professor at the Institute of Technology & Management, Gwalior. With a strong educational background including a B.Sc. in Electronics, M.Sc. in Computer Science, and M.Tech. in Information Technology, Agrawal is currently pursuing his Ph.D. His career spans over two decades, beginning with five years in the industry as an Embedded Engineer before transitioning to academia in 2007. Over the past 17 years, he has progressed from a Lecturer to his current role as Assistant Professor, demonstrating expertise in areas such as vehicular networks, image processing, and cloud computing security. Agrawal's academic contributions include numerous patents, book chapters, publications in international journals and conferences, as well as presentations at IEEE international conferences. His ongoing research and teaching efforts continue to shape the field of computer science and nurture the next generation of technology professionals.

Zain Balfagih holds a PhD in Computer and Information Technology from Universiti Teknologi PETRONAS Malaysia (2013) and specializes in Service Oriented Architecture (SOA), Cloud computing, and wireless networking. His current focus includes Big data and Internet of Things (IoT), particularly in developing mechanisms for utilizing IoT for water and energy applications. Additionally, he conducts research on identifying obstacles hindering the adoption of these technologies in Middle Eastern countries and seeks solutions to address these challenges. Dr. Zain has a background as a researcher in various Malaysian universities before joining Effat University in 2012, where he currently serves as the Chair of the Computer Science Department.

Hemesh Bhardwaj is a IT faculty in NIELIT Shimla. His area of interest is IoT, Machine Learning.

Sagnik Bhattacharyya received his MS(R) degree in wireless communication and signal processing from IIT Ropar in 2021. He is currently working towards his PhD degree with the Department of Electrical Engineering, IIT Ropar. In 2022, he received the prestigious Prime Minister's Research Fellowship (PMRF). His research interests include Cooperative Communication, Cross-Layer Techniques, Intelligent Transportation Systems, and Rate Splitting Multiple Access.

S. Bhavana is an Asst.Professor, CSE-IOT, Department of Computer Science and Engineering-IOT, Sreenidhi Institute of Science and Technology, Hyderabad, Telangana, India.

Saurabh Chandra is an accomplished legal scholar and currently serves as an Associate Professor & Assistant Dean at the School of Law, Bennett University, India. With a robust academic foundation, he earned his Bachelor of Arts and Bachelor of Laws (B.A. LL.B.) from the prestigious Aligarh Muslim University (A.M.U.). He further specialized in Business Laws, completing his Master of Laws (LL.M.) at the renowned National Law School of India University (NLSIU), Bangalore. Dr. Chandra's academic journey culminated in a Doctorate in Law (Ph.D.) from the esteemed National Law University (NLU) Delhi.Dr. Saurabh Chandra also holds a master's degree in Management with specialization in Human Resources in Management and another in Journalism and Mass Communication. With 16 years of teaching experience, he has established himself as an expert in Business Laws, with numerous publications to his credit. Dr. Saurabh Chandra is an expert in Business Laws, with a profound understanding of the intricate legal frameworks that govern the business world.Dr. Chandra brings a wealth of experience in Academic Administration, having successfully held various key administrative positions.

Swetha Chenchu is an Asst.Professor, CSE-IOT, Department of Computer Science and Engineering-IOT, Sreenidhi Institute of Science and Technology, Hyderabad, Telangana, India.

Sam Darshi received the B.Tech. degree (Hons.) in electronics and communication from M. J. P. Rohilkhand University, Bareilly, Uttar Pradesh, and the Ph.D. degree from the Department of Electronics and Electrical Engineering (EEE), IIT Guwahati, India. He is currently working as an Assistant Professor at the IIT Ropar, Punjab. Prior to joining IIT Ropar, he worked as a Faculty Member with the IIIT Guwahati. His research interests include almost all aspects of wireless communication with a special emphasis on advanced underwater communication, cooperative communication, next-generation wireless networks, and intelligent transportation systems.

T Dattaram is a Student CSE-IOT, Sreenidhi Institute of Science and Technology (venkatnarayana.t@sreenidhi.edu.in) YAMNAMPET GHATKESAR Hyderabad, Telangana 501301, IN

Sandip Dey was an Associate Professor and the HOD with the Department of Computer Science and Engineering, Global Institute of Management and Technology, India. He worked as an Assistant Professor in OmDayal Group of Institution and Camellia Institute of Technology for several years. Before that he worked as a Lecturer in Narula Institute of Technology. He is currently an Assistant Professor in the Department of Computer Science, Sukanta Mahavidyalaya, Jalpaiguri. He has coauthored one book titled “Quantum-Inspired Metaheuristics for Image Analysis”; Publisher: John Wiley & Sons, UK, coedited three books, and authored or coauthored about 60 research publications in international journals, book chapters, and conference proceedings. His research interests include soft computing, hybrid intelligence, quantum computing, and image analysis. Dey reviewed several international peer-reviewed journals like Applied Soft Computing, Swarm and Evolutionary Computation, Information Sciences, Journal of Computational Science, Engineering Applications of Artificial Intelligence, Expert Systems With Applications, Advanced Engineering Informatics, IEEE Transactions on Signal Processing and so on.

Deepak Gupta is a distinguished academician with over 20 years of teaching experience and expertise in Computer Science and Engineering. He holds a Ph.D. in Computer Engineering and has an extensive research portfolio, including 42 international journal papers, multiple conference proceedings, and several books. His research interests span machine learning, artificial intelligence, and data mining. Currently, Dr. Gupta serves as an Associate Professor and Academic Coordinator in the Department of Computer Science & Engineering at ITM Gwalior, MP, India. He has also contributed significantly to the field through patents and copyrights and actively engages as a reviewer, conference chair and editor.

Shobhana Kashyap is a Research Scholar in Department of CSE in Dr. B. R. Ambedkar National Institute of Technology, Jalandhar Punjab, India since 2019. She has completed her master's from Thapar University, Patiala India. Her current research is concerned with Machine Learning and Cloud Computing. She has two year teaching experience and written papers for national and international conferences.

Christian Kaunert is Professor of International Security at Dublin City University, Ireland. He is also Professor of Policing and Security, as well as Director of the International Centre for Policing and Security at the University of South Wales. In addition, he is Jean Monnet Chair, Director of the Jean Monnet Centre of Excellence and Director of the Jean Monnet Network on EU Counter-Terrorism.

Pankaj Kumar received his B.Tech. degree (First Division with Honours) in Electronics and Communication Engineering from Uttar Pradesh Technical University, Lucknow, India, and M.Tech. degree (First Division with Honours) with specialisation in digital communication from Dr. A. P. J. Abdul Kalam Technical University, formerly Uttar Pradesh Technical University. He completed his Ph.D. degree from the Indian Institute of Technology Ropar, India. He did his postdoctoral research at the Indian Institute of Science (IISc), Bangalore, India. He is currently working as an Assistant Professor in the Department of Information and Communication Technology, Manipal Institute of Technology, Manipal Academy of Higher Education, Manipal, India. His research interests include, drone assisted cooperative networks, network coding, network coded cooperation, correlation, D2D communication, energy harvesting, and intelligent reflecting surface, NOMA.

Vijaya Kittu Manda has nearly 13+ years of experience in Business Management and Technology. He is a Researcher at PBMEIT, India and works in capital markets, financial planning, and investing. He is an Advocate, a technocrat, an academician, a book writer, and a stock market enthusiast. He has 11 University Postgraduate Degrees in various disciplines. He is a Ph.D. in Management (Finance). His thesis was on Mutual Funds and their Market Competition. His thesis won the prestigious NSE-IEA Best Thesis Award in 2023. He is currently pursuing his second Ph.D. in Computer Science with focus on Blockchain. He contributed over 750 articles to various magazines. He is the Chief Editor for a Management Book Series, is a Peer Review, is a Certified Peer Review Supervisor, is a Session Chair and an Advisor for various academic and industrial conferences. He writes Research Papers and Case Studies, Book Chapters, and sits on Editorial Boards of various publishers. He is a guest speaker for colleges and universities.

Prashant L. Mandale serves as an Assistant Professor at the International Institute of Information Technology (I2IT), Pune. He holds a bachelor's degree from SPPU University and a master's degree from SGGS IE and T, Nanded, Maharashtra, earned in 2012 and 2015, respectively. With over 10 years of teaching experience in Computer Engineering and Information Technology, coupled with 1 years of dedicated research and 1 year in industry, Mr. Mandale is well-versed in the academic and practical facets of his field. Currently pursuing a Ph.D. in Data Science and Cloud, his research interests span across Parallel Computing, Cloud, DevOps, Algorithms, and AI. Mr. Mandale has contributed significantly to the academic community, with 9 published papers in National and International journals and conferences with Scopus indexeding. Additionally, he has authored content for a local publication aligned with the Mumbai University syllabus and another book on DevOps.

Sangeetha Radhakrishnan is an Assistant Professor at Department of Commerce, School of Social Sciences and Languages, Vellore Institute of Technology. She is an M.Phil. Degree in Commerce from Madurai Kamaraj University and a Ph.D. in Commerce from Bharathiar University. She is also a recognized Research Supervisor for Ph.D. & M.Phil. in Bharathiar University. She has more than 12 years of teaching experience in reputed academic institutions. She regularly attends and presents research work in National and International Seminars and Conferences. She has published several articles in International Peer Reviewed Journals and UGC Care Listed journals. She has completed NPTEL courses with All India topper in Financial Accounting Course and MOOC Courses in Coursera, Udemy, Alison etc., She also regularly contributes book chapters. She is a Life Member in IARA and International Association of Academic plus Corporate.

Arunima Shastri is a legal academic with a focus on the intersections of law, technology, and social justice. She holds an LL.M. in Corporate and Commercial Law from NLU Odisha and is currently completing her Ph.D. at Gujarat National Law University, where her research critically explores the interplay between intellectual property rights, visual art, deepfakes, and digital expression. Her teaching and research interests include intellectual property law, data privacy, blockchain regulation, and human rights law, with a particular emphasis on LGBTQ+ rights. Through her work, she bridges doctrinal legal analysis with contemporary technological and cultural developments, contributing to scholarship and curriculum development in emerging areas of law.

Bhupinder Singh working as Professor at Sharda University, India. Also, Honorary Professor in University of South Wales UK and Santo Tomas University

Tunja, Colombia. His areas of publications as Smart Healthcare, Medicines, fuzzy logics, artificial intelligence, robotics, machine learning, deep learning, federated learning, IoT, PV Glasses, metaverse and many more. He has 3 books, 139 paper publications, 163 paper presentations in international/national conferences and seminars, participated in more than 40 workshops/FDP's/QIP's, 25 courses from international universities of repute, organized more than 59 events with international and national academicians and industry people's, editor-in-chief and co-editor in journals, developed new courses. He has given talks at international universities, resource person in international conferences such as in Nanyang Technological University Singapore, Tashkent State University of Law Uzbekistan; KIMEP University Kazakhstan, All'ah meh Tabatabi University Iran, the Iranian Association of International Criminal law, Iran and Hague Center for International Law and Investment, The Netherlands, Northumbria University Newcastle UK,

Venkat T is a Professor and HOD CSE-IOT, Department of Computer Science and Engineering-IOT, Sreenidhi Institute of Science and Technology, Hyderabad, Telangana, India.

Sharda Thete has a Masters In Computer Engineering. 7 Years Of Teaching Experience. Expertise in DSA, Cyber Security.

Deepika Walanjkar is a Ph.D. Pursuing (Symbiosis International University, Lavale Pune) Domain (Blockchain Technology and Artificial Intelligence-Healthcare) M.E. Computer Engineering 6 Years of Teaching Experience

Aqilah Yaacob is a senior lecturer, programme director and certified trainer at Taylor's University. Recognized with the 2023 President Award for Transformational Teaching and Learning, she pioneers innovative educational approaches, fostering transdisciplinary projects with students to address Sustainable Development Goals. Her initiatives have garnered numerous accolades, including innovative teaching and learning awards, and successfully guiding students to win national and international business competitions. Dr Aqilah engages with communities and industries, judging social entrepreneurship programs and securing research grants for sustainability initiatives. With a Ph.D. from Cardiff University, her work appears in prestigious journals, book chapters and conferences worldwide. Passionate about ethical business practices, she shares her insights through media appearances and advisory roles, empowering underprivileged communities globally.

Index

M

O

P

R

S

U

V

www.ingramcontent.com/pod-product-compliance
Lightning Source LLC
LaVergne TN
LVHW080912300926
849162LV00003B/656